Effects and Implications of COVID-19 for the Human Senses, Consumer Preferences, Appetite and Eating Behaviour

Effects and Implications of COVID-19 for the Human Senses, Consumer Preferences, Appetite and Eating Behaviour

Editor

Derek V. Byrne

MDPI • Basel • Beijing • Wuhan • Barcelona • Belgrade • Manchester • Tokyo • Cluj • Tianjin

MDPI

Editor
Derek V. Byrne
Aarhus University
Denmark

Editorial Office
MDPI
St. Alban-Anlage 66
4052 Basel, Switzerland

This is a reprint of articles from the Special Issue published online in the open access journal *Foods* (ISSN 2304-8158) (available at: https://www.mdpi.com/journal/foods/special_issues/covid_human_sense).

For citation purposes, cite each article independently as indicated on the article page online and as indicated below:

LastName, A.A.; LastName, B.B.; LastName, C.C. Article Title. *Journal Name* **Year**, *Volume Number*, Page Range.

ISBN 978-3-0365-5017-6 (Hbk)
ISBN 978-3-0365-5018-3 (PDF)

Cover image courtesy of Derek V. Byrne

Contents

About the Editor

Derek V. Byrne

Derek V. Byrne is Professor of Sensory and Consumer Science and a Science Leader of the Food Quality Perception & Society (FQS) Team at The Department of Food Science, Aarhus University in Denmark. Derek has a PhD in sensory science from the University of Copenhagen and an MSc in Food and Nutritional Science from University College Cork in Ireland. He has over 25 years of experience in sensory and consumer science research, from the senses, through to product quality and design, and on to the importance of sensory science in food industry applicability. Derek's team at Aarhus University focuses on understanding food quality and perception via a cross-disciplinary synergy of multisensory human food analysis, experimental psychology, physiological responses, and cognitive neuroscience in the design and development of high quality, better-tasting, more stimulating, more memorable, and healthier food and drink experiences.

Preface to "Effects and Implications of COVID-19 for the Human Senses, Consumer Preferences, Appetite and Eating Behaviour"

Severe Acute Respiratory Syndrome Coronavirus-2 (SARS-CoV-2) evolved into a global pandemic in 2020. The assessment of coronavirus disease 19 (COVID-19) patients presented health conditions including, in many cases, a mild to severe loss of smell and tasting abilities among patients. Initial work has shown short- and likely longer-term negative effects on the human senses, with some indications for effects on consumer preferences; however, as of yet, very little is known about the impacts on eating behaviours and consequent longer-term effects on appetite.

Food enjoyment for example is a key aspect of people's appetite, and any loss in expected pleasure greatly affects our motivation to eat, potentially leading to persons affected by COVID-19 experiencing core changes in relation to their food intake practices, which may potentially have long-term implications for health and recovery.

The aim of this Special Issue anthology was, for the first time, to bring together researchers with key insights on how COVID-19 has impacted appetite and eating behaviours, from the fundamental to the applicable, as assessed by human sensory perception and behaviour.

Overall, the research included in this collection is diverse and covers a wide range of investigations in relation to the effects and implications of COVID-19 for the human senses, consumer preferences, appetite and eating behaviour. Studies are included that explore issues from the fundamentals of appetite and the senses to real world applicability regarding food choice, safety perception and purchasing behavioural change as a result of the COVID-19 pandemic and its various waves and lockdowns across the world.

The broad nature of the studies included emphasizes the importance and critical nature of the inclusion of the human senses and consumer preference and behaviour in relation to addressing the after effects of the COVID-19 pandemic and its implications.

An overall conclusion with respect to this collection would be that human senses, consumer acceptance, and preferences are core to future food design, with respect to understanding COVID-19's effects on human perception effects on a global scale.

Derek V. Byrne
Editor

foods

MDPI

Editorial

Effects and Implications of COVID-19 for the Human Senses, Consumer Preferences, Appetite and Eating Behaviour: Volume I

Derek Victor Byrne [1,2]

[1] Food Quality Perception and Society Science Team, iSense Lab, Department of Food Science, Faculty of Technical Sciences, Aarhus University, DK-8200 Aarhus, Denmark; derekv.byrne@food.au.dk

[2] Food & Health Research, Sino-Danish Center (SDC), Niels Jensens Vej 2, DK-8000 Aarhus C, Denmark

Citation: Byrne, D.V. Effects and Implications of COVID-19 for the Human Senses, Consumer Preferences, Appetite and Eating Behaviour: Volume I. *Foods* **2022**, *11*, 1738. https://doi.org/10.3390/foods11121738

Received: 27 May 2022
Accepted: 7 June 2022
Published: 14 June 2022

Publisher's Note: MDPI stays neutral with regard to jurisdictional claims in published maps and institutional affiliations.

1. Introduction

Severe Acute Respiratory Syndrome Coronavirus-2 (SARS-CoV-2) evolved into a global pandemic in 2020 [1]. The assessment of coronavirus disease 19 (COVID-19) patients has presented a health condition including, in many cases, a mild to severe loss of smell and tasting abilities among patients, see, e.g., [2,3].

Initial work has shown short- and likely longer-term negative effects on the human senses, with some indications for effects on consumer preferences; however, as of yet, very little is known about the impacts on eating behaviours and consequent longer-term effects on appetite (see [4]).

Food enjoyment is a key aspect of people's appetite, and any loss in expected pleasure greatly affects our motivation to eat, potentially leading to persons affected by COVID-19 to experience core changes in relation to their food intake practices, which may potentially have long-term implications for health and recovery [5].

The aim of this Special Issue was, for the first time, to bring together researchers with key insights on how COVID-19 has impacted appetite and eating behaviours, from the fundamental to the applicable, as assessed by human sensory perception and behaviour [6–17].

Through this call for publications, we wished to document and bring together ongoing key research in order to ensure that this research has a lasting impact regarding our future understanding of the measures developed to help and treat people affected during the ongoing pandemic.

Moreover, we requested the inclusion of a range of research from fundamental effects on the senses, to changes in consumer preferences all the way to how and why COVID-19 has changed consumer behaviours in relation to food and eating in the longer term [6–17].

The articles included have been categorized based on their core aims and findings, and they cover research in relation to COVID-19 and the senses in four key areas, with respect to appetite and eating behaviour [6–8], food choice and preference [9–11], the perception of food risk and safety, [12–14] and finally to the effects on purchasing behaviour during and after the initial waves of the COVID-19 pandemic [15–17]. This collection of articles is, in essence, a time capsule of the wide focus and importance of sensory and consumer science in the COVID-19 space thus far, and its highlighting of effects on societies senses, eating practices and behaviours.

Ultimately, the goal was to publish the Special Issue collection as an open-source book to act as a tool for understanding the long-term effects of COVID-19 on human health related to food and eating issues. This endeavour is now complete with this group of publications, now designated as Volume I on the 'Effects and Implications of COVID-19 for the Human Senses, Consumer Preferences, Appetite and Eating Behaviour'. Due to a continuously growing body of work on COVID-19 being submitted to the Sensory and Consumer area of *Foods*, and to ensure we place a focus on the important work on COVID-19 in the sensory sphere, we have determined that a Volume II of this book will be curated and published in due course.

2. A Synopsis of Special Issues Research

2.1. Appetite and Eating Behaviour Change

Thus, in relation to appetite and eating behaviour change, authors Høier et al. (2021) investigated subjective strategies for maintaining appetite in recovering COVID-19 patients [15]. The study involved in-depth interviews, focusing on patients suffering from the long-term effects of COVID-19. The results were analysed using a thematic analysis for qualitative data. The results on strategies for maintaining appetite included a focus on well-functioning senses, a focus on familiar foods, a focus on the eating environment, and finally a focus on post-ingestive well-being. Høier et al. (2021) found that factors prior to, during and after food intake, as well as the context, could influence both the desire to eat and the pleasure related to food intake [15]. Moreover, the authors indicated that, as ageusia and anosmia make the characterization of food difficult, being able to recognize and memorize its flavour was important to engage in consumption; under normal circumstances, the hedonic value of food relies predominantly on the flavour of the food. When suffering from chemosensory dysfunction, shifting focus towards the texture of food, including trigeminal stimulation during consumption, was beneficial for maintaining appetite and food-related pleasure. Furthermore, a focus on the holistic satisfying feelings of choosing healthy food, as well as a focus on other people's enjoyment during meals, were reported to boost well-being around food intake. Høier et al. (2021) concluded that research elaborated our understanding of the complex consequences of COVID-19 and can be applied in health-promoting initiatives targeting patients recovering from COVID-19 [15].

Furthermore, regarding changes in eating behaviour, Chaaban et al. (2021) investigated the acute and long-term effects of COVID-19 disease on the desire for food, hunger, and satiety sensations; smell, taste, and flavour perception; meals and intake of food types; and the frequency of commonly applied strategies to tackle potential changes in appetite and sensory perception [14]. In this study, an online survey was conducted among Danish adults who had experienced changes in appetite, sensory perception, and/or food-related pleasure due to COVID-19 [14]. The overall results indicated appetite-altering effects at all times during the day when suffering from COVID-19 and were often associated with impaired sensory function. The authors also showed severe sensory perception alterations, namely, for the perception of taste and for the perception of smell. Eating behavioural changes included alteration in quantitative and qualitative aspects of intake. The effects were, in general, more pronounced during the acute phase of disease than during the post-acute phase. Chaaban et al. (2021) concluded that the findings illustrate the complexity by which COVID-19 affects human appetite, sensory perception, and eating behaviour, but also point to strategies to cope with these changes [14].

Finally, in relation to appetite, Parker et al. (2022) looked at parosmia (a distortion in sense of smell) and its effects on food perception as a result of COVID-19. The aim of this study was to identify the key food triggers of parosmic distortions and investigate the relationship between distortion and disgust in order to establish the impact of parosmia on diet and quality of life. The authors indicate that olfactory dysfunction is amongst many symptoms of long COVID [6]. The authors contend that whilst most people that experience smell loss post COVID-19 recover their sense of smell and taste within a few weeks, around 10% of cases experience long-term problems, and their smell recovery journey often begins a few months later [6]. In the authors' cross-sectional study, respondents experiencing smell distortions completed a questionnaire covering aspects of smell loss, parosmia and the associated change in valence of everyday items. It was determined that there was a significant correlation between strength and disgust and when the selected items were reported as distorted—they were described as either unpleasant or gag-inducing 84% of the time. The authors concluded that this shift in valence associated with loss of expected pleasure and the presence of strange tastes and burning sensations could certainly lead to changes in eating behaviours and serious longer-term consequences for mental health and quality of life [6].

2.2. Food Choice and Preference Effects

In relation to food choice and preference changes as a result of COVID-19, Scacchi et al. (2021) investigated how the Italian lockdown affected self-perceived food purchases (FP), occurrence of impulse buying (IB), and household food waste production (HFWP) as well as their respective determinants. A cross-sectional survey was distributed in May 2020, collecting an opportunistic sample of the Italian population. Most of the sample increased overall FP (53.4%) and food consumption (43.4%), and reduced HFWP (53.7%) and halved the prevalence of IB (20.9%) compared to the period before the lockdown (42.5%) [16]. Baking ingredients, fresh vegetables, fresh fruit and chocolate had the largest sales increase by individuals, while bakery products, fresh fish and salted snacks purchases highly decreased. Increased FP were associated with the occurrence of IB and inversely associated with not having worked during lockdown. Multivariable logistic regressions revealed that the occurrence of IB was associated with low perceived dietary quality according to the Emotional Overeating Questionnaire, and inversely associated with decreased HFWP. Reduced HFWP was associated with higher perceived dietary quality and negatively associated with a low score WHO-5 Well-Being Index [16]. Scacchi et al. (2021) concluded that the Italian lockdown highly affected FC behaviours, leading to positive and more sustainable habits towards food purchases and consumption. Public health interventions are needed to keep these new positive effects and avoid negative consequences in case of future lockdowns

In a study investigating trends in coffee and tea consumption during the COVID-19 pandemic, Castellana et al. (2021) mentioned that over the initial two years of the pandemic, many countries have enforced confinement to limit both the spread of COVID-19 and the demand for medical care. Confinement has of course resulted in a disruption of work routines, boredom, depression, and changes in eating habits, as well as changes in the consumption of coffee and tea. The authors indicate that beverage choices also contribute to daily calorie intake and hydration, particularly so-called 'nervine' beverages such as coffee and tea, in view of their purported potential to promote psychological well-being [17]. Castellana et al. (2021) investigated a large body of published studies in a database study, examining articles tracking consumption of tea and coffee. The authors found studies that indicated coffee consumption increased to some degree and tea consumption clearly increased. The authors indicate that the lack of a strong trend in coffee consumption as the result of the COVID-19 pandemic calls for additional investigation. Moreover, the authors state that potential health implications should not be overlooked, especially since caffeine consumption may directly or indirectly promote bronchodilation, interfere in the process of immunomodulation, and hinder viral intracellular transcription while undergoing COVID-19 infection [17]. In relation to tea, reflecting perhaps a discomfited mood and the socially confining setting, the authors found a marked increase in tea consumption. Tea is usually linked to routine and ritualized household consumption. Tea is historically instrumental in bringing the family closer together and provides a platform for sharing. In contrast, coffee consumption needed to be considered in a social, aesthetic and emotional context. Therefore, Castellana et al. (2021) conclude that setting aside the social context, the increased consumption of tea should be understood in emotional and family-related settings. From this perspective, this beverage has long been associated with mood and performance enhancements, such as a greater relaxation and concentration [17].

Finally, in relation to studies focused on food choice and preference changes as a result of the COVID-19 pandemic, Górska et al. (2021) indicated that psychological factors and restrictions imposed due to the pandemic may influence eating behaviours and physical activity. With the above thesis in mind, questionnaire-based surveys were conducted amongst residents of five European countries: Poland, Italy, Spain, Portugal and Great Britain (England and Scotland). A structured questionnaire was used to conduct anonymous internet surveys in 2020. It contained questions pertaining to sociodemographic data, eating behaviours, the impact of the pandemic on the diet and physical activity [11]. The questionnaire was made available to Internet users across the five countries. Górska et al.

(2021) found that age was the parameter that impacted changing eating behaviours to the largest extent during the pandemic. It was also found that during the pandemic, regular consumption of meals was most dependent on various factors. The negative impact of the pandemic within this scope was most profound amongst women, city residents regardless of gender and people over 35 years of age. A change in the frequency of consumption of selected product groups during the pandemic was also observed. A reduced consumption of meat and fish was identified [11], in particular, among people under 35 living in Portugal, where almost half declared lower consumption of meat, and more than half reported lower consumption of fish. In an analysis of the results, the authors also showed an increase in the consumption of products with lower nutritional values, particularly amongst people under 35 years of age and also amongst residents of Great Britain (regardless of age). Moreover, the authors indicated that results showed that the pandemic may have had an impact on weight reduction [11]. Górska et al. (2021) survey results showed that the impact of the pandemic on eating behaviours was particularly profound when it came to meal consumption regularity. Changes to the consumption of products with lower nutritional values, which may decrease immunity, were also found during the pandemic. In conclusion, Górska et al. (2021) showed that the problem associated with consuming products with lower nutritional values was particularly evident amongst people under 35. Considering the global character of SARS-CoV-2 transmission, further research is necessary to determine its impact on the diet, nutritional status and physical activity [11].

2.3. Food Risk and Safety Perception

In a study focusing on food risk and safety perception as a result of COVID-19, Vandenhaute et al. (2022) looked specifically at safety measures in the food service sector and consumers' attitudes and transparency perceptions at three different stages of the pandemic. Vandenhaute et al. (2022) indicate the study aims to examine consumers' attitudes towards, and transparency perceptions of, COVID-19-related safety measures and to identify determinants of consumers' intentions and behaviours regarding visiting restaurants and bars once reopened [7]. By surveying food service businesses in Belgium both during and in between waves of infections, the authors' study allowed for a comparison between both target groups, i.e., 1697 consumers and 780 businesses. Vandenhaute et al. (2022) describe that the findings demonstrate that consumers evaluated safety measures as important when revisiting restaurants and bars, against business owners' expectations [7]. Both consumers' revisit intentions and behaviours are influenced by the perceived importance of hygiene measures (negatively) and past visit frequency (positively). These authors concluded that the study highlights the importance of good compliance with safety measures as a strategy to attract customers during the reopening period. Further, the findings emphasize the importance of transparent communication by food service businesses and the government [7].

Moreover, Cantalapiedra et al. (2022) in their paper, "Facing Food Risk Perception: Influences of Confinement by SARS-CoV-2 Pandemic in Young Population", indicate a new food safety level of trust in food risk perception has been noticed, as a consequence of the SARS-CoV-2 pandemic. The pandemic encouraged a review of nutritional recommendations for the population, mainly for the young population. Here, the results of a survey designed for a young population, from the University of Valencia, Spain, in the health branch, and in charge of carrying out the shopping task for their household, were reported. The study reports three different scenarios and years, as defined by the SARS-CoV-2 pandemic: before the pandemic (the period of January–December 2019), during the pandemic lockdown (the period of March 2020–August 2020), and after the pandemic lockdown (the period of September 2020–June 2021) [8]. Cantalapiedra et al.'s (2022) survey was designed with questions, profiling responses using the best–worst elicitation (BWE) format. Results reported that trust and evaluation of information differed in all three scenarios. Results reported that trust and evaluation of information differed in all three scenarios. In the SARS-CoV-2 pandemic, there was a high increase in trust in the information provided 'inside' by the shopping location, while there were no changes for

the 'outside'; trust in cooperative stakeholders went from a medium-low to medium-high score, while, for individual stakeholders, it was maintained as a medium score, and trust in information on food products was kept at high score [8]. The authors indicate that, regarding the evaluation of the information provided by stakeholders, a tendency towards a maintaining a medium score was seen, while that from the channels of distribution went from medium-low to medium-high for buying on-site. A uniform tendency was observed for online/other distribution channels for all three years and descriptors studied: "Internet", "Farmer on-demand", and "Cooperative consumers" (<50%). In summation, Cantalapiedra et al. (2022) stated that their research provides findings of implications that contribute to the changing the perception of food risk, due to the COVID-19 pandemic, i.e., the adaptation of the young population, trust in safety and quality, and importance of coordination from all communication points to avoid negative or strong consequences, in case of future lockdowns or health crises [8].

Lastly, around food risk and safety perception in this collection, Li et al. (2021) investigated trends in food preferences and sustainable behaviour in Spanish consumers during the COVID-19 lockdown. The authors indicate that the research not only investigates trends in Spanish consumers' general food shopping and consumption habits during the lockdown, but also investigates these trends from the perspective of sustainable purchasing [10]. Specifically, total food consumption (C), food expenditure (E), and purchase of food with sustainable attributes (S) were measured. Data were collected from a semi-structured questionnaire which was distributed online among 1203 participants. Li et al. (2021) describe how logit models showed that gender, age, employment status, and consumers' experiences were associated with total food consumption and expenditure during the lockdown. In addition, Li et al. (2021) state that consumers' risk perceptions, shopping places, trust level in information sources, and risk preference were highly essential factors influencing consumers' preferences and sustainable behaviour. Consumers' objective knowledge regarding COVID-19 was related to expenditure. Furthermore, family structure only affected expenditure, while income and place of residence influenced food consumption. Mood was associated with expenditure and the purchase of sustainable food. Household size affected purchasing behaviour towards food with sustainable attributes. Li et al. (2021) summaries that this research provides references for stakeholders that help them to adapt to the new COVID-19 situation [10].

2.4. Purchasing Behaviour and Decisions

The final grouping of works in the present collection covers purchasing behaviour and decisions as affected by the COVID-19 pandemic. Qi et al. (2021) looked at explaining Chinese consumers' green food purchase intentions during the COVID-19 pandemic via an extended theory of planned behaviour [12]. The authors indicate that as a result of the COVID-19 pandemic, consumers' habits and behaviours have been strongly influenced, potentially creating a more sustainable and healthier era of consumption. Hence, the authors conclude there is a potential for further expanding the green food sector in China [12]. Qi et al. (2021) present that the theory of planned behaviour (TPB) is a widely used framework to explain consumers' food choices. The authors state that considering consumers' internal norms, their perceptions of green food attributes, and the shifting consumer behaviour, their study has extended the TPB framework (E-TPB) by adding constructs of moral attitude, health consciousness, and the impact of COVID-19 (IOC). Qi et al. (2021) analysed the results of structural equation modelling among 360 functional samples, revealing that the E-TPB model has a superior explanatory and predictive power compared with the original TPB model regarding Chinese consumers' green food buying intentions in the current and post-pandemic periods [12]. The authors' path analysis (a form of multiple regression statistical analysis that is used to evaluate causal models by examining the relationships between a dependent variable and two or more independent variables) demonstrated that attitude, perceived behavioural control, moral attitude, health consciousness, and IOC have significant positive effects on green food purchase intentions. In conclusion, however, the authors state that the association between subjective norm and purchase intention varies within the TPB and E-TPB models, which showed a non-significant impact in E-TPB. Overall, Qi et al. (2021) indicate that these findings can generate more suitable managerial implications to promote green food consumption in China during the current and post-pandemic periods [12].

A second investigation in this space looked at COVID-19′s first wave, regarding an examination of the pandemic's impact on food purchasing behaviour in the Eurozone [13]. Gutiérrez-Villar et al. (2021) present that COVID-19 has had a negative impact on the living conditions of people in all countries worldwide. With a devastating economic crisis where many families are finding it difficult to pay bills and make ends meet, increases in the prices of food basket staples can be very worrying. Their study examines the relationship between the incidence of the pandemic during the first wave in 16 Eurozone countries with the variation experienced in food prices [13]. Gutiérrez-Villar et al. (2021) analysed the harmonised index of consumer food prices (included in HICP) and the classification of the degree of pandemic impact by country, the latter established with the index of deaths provided by the Johns Hopkins Center [13]. The procedure the authors used compared actual food prices during the first wave (March to June 2020) with those foreseeable in the absence of the pandemic. Time series analysis was used, dividing the research period into two phases. Gutiérrez-Villar et al. (2021) indicated that in both phases, the Holt–Winters model was applied for estimation and subsequent prediction. After a contrast using Kendall's tau correlation index, Gutiérrez-Villar et al. (2021) concluded that in the countries with the highest death rates during the first wave, there was a higher increase in food prices than in the least affected countries of the Eurozone [13].

Lastly, in this collection, Jun et al. (2022) looked at how customer decisions were influenced to use online food delivery services during the COVID-19 pandemic. The authors contend that despite the popularity of online food delivery systems in the food service industry, there have been few studies into customers' decision-making process to use online food delivery services during the COVID-19 pandemic. Jun et al.'s (2022) study applied the technology acceptance model (TAM) to examine the factors affecting customers' intention to use online food delivery services [9]. Authors results showed that perceived usefulness affects customer's online food delivery usage directly and indirectly through customer attitude; enjoyment and trust are also key factors determining behaviour intention toward customer attitude using online food delivery services, and that there is a positive relationship between social influence and customer attitude in addition to a positive relationship between customer attitude and behaviour intention in the online food delivery service context [9]. Jun et al. (2022) conclude that these findings provide theoretical and managerial implications that contribute to the online food delivery service industry [9].

3. Conclusions

Overall, the research included in this Special Issue collection is diverse and covers a wide range of investigations in relation to the effects and implications of COVID-19 for the human senses, consumer preferences, appetite and eating behaviour. Studies are included from the fundamentals of appetite and the senses on to real world applicability regarding food choice, safety perception and purchasing behavioural change as a result of the COVID-19 pandemic and its various waves and lockdowns across the world.

The diverse nature of the studies included in this Special Issue emphasizes the importance and critical nature of the inclusion of the human senses and consumer preference and behaviour in relation to addressing the after effects of the COVID-19 pandemic and its implications.

An overall conclusion with respect to this article collection would be that the human senses, consumer acceptance, and preferences are core to future food design, with respect to understanding COVID-19's effects on human perception effects on a global scale.

Funding: This work was supported in part by the university partnership Denmark-China, Sino Danish Centre (SDC), (grant number: 30067; DVB) within the 'Food and Health Research Theme', Aarhus, Denmark.

Acknowledgments: The author thanks the Food Quality Perception & Society Team and iSENSE Lab at the Department of Food Science at Aarhus University, Denmark.

Conflicts of Interest: The author declares no conflict of interest.

References

1. Worldometers. COVID-19 Coronavirus Pandemic. 2020. Available online: https://www.worldometers.info/coronavirus/ (accessed on 17 May 2022).
2. Marshall, M. COVID's toll on smell and taste: What scientists do and don't know. *Nature* **2021**, *589*, 342. [CrossRef] [PubMed]
3. Agyeman, A.A.; Chin, K.L.; Landersdorfer, C.B.; Liew, D.; Ofori-Asenso, R. Smell and Taste Dysfunction in Patients With COVID-19: A Systematic Review and Meta-analysis. *Mayo Clin. Proc.* **2020**, *95*, 1621. [CrossRef] [PubMed]
4. Grant, M.C.; Arbyn, M.; Mohammed, Z.; McGuinness, L.; Clarke, E.L.; Wade, R.G. The prevalence of symptoms in 24,410 individuals infected by the novel coronavirus (SARS-CoV-2; COVID-19): A systematic review and meta-analysis of 148 studies from 9 countries. *PLoS ONE* **2020**, *15*, e0234765. [CrossRef]
5. González-Monroy, C.; Gómez-Gómez, I.; Olarte-Sánchez, C.M.; Motrico, E. Eating Behaviour Changes during the COVID-19 Pandemic: A Systematic Review of Longitudinal Studies. *Int. J. Environ. Res. Public Health* **2021**, *18*, 11130. [CrossRef]
6. Parker, J.K.; Methven, L.; Pellegrino, R.; Smith, B.C.; Gane, S.; Kelly, C.E. Emerging Pattern of Post-COVID-19 Parosmia and Its Effect on Food Perception. *Foods* **2022**, *11*, 967. [CrossRef]
7. Vandenhaute, H.; Gellynck, X.; De Steur, H. COVID-19 Safety Measures in the Food Service Sector: Consumers' Attitudes and Transparency Perceptions at Three Different Stages of the Pandemic. *Foods* **2022**, *11*, 810. [CrossRef] [PubMed]
8. Cantalapiedra, F.; Juan, C.; Juan-García, A. Facing Food Risk Perception: Influences of Confinement by SARS-CoV-2 Pandemic in Young Population. *Foods* **2022**, *11*, 662. [CrossRef] [PubMed]
9. Jun, K.; Yoon, B.; Lee, S.; Lee, D.-S. Factors Influencing Customer Decisions to Use Online Food Delivery Service during the COVID-19 Pandemic. *Foods* **2022**, *11*, 64. [CrossRef]
10. Li, S.; Kallas, Z.; Rahmani, D.; Gil, J.M. Trends in Food Preferences and Sustainable Behavior during the COVID-19 Lockdown: Evidence from Spanish Consumers. *Foods* **2021**, *10*, 1898. [CrossRef] [PubMed]
11. Górska, P.; Górna, I.; Miechowicz, I.; Przysławski, J. Changes in Eating Behaviour during SARS-CoV-2 Pandemic among the Inhabitants of Five European Countries. *Foods* **2021**, *10*, 1624. [CrossRef] [PubMed]
12. Qi, X.; Ploeger, A. Explaining Chinese Consumers' Green Food Purchase Intentions during the COVID-19 Pandemic: An Extended Theory of Planned Behaviour. *Foods* **2021**, *10*, 1200. [CrossRef] [PubMed]
13. Gutiérrez-Villar, B.; Melero-Bolaños, R.; Carbonero-Ruz, M. COVID-19's First Wave: Examination of Impact on Food Purchasing Behaviour in the Eurozone. *Foods* **2021**, *10*, 1179. [CrossRef] [PubMed]
14. Chaaban, N.; Høier, A.T.Z.B.; Andersen, B.V. A Detailed Characterisation of Appetite, Sensory Perceptional, and Eating-Behavioural Effects of COVID-19: Self-Reports from the Acute and Post-Acute Phase of Disease. *Foods* **2021**, *10*, 892. [CrossRef] [PubMed]
15. Høier, A.T.Z.B.; Chaaban, N.; Andersen, B.V. Possibilities for Maintaining Appetite in Recovering COVID-19 Patients. *Foods* **2021**, *10*, 464. [CrossRef] [PubMed]
16. Scacchi, A.; Catozzi, D.; Boietti, E.; Bert, F.; Siliquini, R. COVID-19 Lockdown and Self-Perceived Changes of Food Choice, Waste, Impulse Buying and Their Determinants in Italy: QuarantEat, a Cross-Sectional Study. *Foods* **2021**, *10*, 306. [CrossRef] [PubMed]
17. Castellana, F.; De Nucci, S.; De Pergola, G.; Di Chito, M.; Lisco, G.; Triggiani, V.; Sardone, R.; Zupo, R. Trends in Coffee and Tea Consumption during the COVID-19 Pandemic. *Foods* **2021**, *10*, 2458. [CrossRef] [PubMed]

foods

MDPI

Article

Emerging Pattern of Post-COVID-19 Parosmia and Its Effect on Food Perception

Jane K. Parker [1,*], Lisa Methven [1], Robert Pellegrino [2], Barry C. Smith [3], Simon Gane [4] and Christine E. Kelly [1,5]

1 Department of Food and Nutritional Sciences, University of Reading, Reading RG6 6DZ, UK; l.methven@reading.ac.uk (L.M.); chris@abscent.org (C.E.K.)
2 Monell Chemical Senses Center, Philadelphia, PA 19104, USA; rpellegrino@monell.org
3 Centre for the Study of the Senses, Institute of Philosophy, School of Advanced Study, University of London, London WC1E 7HU, UK; barry.smith@sas.ac.uk
4 Royal National Ear, Nose and Throat and Eastman Dental Hospitals, University College London Hospital, 47-49 Huntley St., London WC1E 6DG, UK; simongane@nhs.net
5 AbScent, 14 London Road, Andover SP10 2PA, UK
* Correspondence: j.k.parker@reading.ac.uk

Abstract: Olfactory dysfunction is amongst the many symptoms of Long COVID. Whilst most people that experience smell loss post COVID-19 recover their sense of smell and taste within a few weeks, about 10% of cases experience long-term problems, and their smell recovery journey often begins a few months later when everyday items start to smell distorted. This is known as parosmia. The aim of this study was to identify the key food triggers of parosmic distortions and investigate the relationship between distortion and disgust in order to establish the impact of parosmia on diet and quality of life. In this cross-sectional study (n = 727), respondents experiencing smell distortions completed a questionnaire covering aspects of smell loss, parosmia and the associated change in valence of everyday items. There was a significant correlation between strength and disgust ($p < 0.0001$), and when the selected items were reported as distorted, they were described as either unpleasant or gag-inducing 84% of the time. This change in valence associated with loss of expected pleasure and the presence of strange tastes and burning sensations must certainly lead to changes in eating behaviours and serious longer-term consequences for mental health and quality of life.

Keywords: COVID-19; olfactory distortions; parosmia; trigger foods; disgust; valence

Citation: Parker, J.K.; Methven, L.; Pellegrino, R.; Smith, B.C.; Gane, S.; Kelly, C.E. Emerging Pattern of Post-COVID-19 Parosmia and Its Effect on Food Perception. *Foods* **2022**, *11*, 967. https://doi.org/10.3390/foods11070967

Academic Editor: Derek V. Byrne

Received: 16 February 2022
Accepted: 24 March 2022
Published: 27 March 2022

1. Introduction

Sense of smell guides our selection and appreciation of food and plays a dominant role in flavour perception [1]. When that sense is missing or impaired, the consequences are far-reaching. The impact of olfactory dysfunction on diet, quality of life, and interpersonal relationships is well-documented [2]; never more so as witnessed poignantly during the COVID-19 pandemic, with olfactory loss recognised worldwide as one of the official symptoms of COVID-19. Burges Watson et al. identified from a co-created study, based on social media posts of those with post-COVID-19 alterations in taste and smell, three broad concerns: (i) a radically altered experience of food and eating, (ii) difficulty in making sense of the altered experience, and (iii) altered relationships to the world [3]. They concluded that in cases where the sense of smell is not recovered within 2–3 weeks, the effect is not mild, given that it may last for months, and it has "serious implications for food, eating, health, work and well-being".

Prior to the COVID-19 pandemic, olfactory disorders were a largely unrecognised problem, even though they prevailed in up to 23% of the population [4]. Such disorders are a known consequence of viral illness or infection [5] and are often referred to as post-infectious or post-viral olfactory dysfunction. In long-term cases (experienced by about 10% of all COVID-19 cases), there is an initial loss of sense of smell (anosmia), and as the recovery process begins, typically 2–3 months after the initial loss [6], many experience qualitative olfactory disorders. Parosmia is one such qualitative disorder which alters an individual's perception of odours in such a way that every day smells are commonly described as "distorted". These distortions are often associated with strong dislike or disgust and can persist in some cases for up to 10 years [7]. In extreme cases of parosmia, some triggers can provoke nausea and vomiting [8]. However, it is generally recognised as a sign of recovery and has been identified recently as an independent predictor for complete recovery [9].

We present here the impact of this condition on the perception of food within the greater context of the onset, symptoms, and duration of COVID-19-related olfactory disorders. The main aim of this investigation is to examine the late-emerging pattern of qualitative olfactory dysfunction and its effect on the perception of common foods and beverages. Our focus in this paper is on parosmia, the foods that trigger the distortions associated with parosmia and, in particular, on trying to understand the relationships between distortion and disgust. We explore differences between cases which are post-COVID-19 and cases which were attributed to other viral infections, and investigate characteristics of parosmia. An explanation of the aetiology of the disease is important in understanding the impact on food; hence, the first part of this paper will address these biomedical issues.

2. Materials and Methods

2.1. Ethics and Recruitment

All subjects gave their informed consent for inclusion before they participated in the study. The investigations were carried out following the rules of the Declaration of Helsinki of 1975, and the protocol was approved by the School of Chemistry, Food and Pharmacy Research ethics committee of the University of Reading on 10 June 2020 (study number 29.2020). It was registered under the US Library of Medicine as trial NCT04868435.

This is a cross-sectional study. Participants were recruited through ENT clinics, Facebook (AbScent Parosmia and Phantosmia Support group and personal accounts), and Twitter between 19 June 2020 and 5 September 2021. Volunteers aged 18 or over who were experiencing smell distortions or for whom everyday things smelled different, odd, or disgusting were invited to participate in the fully anonymised survey. Entry into the study was dependent on the participants completing a standard unlinked online consent form which then took them to the survey landing page. Participation in this online study was voluntary, and respondents received no remuneration. The survey was carried out on Compusense (West Guelph, ON, Canada).

2.2. The Questionnaire

After completing the consent form, respondents completed the six-part survey. Demographic data (age, gender, country of residence, ethnic group, and smoking status) were collected in Section 1, whilst Section 2 asked questions about the speed, timings, and aetiology of the respondents' initial loss of smell (anosmia). They were asked when they lost their sense of smell (date) and the likely aetiology of their symptoms (COVID-19, other viral illness, accident (including head or brain injury), unexplained (or idiopathic), or other/do not know). If the cause was COVID-19 or another viral infection, the speed of loss was reported as one of 4 categories: very suddenly before the onset of other symptoms of infection, very suddenly during infection, very suddenly after infection, or gradually. In cases where it was attributed to COVID-19, further questions were asked about diagnosis (PCR, antibody test, or no test) and severity. Section 3 asked about the onset of parosmia, whether it had been preceded by any partial or full recovery of the sense of smell (none, a few hints, any partial or full recovery), and whether the symptoms fluctuated significantly (four categories: no fluctuation, infrequent or minor fluctuations, significant daily random fluctuations, significant but generally get better during the day). In Section 4, respondents were asked to indicate whether they could taste salt and sugar (three categories: taste as normal, taste weaker, or cannot taste), whether they could detect heat in spices (yes/no), and whether they had experienced any metallic taste (yes/no) or burning sensations in their nose or throat (yes/no).

The core survey (Section 5) concerned the respondents' perception of 14 foods (referred to as the 14 "set triggers") which had been pre-selected based on data from other studies [8,10]. Onion, meat, coffee, and eggs were selected on the basis of being among the most commonly reported trigger foods for parosmia. Chocolate, peanuts, bacon, fried foods, peppers, cucumber, and melon were selected because they were known but less frequent triggers, whereas butter, apple, and rose were selected as examples of "safe" foods and smells that were less likely to trigger parosmia.

For each item, the respondents were asked to record whether they perceived the smell of that item as distorted (four categories: smells like it did before, smells distorted, I cannot smell it at all, or I am not familiar with this food/smell so cannot answer). If either of the last two answers was selected, the survey skipped to the next item. Those who selected "distorted" were asked to provide two to three words to describe the distortion. Next, respondents were asked to rate their hedonic assessment of the smell as pleasant (score = 1), neither pleasant nor unpleasant (2), unpleasant (3), or so bad I want to gag/vomit/leave the room (4). An additional option had been provided in the questionnaire, "This food has always smelt unpleasant", to allow for those where there was no change in hedonic rating because the item had always been perceived as unpleasant. This option was used in 160 out of 6447 observations, and these data were excluded on the grounds that there had been no change in hedonic valence due to parosmia, but we suspect that this answer may have been misinterpreted by many of the respondents. Lastly, respondents were asked to record the strength of the smell now in comparison to before smell loss: weaker than before (score = 1), same as before (2), or stronger than before (3) with the additional option to report that the intensity fluctuated (not associated with a score).

Faecal odour had been highlighted in a previous publication based on social media [8] as being less unpleasant and more tolerable for some, whilst for others, there was a switch in hedonic valence from repulsive to pleasant. This was explored further by asking about the distortion of faecal odour on a three-category scale (same as before, distorted, or cannot smell) and asking about the hedonic quality on a two-category scale (no longer unpleasant, just as unpleasant as before).

Section 6 of the survey involved a check-all-that-apply (CATA) question covering an additional 20 possible triggers selected to cover a wide range of food, drink and some environmental or personal care items, with the opportunity to add further triggers as free text. The final question gave the respondents the opportunity to add any further comments.

2.3. Data Analysis

Data relating to respondents' demographics, aetiology, onset, and recovery were expressed as total count (n) and proportion (%). To investigate associations between the different aetiologies and onset, partial recovery and frequency of fluctuation, contingency tables were prepared on the counts and analysed using Fisher's Exact test ($\alpha = 0.05$). To determine whether there were significant differences in taste loss between respondents that had suffered COVID-19 versus other viral infections, the count data were similarly analysed by Fisher's Exact test ($\alpha = 0.05$).

The Kruskal–Wallis two-tailed test with multiple pairwise comparisons with a Bonferroni correction was used to determine whether disgust was significantly different between the set triggers. The Kruskal–Wallis two-tailed test with multiple pairwise comparisons using Dunn's procedure was used to determine whether strength was significantly different between the set triggers (from Section 5). Kruskal–Wallis and Dunn's procedure were similarly used to determine whether there was a relationship between distortion and both hedonic valence and strength and between strength and hedonic valence. Statistical significance was considered at the 5% level ($p \leq 0.05$).

Descriptions of distorted food items (including faecal odour) were cleaned, and words were spell-checked with Hunspell using a large English dictionary [11]. For the word clouds, single word or compound adjectives were extracted from the descriptions and qualifiers suggesting qualitative changes (weaker, stronger, faint, etc.) were removed. Obvious synonyms were combined (e.g., gasoline/petrol, garbage/trash/bin, toxic/poisonous, poo/poop/faeces/feces/faecal/fecal odour, synthetic/artificial, cat food/dog food) and words with the same root were combined under one term (e.g., chemical/chemically, earth/earthy, burnt/burning rotten/rotting, but sick and sickly for example were deemed to relate to different smells). The frequencies of the words reported for each trigger were calculated, and words where the frequency per item was never more than 1 were removed. This was carried out for the 14 set triggers as well as for the answers to the question on faecal odour. Words for each were visually represented in word clouds with the size representing the frequency using ggwordcloud [12]. Descriptions of distortions next underwent sentiment analysis using the sentimentR package [13] for each item, were averaged and then compared using ANOVA with post hoc analysis and Tukey's honest significant difference (HSD).

The 120 words were further split into descriptive words where there was a true description of aroma character (79), hedonic words where there was a clear valence attributed to the word (30), and the remaining words (11) (e.g., indescribable, different, funky, unusual). Principal component analysis (PCA) using covariance was carried out on the frequencies of the descriptive words and on the frequencies of the hedonic words.

Manual counts of the items mentioned in the free text were performed in order to identify the most frequently reported triggers. All those items previously assessed in either the set triggers (14) or the CATA (20) were disregarded, as were complex dishes that contained a number of potential triggers (e.g., curry, falafel, pasta sauce, baked beans), and the focus was on simple ingredients and personal care, home care, or environmental odours. These were counted in a word search, using the word or the root of the word where the word was commonly misspelt or had regional variations. However, each incidence was verified since the contextual significance in which these words were mentioned varied: many people chose to tell us which items were not distorted, or which items came back first, or some words were used as descriptors for others (e.g., coffee smells like bleach, so bleach was disregarded on that occasion).

3. Results

3.1. Demographic Characteristics

The questionnaire was started by 945 people, 17 preliminary practice runs were removed, as were a further 201 non-completers, leaving 727 respondents who completed the whole survey (78%). All demographic data are shown in Table 1. We note that the demographics are skewed towards white (87%), females (90%), and those living predominantly in the UK (45%) or the USA (41%). These demographics reflect those of the AbScent Facebook groups, which were the major source for recruitment, where the proportion of women responding to the survey reported in [14] was 76%, and 76% were residents in either the UK or the USA.

3.2. Origin and Progression of Olfactory Dysfunction

3.2.1. Aetiology

In this study, 92% of the cases of smell loss were attributed to a viral infection which is consistent with data reported in a similar but larger self-selecting cross-sectional study carried out at a similar time showing that for those respondents with parosmia, 89% were post-viral cases [15]. Table 1 shows 83% of the respondents had lost their sense of smell due to COVID-19, whereas only 9% had lost their sense of smell from non-COVID-19 infections. This is not surprising as there has been a surge in cases of post-COVID-19 olfactory dysfunction since the start of the pandemic in January 2020.

Table 1. Demographics and Aetiology of Parosmia.

Statistic		Count	Percent
Total Respondents		727	
Sex	Male	76	10%
	Female	651	90%
Age	Range (years)	18–75	
	Mean (years)	43	
Country of Residence	UK	330	45%
	USA	297	41%
	Canada	15	2%
	Spain	9	1%
	The Netherlands	8	1%
	Others (<1%)	68	9%
Ethnicity	White European	396	54%
	White North American	205	28%
	White Other	30	4%
	Other Ethnic Group	20	3%
	White South American	17	2%
	South Asian	11	2%
	Prefer not to say	10	1%
	Others (<1%)	38	5%
Smoking status	Smoker	52	7%
	Non-smoker	531	73%
	Ex-smoker	144	20%
Aetiology	COVID-19 (diagnosed)	367	50%
	COVID-19 (self-diagnosed)	239	33%
	Viral non-COVID-19 pre-Dec 2019	58	8%
	Viral non-COVID-19 post-Dec 2019	5	1%
	Accident, head/brain injury	14	2%
	Unexplained (idiopathic)	12	2%
	Other	7	1%
	Do not know	25	3%
Severity of COVID-19	Loss of smell only	66	11%
	Mild	194	32%
	Moderate	226	37%
	Severe	117	19%
	Very severe (hospitalised)	3	0.5%

3.2.2. Timings of Smell Loss

Most of the respondents (98%) reported the onset of distortions within the past 10 years and 88% within the last 2 years. Most (95%) experienced the onset of parosmia less than 6 months after their initial loss of smell with a mean time of 4.4 months and a median time of 2.8 months, similar to other pre- or post-COVID-19 cases ([16,17], respectively) and consistent with the peak survey response rate in August 2020, just 4 months after the peak in cases of COVID-19 in the UK. The timescale for the remaining 5% spread between 6 months and 24 years.

In all post-viral cases, about half of the respondents (56%) reported their loss of sense of smell as concomitant with other symptoms. However, 21% of post-COVID-19 respondents lost their sense of smell very suddenly preceding onset of other symptoms (Table 2), consistent with data from Borsetto et al. [18], who reported in a systematic review that typically 20% of post-COVID-19 cases experienced a loss of sense of smell as the first symptom. This was rarely the case (2%) for non-COVID-19 post-viral respondents. The use of Fisher's exact test, which takes into account groups of different sizes, showed this difference was significant at $p < 0.0001$. This early onset was noted by Gane et al. [19], who identified the Isolated Sudden Onset of Anosmia as a novel post-COVID-19 syndrome. Although the difference in the size of the two groups does place some limitations on the conclusions we have drawn, the results are fully in line with other studies. Early onset is the reason why it became so important to recognise loss of sense of smell as an official symptom of COVID-19 to minimise further spread of the disease.

Table 2. Timings of Smell Loss for All Post-Viral Cases.

	COVID-19 *n* = 606	**Non-COVID-19** *n* = 63
very suddenly, before the other symptoms of infection appeared	127 (21%)	1 (2%)
very suddenly, during the infection	344 (57%)	29 (46%)
very suddenly, after the infection	69 (11%)	9 (14%)
gradually, I only noticed it was gone when I was recovering from the infection	66 (11%)	24 (28%)

3.2.3. Severity of COVID-19

The majority of post-COVID-19 cases were self-reported as either mild, moderate or asymptomatic, with 19% of respondents reporting severe symptoms and three respondents having been hospitalised (Table 1). The current literature shows that olfactory dysfunction is more prevalent in mild COVID-19, with 86% of patients in a cohort of mild to medium cases reporting olfactory dysfunction [20–22], but this is difficult to assess since smell and taste checks were rarely performed on those hospitalised with severe respiratory conditions.

3.2.4. Intermittent Recovery of Olfactory Function

Most of the post-COVID-19 respondents (82%) reported some recovery of their olfactory function prior to the onset of parosmia (Table 3) compared to 27% of non-COVID-19 post-viral respondents. The use of Fisher's exact test, which takes into account groups of different sizes, showed this difference in pre-parosmia recovery is significantly more prevalent in post-COVID-19 respondents ($p < 0.0001$). However, Borsetto et al. [18] showed the disparity between self-reported olfactory function and objective testing, with examples of significant recovery when objective tests showed little improvement and vice versa. Thus, the terms "full recovery" and "no recovery" need to be treated with caution, but undoubtedly, respondents were acutely aware of changes in their olfactory function. Although these differences in recovery are quite significant, there are limitations in the conclusions drawn arising from the fact that those with non-COVID-19 are not reporting recent changes, leading to a potential memory bias between the two groups.

Table 3. Intermittent Recovery of Olfactory Function.

	COVID-19	**Non-COVID-19**
no recovery of normal sense of smell	108 (18%)	46 (73%)
just a few hints that a sense of smell was returning and nothing else	171 (28%)	11 (17%)
partial recovery of a normal sense of smell	206 (34%)	5 (8%)
full recovery of a normal sense of smell	121 (20%)	1 (2%)

Furthermore, over half of respondents reported that their symptoms fluctuated, as had been highlighted in the thematic analysis of social media posts [8]. Minor or infrequent fluctuations in symptoms were reported significantly more frequently for post-COVID-19 cases ($p < 0.001$), whereas the lack of recovery and lack of fluctuations were both reported significantly more frequently for non-COVID-19 post-viral cases (both $p < 0.0001$).

Those with hints of recovery were significantly more associated with infrequent minor fluctuations ($p < 0.0001$), and those with partial recovery were significantly more associated with any level of fluctuation than no fluctuation ($p < 0.0001$); thus, the partial recovery is likely to be intermittent, as opposed to a stable increase in olfactory function. This notion of partial but sporadic recovery seems to be more prevalent post-COVID-19 and is characterised by fluctuations in the olfactory dysfunction prior to the full onset of parosmia. Such apparently random changes in olfactory (dys)function remain a puzzle for those who are interested in understanding the underlying mechanisms of parosmia as it is difficult to rationalise with the prevailing theory of widespread destruction and slow (misguided) regeneration of olfactory sensory neurons [16,17,23].

3.2.5. Changes in Taste and Chemesthesis

Parma et al. [24] demonstrated that loss of smell, taste, and trigeminal sensation were all compromised post-COVID-19 with self-reported decreases of 80, 69, and 37%, respectively within 2 weeks of a respiratory illness. However, our data, which, on average, were collected 3 months after smell loss, show very little evidence of (residual) loss of the taste of sugar, salt, and the heat in spices in post-COVID-19 respondents with only 3.5%, 1.3%, and 9.6% of cases, respectively, reporting a loss of these senses. This concurs with previous literature on smell and taste loss post-COVID-19, which shows that post-COVID-19 ageusia improves in most cases after 10 days [25] and taste often improves, whilst olfaction does not [26]. However, taste was weaker for 31% of post-COVID-19 respondents for sweet and 27% for salt. This may be partial recovery or due to the unfamiliarity of experiencing those sensations without the co-presence of smell. The recovery of taste and trigeminal sensation is good news for those struggling to eat, enabling them to take an interest and explore a greater variety in the gustatory and somatosensory properties of foods. However, 45% of all respondents reported the presence of a metallic taste in the mouth, 31% reported burning nasal passages, and 14% reported a burning sensation in the mouth. No significant association was found ($p > 0.05$) for any of these tastes or trigeminal sensations with non-COVID-19 or post-COVID-19 aetiologies. Such distortions of the gustatory and trigeminal senses reinforce the aversion to many foods created by parosmic distortions, further restricting food selection and adversely affecting the eating behaviour and nutritional quality of the diet.

3.3. Distortion, Disgust, and Strength of Triggers Foods

3.3.1. Key Trigger Foods

Of the 14 set triggers in the questionnaire, coffee, meat, and onion were the most frequently distorted and least likely to be undetected or normal amongst those who were familiar with the items (Table 4).

Table 4. Frequency of Food Items being Reported as (a) Distorted, (b) Not Detected, or (c) Normal.

Item	Reported as Distorted Freq (%) [1]	Item	Reported as Not Detected Freq (%) [1]	Item	Reported as Normal Freq (%) [1]
Butter	18	Coffee	7	Meat	11
Apple	23	Onion	18	Coffee	11
Rose	25	Meat	18	Onion	12
Cucumber	29	Fried foods	18	Egg	18
Melon	32	Bacon	21	Fried foods	20
Peppers	36	Peanuts	24	Bacon	22
Chocolate	43	Chocolate	27	Peanuts	26
Peanuts	50	Egg	29	Chocolate	30
Egg	53	Cucumber	30	Peppers	31
Bacon	57	Peppers	33	Melon	34
Fried foods	61	Melon	33	Rose	36
Onion	70	Apple	36	Butter	39
Meat	71	Rose	39	Cucumber	40
Coffee	82	Butter	44	Apple	41

[1] Percentage frequency excluding those who are not familiar with or do not consume the item.

3.3.2. Characterisation of Distortions

The three words used most frequently to describe the distortions were "rotten"/"rotting", "chemical", and "burnt" used in 14, 11, and 7% of all descriptions. "Burnt" was used more frequently for coffee, whereas "rotten"/"rotting" were used more frequently to describe meat and onions. "Chemical", which was used across the range of trigger foods, is often a catch-all term discouraged in sensory profiling as, without further definition, it can relate to any number of chemical odours and tends to cover a range of otherwise unidentifiable odours. "Sweet" was the next most frequent (5%) and was the term most frequently reported for faecal odour. The words "indescribable"/"cannot describe" were used 94 times, reflecting the difficulty respondents had in finding appropriate words for smells they consider novel. Indeed, given that most of the support provided for those with parosmia is through online support groups, the selection of some of these descriptors may have been influenced by online discussions. The word frequencies are represented in the word clouds in Figure A1. It is worth noting that many of these terms offered by respondents seem to be chosen for their negative hedonic value ("rotting") rather than for their a priori resemblance to the actual descriptor. Comments about the indescribability of these parosmic odours support the argument that, often, descriptors used are shorthand for the level of disgust felt. Indeed, previous work has shown that disgust is the highest emotion expressed in descriptions of parosmic triggers [15].

3.3.3. Valence of Distortions

The overall hedonic ratings for all 14 set trigger foods that were described as distorted are shown in Figure 1. In total, 84% of all hedonic responses were unpleasant or gag-inducing, and 16% were rated as pleasant or neutral. Although coffee is most frequently reported as distorted, meat and onions had significantly higher scores for disgust than coffee ($p = 0.007$ and 0.009 respectively). Rose had a significantly lower mean disgust score than any of the other items (p ranges from 0.025 (apple) to 0.0001), with about half of those finding it distorted, scoring it as pleasant or neutral.

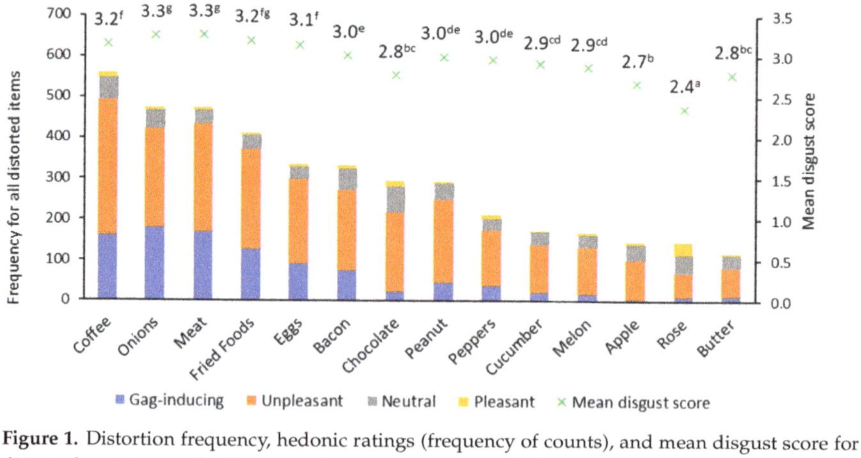

Figure 1. Distortion frequency, hedonic ratings (frequency of counts), and mean disgust score for distorted set triggers. For the mean disgust score, items with the same letters in the superscript are not significantly different from each other using Kruskal–Wallis with Bonferroni correction.

3.3.4. Strength of Distortions

Coffee, egg, fried food, onion, and meat were rated as significantly stronger than all other triggers except bacon (which was scored separately from meat, originally based on the different molecular composition of its aroma) ($p < 0.001$). Distorted smells were significantly more unpleasant and significantly stronger than non-distorted smells (both $p < 0.0001$). The strongest aromas were significantly less liked (or more disgusting) overall ($p < 0.0001$).

3.3.5. Faecal Odour

Over half of the respondents (60%) reported distortion of faecal odour, whilst 34% reported no smell at all (similar order of magnitude to those who could not smell apple or rose), and only 6% reported no change to faecal odour. Of those that could perceive faecal odour (66%), 30% reported it as not unpleasant anymore.

The relationships between the 14 main triggers and faecal odour (denoted as "poo") were investigated further using PCA. In Figure 2A, the PCA carried out on just the descriptive word frequencies shows a clear separation along PC1 of the common triggers (coffee, fried foods, poo, bacon, meat, onion) from the less common triggers. They are all associated with the descriptor "chemical" (as also shown in the word clouds in Figure A1), but coffee is separated from meat and onions on PC2, being associated with burnt notes, whereas meat and onion are associated with rotting/rotten notes, and fried foods and poo are positioned between the two providing elements of both. In Figure 2B, which shows a PCA based on just the hedonic word frequencies, the separation is quite different with both coffee and onion associated with PC1 and words such as "disgusting", "horrible", "gross" and "unpleasant" confirming the strong negative valence. Meat, egg, and fried foods were closer to the origin, and the other less frequent triggers were at the negative extreme of PC1. Faecal odour ("poo"), however, is separated on PC2, associated with terms such as "weird", "less unpleasant", and "not bad" (and "better" and "pleasant" not shown), which have significant components on PC2.

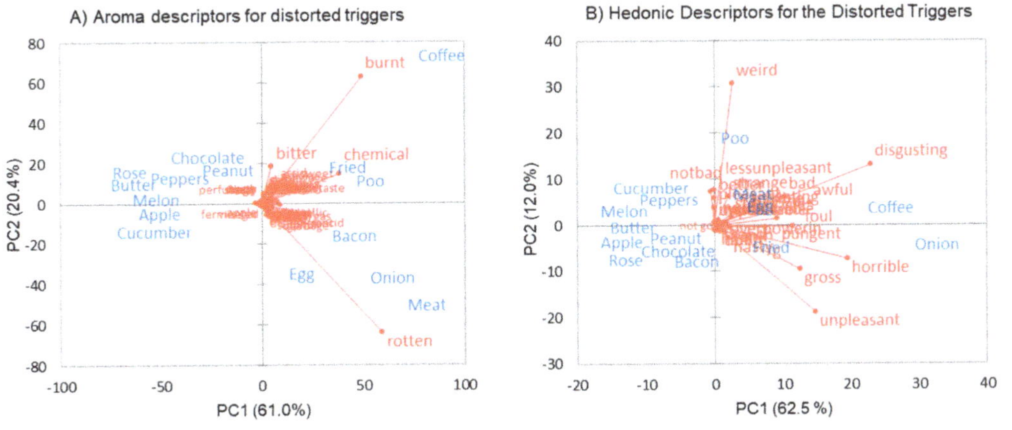

Figure 2. Frequencies of words used to describe the set triggers and faecal odour; PC1 vs. PC2 (**A**) = aroma descriptors, (**B**) = hedonic descriptors. Blue denotes set triggers; red denotes descriptors. Raw data are provided in Supplementary Table S1.

This reverse change in valence has been discussed before [8,15], but here, we can show some objective evidence from a larger survey where we have directly asked respondents to rate the distortion and disgust of faecal odour alongside common food triggers. Sentiment analysis of the descriptors generated for the distortions is presented in Figure 3. Faecal odour ranks as more positive than any of the other 14 triggers, akin to the sentiments expressed for apple, melon, and chocolate, whereas onion, meat, and coffee rank as the most negative.

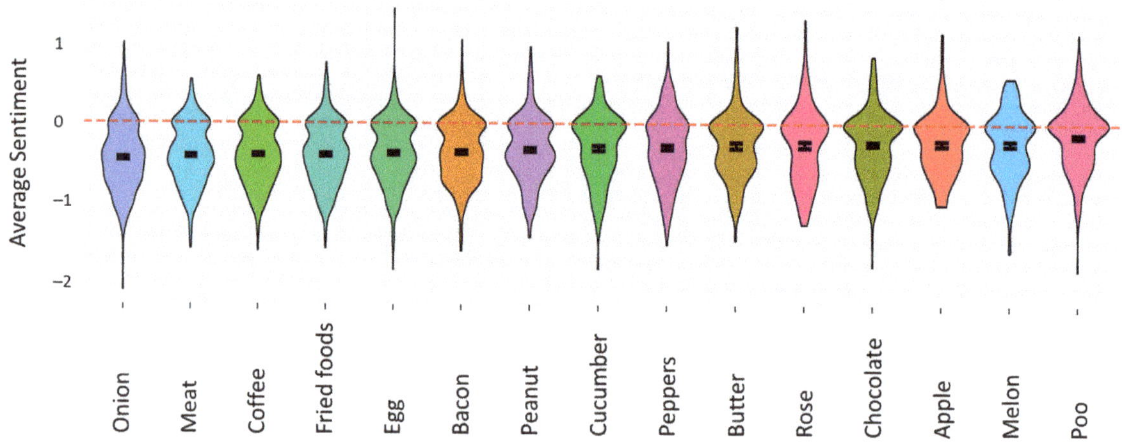

Figure 3. Sentiment analysis of the 14 set triggers and faecal odour. Distribution of sentiment for triggers is ordered from low to high sentiment. All triggers are below neutral sentiment (zero or red dashed line), with some (left side) being more negative than others (right side). Error bars represent standard errors.

3.3.6. Additional Triggers of Distortion (CATA)

The data from the CATA question showed that the 20 pre-selected items were each identified as triggers at least 80 times, so even those that had been selected as safe foods were reported as distorted in at least 10% of cases. These data, which were combined with the data from the 14 set triggers, and the frequency of distortions (Table 5), demonstrate that triggers are by no means universal, and therefore, our understanding of food consumption by those with parosmia cannot be considered straightforward. Unsurprisingly, garlic ranks high up the list with onion, but room freshener, cola drinks, and petrol are also high on the list, indicating it is not just the aroma molecules present in food (particularly cooked foods) that trigger the distortions. Although specific aroma compounds generated via the Maillard reaction during cooking are known to consistently trigger distortions in those suffering from parosmia [8], Table 5 suggests that there are other trigger molecules to be found. The bottom of the table is dominated by fresh fruit and nuts, indicating that for many, these may be safe foods.

Table 5. Frequency of Items being Reported as Distorted (20 from the CATA question with the 14 preset triggers in italics).

Distorted Aromas	Count [1]	Distorted Aromas (Contd)	Count [1]	Distorted Aromas (Contd)	Count [1]
Coffee	570	*Peanuts*	301	Celery	142
Garlic	496	Bananas	289	Carrots	131
Onion	495	Popcorn	288	Peaches	130
Meat	490	Cigarettes	277	Vanilla	130
Room freshener	435	*Peppers*	217	Mango	122
Fried foods	424	Tomatoes	204	Hazelnuts	122
Cola drinks	356	*Cucumber*	178	*Butter*	120
Egg	350	*Melon*	171	Walnuts	115
Bacon	340	*Rose*	148	Grapefruit	112
Toast	336	*Apple*	147	Passionfruit	84
Petrol	318	Raspberries	146	Honey	80
Chocolate	302				

[1] Count = number of times scored as distorted.

3.3.7. Additional Triggers (Free Text)

Analysis of the free choice answers reported well over 220 additional triggers, almost to the point that no food remained unmentioned. These were counted, and the 87 most frequently reported are listed in Table A1. What was clear from this section was the impact on people's experience of non-food items, particularly those related to daily personal care activities such as showering and oral care. Water was mentioned as a trigger by 44 people and air by 14. For 41, who reported that everything was distorted, there was no respite from parosmia.

3.4. Analysis of the Free Text Question

Free text answers to the survey question "Do you have anything else you would like to tell us" gave us more detail of the impact of living with parosmia. Several themes reccurred, and a selection of poignant comments is quoted below. A number of comments related to the difficulty in describing the experience of parosmia were found in the free text:

> *"It is really hard to find words to describe the new smells. I talked to my doctor three times since March, and his suggestions underlines the need for more knowledge and awareness. He answered me as if he really did not understand that the smells are different now . . . ".*

Several comments referred to "the COVID smell" or "Parosmia smell". At onset, this seemed to be one single smell, and then with time, this experience diverged into two or more different "COVID smells" that could be roughly grouped together based on food type:

> *"I only sense one unpleasant smell. Everything that has distorted smell, smells almost the same".*

> *"I seem to have two types of distortion 'categories'—coffee, chocolate, onions etc taste like a musky, nutty, rancid, earthy taste. And things like peppers and melon taste more chemical, like something that would be flammable".*

Consternation over the perceptions of disgust in the face of food was a common theme in the free text. This also extended to body odours, with a reversal of valence. What once smelled good (food) began to smell like body waste, and body waste became less offensive. This was hard for respondents to rationalise:

> *"For me, my feaces [sic], urine and sweat have the same bad smell like the other distorted smells (like onion and fried meat). The bad smell is not familiar to me from before and it messes my head that my food smells like my body waste".*

Burning nose and throat, and particularly nasty (metallic) tastes, were described in more detail:

> *"Shampoos, cleaning products, soaps, perfumes, etc.,smell so strong I feel like my nose burns".*

> *"I have a weird constant taste in my mouth which matches the weird smell I keep smelling in everything".*

A burning mouth has previously been associated with low body weight [27] and higher olfactory thresholds [28]. With aroma either absent or distorted, these additional sensations compound the problems associated with eating. With so many foods unbearably distorted, some commented on how their diet had changed:

> *"My diet is quite limited at present so there may be other foods that smell distorted but I haven't tried to eat them in the last week".*

> *"I am on a strict diet of fresh corn [. . .], apricots, peaches, plums, grapes, cherries, cucumber, Fairlife protein shakes, diet Dr Pepper (coke and other sodas taste like pure dirt/mud), fresh mozzarella (all other hard cheeses, especially yellow cheddar, are rancid)".*

As a result, those with long term parosmia may be subject to weight gain or weight loss depending on the severity of their symptoms and the range of foods that are tolerable:

"Weight is increasing as the only tolerable foods are all sugary, stodgy, high fat, high calorie".

"I am very sad and losing weight it feels worse every day that passes".

Fluctuating symptoms were one of the characteristics of post-COVID-19 olfactory dysfunction. Respondents to the free text question reported that fluctuations in symptoms caused considerable anxiety:

"Distortion of smells fluctuates. Sometimes I can smell something and it smells normal, then I smell it again and it is distorted or I can't smell it at all".

Another observation made by a number of respondents was that a "parosmic" smell, smelled once, might linger long after exposure to the item for hours or longer. This olfactory perseveration seems to suggest a malfunction of the attenuation system. It was not interrogated in the questionnaire but has been observed in social media threads for those with olfactory disorders [8].

"The distorted odours fluctuate in strength but are never totally absent. Sometimes they linger even after the source is removed, and the memory of the odour alone can be enough to make me conscious of it, as though I can smell it, even if there is no odour source present".

Additionally, possibly associated with a malfunction of the attenuation system, "fleeting whiffs" were experienced. Unlike the "smell lock", this was a quick flash of a perception that then disappeared, leaving people frustrated.

"I still smell very little but sometimes have an initial weak or distorted 'whiff' of an item but on a second sniff I cannot smell anything. This has been happening since about 3 months after losing sense of smell".

Five respondents reported experiencing parosmia without associated anosmia or hyposmia.

"I never lost my sense of smell. Parosmia came on suddenly. There was no absence of smell transition as implied in many questions".

Other studies have looked in detail at the emotional impact of parosmia [3,8,15]. The following quotes demonstrate the severity of the impact on quality of life:

"Parosmia really affects mental health. Cannot eat out or socialise. . . . It is the weirdest thing but nobody believes you if you try to explain it. Cannot face the not knowing of how long it might last".

"This needs to be over, it's ruining my life and not worth getting up for this is severely effecting my way of life and it's nice to know others care".

4. Discussion

The list of food ingredients that trigger distortions is long, almost to the point that no food is unmentioned. However, there are clear trends: The worst items are coffee, onion, and meat (worst being loosely defined as a combination of how frequently they were detected, how frequently they were distorted, how frequently they invoked disgust, and how intensely they were perceived). We know that coffee, roasted meat and fried foods have many molecules in common that trigger distortions and have similar formation pathways [10]. Certain pyrazines and certain sulphur-compounds which are formed in the Maillard reaction during the processing/cooking of food have been shown to elicit a "parosmia-like" smell. Since these Maillard reaction products are formed during roasting, frying, grilling, or baking of almost any food, those sensitive to coffee, meat and fried foods are likely to find many cooked foods unpalatable. Milder cooking processes may mitigate the distortions to a certain extent. Additionally, many recipes include onion and/or garlic which contain a different volatile profile to those above, producing thiols and disulphides that are also likely to trigger distortions. Coffee, onions, garlic, fried foods, eggs, and (in a non-vegetarian lifestyle) meat constitute a major part of a typical weekly diet, certainly in western cultures, and it is clear how distortions and sense of disgust in these key items could have a serious impact on diet and nutrition.

Changes in diet with olfactory dysfunction have been discussed by several authors [29–32] and discussed in detail by Chaaban et al. [33]. The relationship is complex depending on the nature, duration, and aetiology of the olfactory dysfunction [29] and can lead to the adoption of both healthier diets of better nutritional quality and diets high in sugar, fat, and salt. More recently, a paper addressing olfactory dysfunction as a result of COVID-19 showed a tendency for the diets to be higher in energy, fat, and sugar [34], but this is not necessarily the case for those with parosmia. The literature on the impact of parosmia on diet is scarce. Burges Watson et al. [3] reported a shift in appetite and intake in both directions: those with olfactory loss having a tendency for high energy diets and increased intake in search of the hedonic pleasure normally associated with food, but for those with parosmia, there was a tendency to avoid eating, leading to dramatic weight loss and further impacts on quality of life and mental health. The fact that key proteins such as meat (including bacon) and eggs are such strong triggers can result in a poor low-protein diet for those with parosmia unless suitable alternatives are sought. For some, safe foods are the less frequent triggers, such as fresh fruit and vegetables. However, safe foods vary from person to person and for some, diets consisting of relatively safe foods mean a diet of "plain potatoes, yogurt and cheese" or "bread, cheese, chips and cake" as reported in [8], resulting in weight gain. The added impact of a continuous metallic taste in the mouth, nose burn, and throat burn only exacerbates these problems.

However, triggers extend beyond food and the kitchen to homecare or personal care products and environmental odours, even water and air, contributing relentlessly to the misery experienced by those with parosmia, to such an extent that for some, there is no safe space either inside or outside of the home. The inability to describe the experience contributes to the frustration of having parosmia. Not only do patients suffer a daily onslaught of relentless and disagreeable smells, but they are not able to summon up enough descriptors to engage their doctors and support circles on the subject. In fact, a constant reminder of being unwell through triggers is a major reason individuals with parosmia may suffer more than those with a simple loss [15,35].

There is mounting literature in which the mechanisms by which SARS-CoV-2 may affect olfactory function are discussed [23,36,37], but few address the underlying mechanism of parosmia. Parosmia has sometimes been loosely characterised as a cross-wiring between the regenerating olfactory receptors and the glomeruli or olfactory bulb [23]. However, many have asked why these distortions are predominantly unpleasant, leading some to conjecture that the distortions of smell may be unpleasant because of a violation of expectations of how the odour of a familiar food or household item is usually perceived. However, this study shows that not all distorted and hence unexpected smells are found to be unpleasant. Thus, merely having an unexpected smell is not sufficient to cause disgust. Since the main triggers are experienced as having a disgusting odour, we need to account for what it is about the nature of the distorted odours that makes them disgusting. Analogies are made to the rotten, earthy, burnt, or chemical smells, but many participants in the study talk about a novel odour, or "that parosmia smell". What is noticeable from the results is the high-intensity scores for those items found to be most disgusting; although this can fluctuate, the intensity and the disgust are correlated. Whatever way the participants are perceiving distorted odours, when they are found to be disgusting, they are perceived to be intense odours. The question remains as to whether their intensity is part of what makes them disgusting or whether it is because people find the novel or distorted odour to be disgusting that they also find it intense. Another hypothesis is that people with parosmia are becoming especially sensitive to particular molecules in a mixture, which means those compounds stand out and are therefore perceived as intense, or that the absence of other aromas that usually mask or round them out leaves the remaining compounds to smell more intense than they would in a mixture.

The reversal in valence described for faecal matter/body waste is disturbing. It has been explained by us previously [10] in terms of molecular triggers. Limited studies demonstrated that those with parosmia did not perceive the normally overpowering smell of indole, skatole, or cresol in faecal matter and therefore perceived only the aroma compounds which are normally masked—these could be pleasant aroma molecules or other trigger molecules.

It is also fascinating to note that some of the distorted items were still rated as pleasant despite the distortion: rose (50% of the time), apple (31%), and butter (29%). In cases where rose was rated as pleasant, it was rated as weaker in 50% of responses and only stronger in 6% of cases. In such cases, it may be that as a result of only partial regeneration of the olfactory sensory neurons, and consequently higher odour thresholds, most of the constituent aroma compounds are present at concentrations closer to the threshold than usual, with some dropping below the threshold, changing the balance and providing the distortion. This may be an instance of incomplete odour characterisation, as proposed by Leopold [35], where there is an imbalance in the aroma profile, which leads to the perceived distortion, but no single molecules trigger the switch in valence.

Rose may be an example of what can be termed "eusomia", which has in the past been used to describe distortions which are pleasant, as opposed to cacosmia where olfactory distortions are negatively perceived [4]. Indeed, 56 of the 727 participants reported distorted items but no change in valence in any of them. Currently, the working definition of parosmia is generally inclusive of both eusomia and cacosmia, describing the twisting and warping of the sense of smell, without mention of the hedonic aspect [4], but there may be instances where the use of more precise terminology could be useful.

5. Conclusions

Here, we show that post-infectious olfactory dysfunction leads to significant distortion and hedonic change in key food items, predominantly but not exclusively towards a negative valence. Such intense distortions, the associated change in valence, loss of expected pleasure, and the presence of strange tastes and burning sensations certainly lead to changes in eating behaviours and serious longer-term consequences for mental health and quality of life. It remains to be seen whether there are any changes in the prevalence and trajectory of parosmia arising from the newer variants of COVID-19.

In subsequent work, we have looked in detail at the individual molecules which trigger distortions for a wide range of food [10], and we discuss in more detail how this can be rationalised with prevailing mechanistic theories. Further work is underway to test our hypothesis that some distortions, particularly those that do not elicit a sense of disgust, may be due to incomplete odour characterisation, as initially proposed by Leopold [35], whereas those that are associated with severe disgust are triggered by individual molecules. Future work should be targeted at understanding how individual molecules relay such aversions to the integrative centres in the brain.

Supplementary Materials: The following supporting information can be downloaded at: https://www.mdpi.com/article/10.3390/foods11070967/s1, Table S1: Descriptors for 14 Set Triggers.

Author Contributions: Conceptualisation, J.K.P., L.M., C.E.K., S.G. and B.C.S.; data curation, L.M., J.K.P. and C.E.K.; formal analysis, L.M. and R.P.; interpretation, J.K.P., R.P., B.C.S., C.E.K. and S.G.; writing—original draft preparation, J.K.P.; writing—review and editing, J.K.P., L.M., R.P., B.C.S., C.E.K. and S.G. All authors have read and agreed to the published version of the manuscript.

Funding: This research received no external funding.

Institutional Review Board Statement: The study was conducted in accordance with the Declaration of Helsinki and the protocol was approved by the School of Chemistry, Food and Pharmacy Research ethics committee of the University of Reading on 10 June 2020 (study number 29.2020).

Informed Consent Statement: Informed consent was obtained from all subjects involved in the study.

Data Availability Statement: The data presented in this study are openly available in the University of Reading Research Data Archive at http://doi.org/10.17864/1947.000367, last accessed on 15 February 2022.

Acknowledgments: The authors would like to thank Anne Hasted of QI Statistics for helpful guidance on PCA.

Conflicts of Interest: The authors declare no conflict of interest.

Appendix A

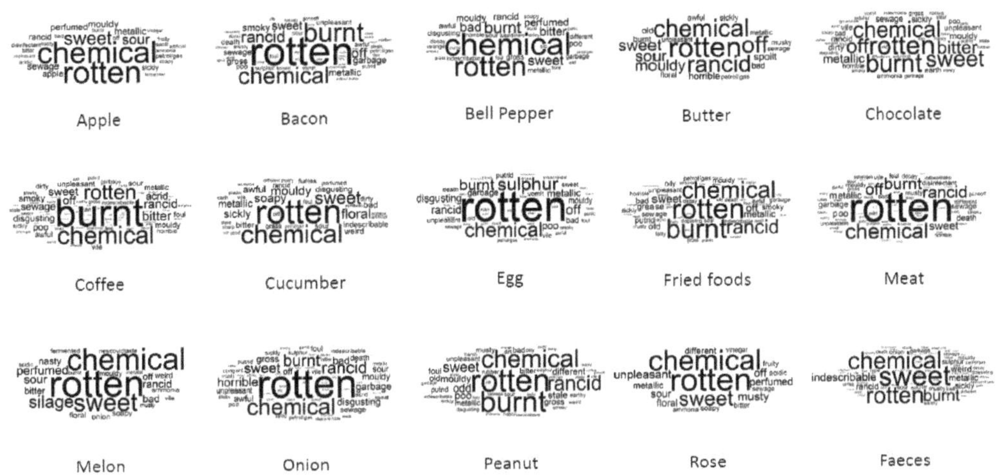

Figure A1. Word clouds representing words used to describe the 14 set triggers and faecal odour. Raw data are provided in Supplementary Table S1.

Table A1. List of triggers found in personal care and homecare products, environment, foods, and beverages.

Other Distorted Smells Personal Care	Count [1]	Other Distorted Smells Home and Environment	Count [1]	Other Distorted Smells Foods and Beverages	Count [1]
Daily washing/ grooming (445)		*Home cleaning (223)*		*Herbs (119)*	
soap	132	cleaning products	110	mint	81
deodorants/antiperspirant	106	detergents	75	herbs	15
shower gel	91	fabric softner	38	basil	12
personal care products	63	*Sanitisers (98)*		rosemary	11
cosmetics and toiletries	35	sanitisers	46	*Fruit (114)*	
nail care	12	bleach	39	citrus (lemon, orange, lime)	83
sun cream	6	wipes	13	strawberries	31
Personal Fragrances (272)		*Environment (96)*		*Alcoholic beverages (103)*	
perfume	223	water	44	wine	63
candles	30	air	14	beer	22
aftershave	19	garbage	10	rubbing alcohol	8
Hair care (247)		chlorine	9	gin	7
shampoo	178	coast	8	cider	3
conditioner	44	rain/petrichor	6	*Carbohydrates (101)*	
hair products	25	manure	4	bread	46
Oral care (146)		tarmac	1	cookies biscuits	15
toothpaste	133	*Essential oils (95)*		baked goods	14
mouthwash	13	other essential oils	39	rice, rice cakes	12
Body aromas (132)		eucalyptus/Vicks	27	pasta	11
other body odour	76	lavender	27	tortilla	3
urine	44	rose essential oil	2	*Dairy (55)*	
breath	7	*Garden (90)*		cheese	24
new born baby/husband	5	grass	47	tea	15

Table A1. *Cont.*

Other Distorted Smells Personal Care	**Count** [1]	**Other Distorted Smells Home and Environment**	**Count** [1]	**Other Distorted Smells Foods and Beverages**	**Count** [1]
		flowers	20	milk	11
		plants, leaves	12	dairy	5
		soil, earth	6	***Vegetables and pulses (54)***	
		trees	5	brassica	21
		Miscellaneous (58)		salad	15
		everything	41	peas and beans, pinto	14
		marijuana	17	chickpeas	2
		Car (43)		soya	2
		petrol/diesel fumes	42	***Spices (48)***	
		plastic/interior	1	coriander, cilantro	19
		Pets (32)		ginger (beer, tea, lotion)	9
		petfood	16	mustard	5
		pet	13	clove	4
		cat litter	3	cinnamon	4
		Smoke (28)		cumin	4
		smoke	22	paprika	3
		bonfire	6	***Fish (18)***	
		Home (4)		fish, tuna	18
		pens/crayons	2		

[1] Count = number of times scored as distorted.

References

1. Spence, C. Just how much of what we taste derives from the sense of smell? *Flavour* **2015**, *4*, 30. [CrossRef]
2. Croy, I.; Nordin, S.; Hummel, T. Olfactory Disorders and Quality of Life—An Updated Review. *Chem. Senses* **2014**, *39*, 185–194. [CrossRef] [PubMed]
3. Burges Watson, D.L.; Campbell, M.; Hopkins, C.; Smith, B.; Kelly, C.; Deary, V. Altered Smell and Taste: Anosmia, parosmia and the impact of long COVID-19. *PLoS ONE* **2020**, *16*, e0256998. [CrossRef] [PubMed]
4. Hummel, T.; Whitcroft, K.L.; Andrews, P.; Altundags, A.; Cinghi, C.; Costanzo, R.M.; Damm, M.; Frasnelli, J.; Gudziol, H.; Gupta, N.; et al. Position paper on olfactory dysfunction. *Rhinol. Suppl.* **2017**, *54*, 1–30. [CrossRef] [PubMed]
5. Keller, A.; Malaspina, D. Hidden consequences of olfactory dysfunction: A patient report series. *BMC Ear Nose Throat Disord.* **2013**, *13*, 8. [CrossRef]
6. Hopkins, C.; Surda, P.; Vaira, L.A.; Lechien, J.R.; Safarian, M.; Saussez, S.; Kumar, N. Six month follow-up of self-reported loss of smell during the COVID-19 pandemic. *Rhinology* **2020**, *59*, 26–31. [CrossRef]
7. Olofsson, J.K.; Ekesten, F.; Nordin, S. Smell distortions: Prevalence, longevity and impact of parosmia in a population-based, longitudinal study spanning 10 years. *PsyArXiv* **2021**. [CrossRef]
8. Parker, J.K.; Kelly, C.E.; Kirkwood, A.F.; Smith, B.C.; Hopkins, C.; Gane, S. Patients' Perspectives on Qualitative Olfactory Dysfunction: Thematic Analysis of Social Media Posts. *JMIR Form. Res.* **2021**, *5*, e29086. [CrossRef]
9. Teaima, A.A.; Salem, O.M.; Teama, M.; Mansour, O.I.; Taha, M.S.; Badr, F.M.; Khater, S.S.; Abdou, K.; Mahmoud, M.S. Patterns and clinical outcomes of olfactory and gustatory disorders in six months: Prospective study of 1031 COVID-19 patients. *Am. J. Otolaryngol.* **2022**, *43*, 103259. [CrossRef]
10. Parker, J.K.; Kelly, C.E.; Gane, S.B. Molecular Mechanism of Parosmia. *medRxiv* **2021**. [CrossRef]
11. Ooms, J. Hunspell: High-Performance Stemmer, Tokenizer, and Spell Checker (3.0.1) [Computer Software]. 2020. Available online: https://cran.r-project.org/web/packages/hunspell/index.html (accessed on 15 February 2022).
12. Pennec, E.; Slowikowski, K. Ggwordcloud: A Word Cloud Geom for 'ggplot2', 0.3.0 [Computer Software]. 2018. Available online: https://cran.r-project.org/web/packages/ggwordcloud/vignettes/ggwordcloud.html (accessed on 15 February 2022).
13. Rinker, T.W. Sentimentr: Calculate Text Polarity Sentiment, (2.7.1) [R Package]. 2019. Available online: https://github.com/trinker/sentimentr (accessed on 15 February 2022).
14. Koyama, S.; Ueha, R.; Kondo, K. Loss of Smell and Taste in Patients with Suspected COVID-19: Analyses of Patients' Reports on Social Media. *J. Med. Internet Res.* **2021**, *23*, e26459. [CrossRef] [PubMed]
15. Pellegrino, R.; Mainland, J.D.; Kelly, C.E.; Parker, J.K.; Hummel, T. Prevalence and Correlates of Parosmia and Phantosmia among Smell Disorders. *Chem. Senses* **2021**, *46*, bjab046. [CrossRef] [PubMed]
16. Hong, S.-C.; Holbrook, E.H.; Leopold, D.A.; Hummel, T. Distorted olfactory perception: A systematic review. *Acta Oto-Laryngol.* **2012**, *132*, S27–S31. [CrossRef] [PubMed]
17. Vaira, L.A.; Hopkins, C.; Sandison, A.; Manca, A.; Machouchas, N.; Turilli, D.; Lechien, J.R.; Barillari, M.R.; Salzano, G.; Cossu, A.; et al. Olfactory epithelium histopathological findings in long-term coronavirus disease 2019 related anosmia. *J. Laryngol. Otol.* **2020**, *134*, 1123–1127. [CrossRef] [PubMed]
18. Borsetto, D.; Hopkins, C.; Philips, V.; Obholzer, R.; Tirelli, G.; Polesel, J.; Boscolo-Rizzo, P. Self-reported alteration of sense of smell or taste in patients with COVID-19: A systematic review and meta-analysis on 3563 patients. *Rhinology* **2020**, *58*, 430–436. [CrossRef] [PubMed]

19. Gane, S.B.; Kelly, C.; Hopkins, C. Isolated sudden onset anosmia in COVID-19 infection. A novel syndrome? *Rhinology* **2020**, *58*, 299–301. [CrossRef]
20. Lechien, J.R.; Chiesa-Estomba, C.M.; Beckers, E.; Mustin, V.; Ducarme, M.; Journe, F.; Marchant, A.; Jouffe, L.; Barillari, M.R.; Cammaroto, G.; et al. Prevalence and 6-month recovery of olfactory dysfunction: A multicentre study of 1363 COVID-19 patients. *J. Intern. Med.* **2021**, *290*, 451–461. [CrossRef]
21. Lechien, J.R.; Chiesa-Estomba, C.M.; De Siati, D.R.; Horoi, M.; Le Bon, S.D.; Rodriguez, A.; Dequanter, D.; Blecic, S.; El Afia, F.; Distinguin, L.; et al. Olfactory and gustatory dysfunctions as a clinical presentation of mild-to-moderate forms of the coronavirus disease (COVID-19): A multicenter European study. *Eur. Arch. Oto-Rhino-L* **2020**, *277*, 2251–2261. [CrossRef]
22. Yan, C.H.; Faraji, F.; Prajapati, D.P.; Ostrander, B.T.; DeConde, A.S. Self-reported olfactory loss associates with outpatient clinical course in COVID-19. *Int. Forum Allergy Rhinol.* **2020**, *10*, 821–831. [CrossRef]
23. Cooper, K.W.; Brann, D.H.; Farruggia, M.C.; Bhutani, S.; Pellegrino, R.; Tsukahara, T.; Weinreb, C.; Joseph, P.V.; Larson, E.D.; Parma, V.; et al. COVID-19 and the Chemical Senses: Supporting Players Take Center Stage. *Neuron* **2020**, *107*, 219–233. [CrossRef]
24. Parma, V.; Ohla, K.; Veldhuizen, M.G.; Niv, M.Y.; Kelly, C.E.; Bakke, A.J.; Cooper, K.W.; Bouysset, C.; Pirastu, N.; Dibattista, M.; et al. More than smell—COVID-19 is associated with severe impairment of smell, taste, and chemesthesis. *Chem. Senses* **2020**, *45*, 609–622. [CrossRef] [PubMed]
25. Alon, E.E.; Glikson, E.; Shoshani, Y.; Dobriyan, A.; Yahalom, R.; Yakirevitch, A. Six-month smell and taste recovery rates in coronavirus disease 2019 patients: A prospective psychophysical study. *J. Laryngol. Oto.* **2021**, *135*, 448–451. [CrossRef] [PubMed]
26. Ohla, K.; Veldhuizen, M.G.; Green, T.; Hannum, M.E.; Bakke, A.J.; Moein, S.T.; Tognetti, A.; Postma, E.M.; Pellegrino, R.; Hwang, L.-D.; et al. Increasing incidence of parosmia and phantosmia in patients recovering from COVID-19 smell loss. *Rhinology*, 2021; *in press*.
27. Deems, D.A.; Doty, R.L.; Settle, R.G.; Moore-Gillon, V.; Shaman, P.; Mester, A.F.; Kimmelman, C.P.; Brightman, V.J.; Snow, J.B., Jr. Smell and Taste Disorders, A Study of 750 Patients from the University of Pennsylvania Smell and Taste Center. *Arch. Otorhinolaryngol. Head Neck Surg.* **1991**, *117*, 519–528. [CrossRef] [PubMed]
28. Siviero, M.; Teixeira, M.J.; Siqueira, J.T.; Siqueira, S.R. Central mechanisms in burning mouth syndrome involving the olfactory nerve: A preliminary study. *Clinics* **2011**, *66*, 509–512. [CrossRef] [PubMed]
29. Stevenson, R.J.; Mahmut, M.K.; Horstmann, A.; Hummel, T. The Aetiology of Olfactory Dysfunction and Its Relationship to Diet Quality. *Brain Sci.* **2020**, *10*, 769. [CrossRef] [PubMed]
30. Zang, Y.; Han, P.; Burghardt, S.; Knaapila, A.; Schriever, V.; Hummel, T. Influence of olfactory dysfunction on the perception of food. *Eur. Arch. Oto-Rhino-L* **2019**, *276*, 2811–2817. [CrossRef]
31. Kershaw, J.C.; Mattes, R.D. Nutrition and taste and smell dysfunction. *World J. Otorhinolaryngol. Head Neck Surg.* **2018**, *4*, 3–10. [CrossRef]
32. Pellegrino, R.; Hummel, T.; Emrich, R.; Chandra, R.; Turner, J.; Trone, T.; Dorminy, C.; Luckett, C.R. Cultural determinants of food attitudes in anosmic patients. *Appetite* **2020**, *147*, 104563. [CrossRef]
33. Chaaban, N.; Hoier, A.T.Z.B.; Andersen, B.V. A Detailed Characterisation of Appetite, Sensory Perceptional, and Eating-Behavioural Effects of COVID-19: Self-Reports from the Acute and Post-Acute Phase of Disease. *Foods* **2021**, *10*, 892. [CrossRef]
34. Rawal, S.; Duffy, V.B.; Berube, L.; Hayes, J.E.; Kant, A.K.; Li, C.-M.; Graubard, B.I.; Hoffman, H.J. Self-Reported Olfactory Dysfunction and Diet Quality: Findings from the 2011-2014 National Health and Nutrition Examination Survey (NHANES). *Nutrients* **2021**, *13*, 4561. [CrossRef]
35. Leopold, D. Distortion of olfactory perception: Diagnosis and treatment. *Chem. Senses* **2002**, *27*, 611–615. [CrossRef] [PubMed]
36. Doty, R.L. The mechanisms of smell loss after SARS-CoV-2 infection. *Lancet Neurol.* **2021**, *20*, 693–695. [CrossRef]
37. Brann, D.H.; Tsukahara, T.; Weinreb, C.; Lipovsek, M.; Van den Berge, K.; Gong, B.; Chance, R.; Macaulay, I.C.; Chou, H.J.; Fletcher, R.B.; et al. Non-neuronal expression of SARS-CoV-2 entry genes in the olfactory system suggests mechanisms underlying COVID-19-associated anosmia. *Sci. Adv.* **2020**, *6*, eabc5801. [CrossRef] [PubMed]

foods

MDPI

Article

COVID-19 Safety Measures in the Food Service Sector: Consumers' Attitudes and Transparency Perceptions at Three Different Stages of the Pandemic

Heidi Vandenhaute, Xavier Gellynck and Hans De Steur *

Division of Agri-Food Marketing & Chain Management, Department of Agricultural Economics, Ghent University, Coupure Links 653, 9000 Ghent, Belgium; heidi.vandenhaute@ugent.be (H.V.); xavier.gellynck@ugent.be (X.G.)
* Correspondence: hans.desteur@ugent.be

Abstract: The food service sector was among the hardest hit by the COVID-19 pandemic. This study aims to examine consumers' attitudes towards and transparency perceptions of COVID-19-related safety measures and to identify determinants of consumers' intentions and behaviour regarding visiting restaurants and bars once reopened. By also surveying food service businesses, this study allows for comparison between both target groups. A total of 1697 consumers and 780 businesses participated in this study, conducted in Belgium both during and in between waves of infections. The findings demonstrate that consumers evaluated safety measures as important when revisiting restaurants and bars, against business owners' expectations. Both consumers' revisit intentions and behaviours are influenced by the perceived importance of hygiene measures (negatively) and past visit frequency (positively). This study highlights the importance of good compliance with safety measures as a strategy to attract customers during the reopening period. Further, our findings emphasize the importance of transparent communication by food service businesses and the government.

Keywords: COVID-19; Belgium; consumer behaviour; food service sector; safety measures; transparency

Citation: Vandenhaute, H.; Gellynck, X.; De Steur, H. COVID-19 Safety Measures in the Food Service Sector: Consumers' Attitudes and Transparency Perceptions at Three Different Stages of the Pandemic. *Foods* **2022**, *11*, 810. https://doi.org/10.3390/foods11060810

Academic Editor: Derek V. Byrne

Received: 31 January 2022
Accepted: 8 March 2022
Published: 11 March 2022

Publisher's Note: MDPI stays neutral with regard to jurisdictional claims in published maps and institutional affiliations.

1. Introduction

The COVID-19 (COrona VIrus Disease-2019) pandemic left behind a trail of unprecedented consequences on everyday life worldwide. Over 9.8 billion vaccine doses have already been administered [1] and enabled many countries to gradually return to normalcy. However, with the arrival of new variants, the pandemic remains pervasive with over 323.6 million reported cases and 5.5 million confirmed deaths as of 16 January 2022 [2]. To slow down the spread of the virus in order to protect their healthcare systems from overloading, governments worldwide implemented strict health measures, imposing major restrictions on all aspects of daily life. Social distancing measures have been prioritized in many countries and this has necessitated the closure of non-essential services, including schools, offices, restaurants, and hotels [3,4]. To reduce interactions and, thus, virus transmission between people, public gatherings were banned, teleworking became the norm and travelling was suspended [5,6]. As the number of positive cases, hospitalisations, and deaths slowly decreased, mitigation measures were relaxed or lifted, and social bubbles were expanded. Nevertheless, many European countries faced a second, third and even fourth wave of infections, along with waves of gradually tightening and relaxing original measures.

Disease outbreaks have the potential to disrupt existing food systems and create food crises [7,8]. Due to the COVID-19 pandemic, food supply chains were confronted with demand shocks, including changing purchasing and consumption patterns [9,10]. Over the last decades, an increasing number of people depend significantly on out-of-home eating services for their daily dietary intake [11,12]. With the closure of restaurant, bars, schools, and offices, meals prepared and consumed at home have replaced the food service sector, exerting additional pressure on the food retailing sector [9]. While out-of-home food consumers are forced to adopt a less convenient lifestyle, catering services see both their revenues and chances of survival either drop or disappear.

2. Literature Review and Aims

2.1. COVID-19 and (Out-Of-Home) Food Consumption Behaviour

Although it is clear that COVID-19 has impacted food consumption behaviours, preliminary results are diverse and contradictory. Shifts towards both healthier and unhealthier diets were identified during the COVID-19 pandemic and lockdowns. On the one hand, lockdown restrictions led to the adoption of a healthier diet and a reduced consumption of unhealthy foods [10,13]. On the other hand, an increase in 'comfort food' consumption and snacking was reported, leading to a decrease in dietary nutritional quality [14–16]. Several reasons may account for these mixed effects. A balanced and diversified diet can help in maintaining and strengthening immunity, which is essential when dealing with viral threats [13,17]. Moreover, with the closure of restaurants, out-of-home eating was mostly substituted by home-cooked meals, which is typically considered to be a healthier choice [18]. As out-of-home food consumption is associated with a higher intake of energy and fat [11,19], people who used to have more meals out-of-home before the lockdown showed an increased adherence to healthier dietary habits during the lockdown [13]. The increased intake of unhealthy food products can be attributed to panic buying and negative emotions, such as anxiety, stress, and boredom, related to the pandemic and the consequential confinement [10,14,20].

Research on the impact of the COVID-19 pandemic on food purchase behaviour, and takeaway and meal delivery in particular, provides inconclusive results. During lockdown periods, the use of food delivery and takeaway was found to both decrease [10,21,22] and increase [23]. Although there is currently no evidence of COVID-19 transmission through food and food packaging [24], fear for unnecessary exposure might explain why people reduced their frequency of ordering food from restaurants [10]. Perceived risks related to COVID-19 negatively affected the intention to buy food through online food delivery services [25,26]. In contrast, purchase intentions during lockdown were positively influenced by the frequency of online food ordering before COVID-19 [26]. According to Poelman, et al. [23], 30% of those who previously used meal delivery services, did so more frequently during lockdown, especially for meals from local restaurants, with a bias towards highly educated and young consumers. As a higher education level is also associated with eating out-of-home [11,27], the increased use of meal delivery and takeaway might be seen as a means to recreate the restaurant experience at home [23].

It is largely unclear how people's consumption and purchasing behaviour evolved as lockdown restrictions were lifted and countries moved towards a 'new normal'. Undoubtedly, COVID-19 will have changed consumers' out-of-home food consumption behaviour in the short and long term, but research focusing on this topic is still scarce [28]. As dining out implies a setting that involves a large number of people in close proximity to one another for an extended period of time, human interaction, and thus risk of infection, is inherent in visiting food service businesses [28]. Further, high-touch surfaces outside the food preparation areas, e.g., restaurant menus, represent a potential risk of cross-contamination [29,30]. Therefore, many consumers did not feel comfortable and were reluctant to revisit restaurants and bars upon reopening [31–33]. In the United States, for instance, Gursoy and Chi [31] found that more than half of consumers were not willing to revisit food service businesses immediately and of those who already had the opportunity to return, only one in four did. This is consistent with the study by Taylor [32], where 50% of respondents had dined in at a restaurant three weeks after lockdown restrictions were lifted across the US. Consumers' intentions towards dining out were negatively affected by a high-risk perception of COVID-19. The more people are concerned about eating

out-of-home, the more likely they are to avoid it [28], while trust in the ability of restaurants and bars to handle COVID-19 positively affects consumers' intention to visit them during the pandemic [34].

2.2. Safety Measures in the Food Service Sector

As it becomes clear that not everyone will be rushing back to restaurants and bars in the short term, it is of utmost importance that business owners do everything they can to improve consumers' willingness to revisit them. Many countries have published policy documents with regulations for the safe reopening of food service businesses [35]. Safety measures in restaurants and bars reduce the possibility of infection and consumers' perceived risk. By thus ensuring customers' health, customers are motivated to dine out, indicating the importance of safety measures [28,34]. Key safety measures customers expect from food service businesses include visible sanitation efforts, social distancing, limited number of customers, more thorough and frequent cleaning of high-touch surfaces and employee training of health and safety protocols [31].

Social distancing and safety measures can be implemented in different ways, which may elicit different attitudes towards dining out. Research has indicated that consumers prefer restaurants that use partitions to ensure social distancing between different parties [32]. Partitioned restaurants, where physical barriers create individual spaces within a larger room, were considered safer, cleaner, and more sanitary. Similarly, perceived threat of COVID-19 increased preferences for restaurant set-ups with private dining tables or rooms [33]. Both cleanliness and customers' cleanliness perceptions have become increasingly important since the COVID-19 outbreak [30,32]. Perceived restaurant cleanliness has a positive effect on customers' satisfaction, which in turn positively impacts revisit intention [32,36,37], whereas cleanliness of restaurants is a key determinant of consumers' decision to select or return to a restaurant [38,39]. By prioritizing cleanliness, food service businesses may not only ensure the health of customers and employees but also attract customers by meeting their cleanliness expectations [30,32]. Adequate implementation as well as communication of the measures were considered important for consumers to enjoy eating out with lower perceived risk of becoming infected [28,34]. As part of their recovery strategy in response to SARS, restaurants also used cleanliness as a selling proposition [40]. By communicating the safety measures taken to ensure customers' health, perceived risk of dining out is expected to decrease, while customers will feel confident to come back [40].

Food service businesses, among the hardest hit by the COVID-19 pandemic, are confronted with numerous uncertainties while facing difficult circumstances. As lockdown periods of mandatory closure (except for takeaway or delivery) were alternated with periods of reopening under strict conditions, restaurants and bars had to drastically change the way they operate to ensure compliance with the imposed safety and social distancing guidelines. Doing so safeguards the health of both customers and employees and encourages customers to revisit their businesses [31]. However, after having endured months of closure, such safety measures further challenge companies' chances of survival. The enforced measures result in additional expenses, e.g., for disinfecting and protective materials, as well as reduced revenues, e.g., due to limited capacity and imposed curfew. Revenues are further diminished by consumers' reduced demand for restaurant services due to the risk involved and their avoidance of eating out [28,41].

2.3. Transparency of Government and Businesses' Communications

A government's control strategies can only be considered successful when its measures are broadly accepted. Non-compliance of safety measures renders them ineffective. Trust in government is key for effective implementation of policy measures that rely on behaviour [42,43]. Previous studies have demonstrated a positive correlation between adoption of recommended health precautions and trust in authorities. People are more likely to comply with health-related recommendations when they trust their efficacy and the institutions issuing them, along with the latter's competence to contain the pandemic [42–48].

Moreover, the adoption of policy recommendations is related to the communication strategy used. People are more likely to undertake recommended precautionary behaviours if the information communicated by the authorities is perceived clear, consistent, sufficient and helpful [46]. Policy communications should be credible and coherent; measures that are inconsistent, unclear, and open to interpretation will cause confusion and undermine compliance [44,49]. A good level of understanding of the measures and the rationale behind them is positively correlated with higher acceptance and more compliant behaviour [47,49].

In addition, clear and unambiguous health information are considered essential in maintaining trust in authorities. A policy of open and transparent communication, providing all necessary information, leads to more public trust [50]. A positive relationship exists between perceived government transparency and trust in government [51]. The same applies at the level of businesses; the food service sector can rebuild consumers' trust by being transparent, appearing credible and sharing timely, accurate, consistent and reliable information [34,52]. Yost and Cheng [53] suggest the importance of restaurant transparency to regain consumers' trust, which may motivate them to resume dining out during the pandemic. Trust in government is also positively associated with the willingness to engage in prosocial behaviours, e.g., making donations to help those who suffer from the pandemic [43]. Further, solidarity with the food service sector, i.e., visiting bars and restaurants or using online food service applications to protect them from bankruptcy, unemployment and liquidity shortage, influences consumers' visit intention during the COVID-19 pandemic [34,54].

2.4. Aims

It is unclear to what extent food service businesses will (have to) change and adapt to the new reality and how this relates to their customers' attitudes towards safety when dining out. More research is needed to assess how safety measures will influence consumers' dining out intentions and behaviours. For food service businesses, this is crucial to restore the demand for their services that is required to survive and recover from this crisis. Therefore, this study examines attitudes, perceptions, and behaviour from the perspective of both consumers and businesses. At the consumer level, this study aims to evaluate their attitudes towards expected and imposed safety measures and to gain insight into consumers' decisions to either visit restaurants and bars as soon as they reopen or postpone their visits. In addition, by investigating food service businesses' perceptions, this study allows for comparison between both target groups. As such, this study attempts to answer the following research questions: (RQ1a) how important do consumers consider safety measures when revisiting food service businesses?; (RQ1b) how does this relate to businesses' perceptions of safety measures and expectations of their customers' attitudes?; and (RQ2) what are the determinants of consumers' intentions and behaviour regarding out-of-home consumption in (post) pandemic times? For RQ2, following prior research findings, we hypothesised that consumers' attitudes towards sanitation negatively influence revisit intention and behaviour, whereas their past visit frequency has a positive influence.

Furthermore, it is highly relevant to explore how perceptions of transparency of COVID-19 (safety) measures link with attitudes and behaviour towards them and what solidarity intentions and expectations consumers and businesses respectively have. Therefore, this study also investigates the following research questions: (RQ3a) to what extent are consumers willing to financially support food service businesses?; (RQ3b) how does this relate to businesses' expectations of their customers' willingness?; and (RQ4) to what extent do consumers' transparency perceptions of communications by food service businesses and the government correlate with their attitudes? In relation to RQ4, we hypothesised that perceived business and government transparency is positively correlated with consumers' perceptions when dining out and acceptance of policy decisions, respectively.

3. Materials and Methods

3.1. Survey Design

Three different cross-sectional studies were conducted at three different stages of the COVID-19 pandemic. All studies addressed both consumers and the food service sector (restaurants and bars). Three standardized surveys were developed, one per study, and were each divided into two sub-surveys for the different target groups. Although the studies were targeting different stages and measures during the pandemic, all surveys were structured in a similar way.

The consumer questionnaires consisted of three parts. The first part contained behavioural questions regarding out-of-home food consumption and takeaway, before and since COVID-19. Past visit (study 1, 2) and takeaway (study 3) frequencies were measured by recoding a 10-point scale, ranging from 1—'never' to 10—'daily', into frequencies per week. Consumers' intentions to revisit food service businesses were evaluated in the first study, while the second study assessed consumers' revisit behaviours. Revisit intention and behaviour were both recoded into dummy variables (yes/no).

The second part measured attitudes towards and perceived transparency of (safety) measures and decisions issued by the government. Safety measures, aimed at preventing virus transmission during food service visits, were developed through consultation of experts (study 1) and government documents (study 2) [55] in light of the current policy. The measures included items such as "Service is performed with mouth mask"; "No possibility of self-service or buffet" and "Clients can only consume while seated". Attitudes were evaluated based on 5-point importance scales, with values ranging from 1—"not at all important" to 5—"very important". While the first study dealt with 21 safety measures expected to be imposed when food service businesses reopen, the second study focused on 14 actually imposed safety measures, extended with perceived compliance of the measures and perceived safety when revisiting. Imposed measures were slightly different from what was expected a priori. This discrepancy was a result of the growing body of knowledge on the virus and the continuously evolving epidemic situation. Perceived compliance refers to the extent to which bars and restaurants adhere to imposed safety measures, while perceived safety relates to the extent to which consumers felt safe during their visit. Both variables were measured using a self-constructed item on a 5-point Likert scale. The third study focused on consumers' attitudes towards government decisions (5-point Likert scale, "I support the government's decision to close food service businesses") and on consumers' willingness to financially support the sector. Five support actions were evaluated, namely: "Extra use of takeaway/delivery of meals"; "Extra tip when using takeaway/delivery of meals"; "Purchasing vouchers"; "Support crowdfunding campaign" and "Paying extra corona contribution at next visit", on a 5-point willingness scale.

Perceived transparency regarding businesses' (study 2) and government (study 3) communication was assessed by using items from Rawlins [56] and measured on a 5-point Likert scale. The subscale substantial information (7 items, namely: "The information communicated by the business/government about the measures is timely; relevant; consistent; complete; easy to understand; accurate; reliable.") was supplemented by one self-constructed item ("The information communicated by the business/government about the measures explains the rationale."). The questionnaires concluded with a set of profiling variables related to socio-demographic characteristics, such as age, gender, and education level.

The surveys addressing food service business owners were composed of two parts. The first part measured attitudes towards government safety measures, reviewed by experts, on 5-point scales. Businesses evaluated the same set of safety measures as consumers, both expected (study 1) and imposed (study 2), though with the option to indicate "not applicable to my business". The first study assessed businesses' expectations of their customers' attitudes towards the measures (5-point importance scale), while the second study focused on the perceived impact of the measures on businesses' profitability, using a scale ranging from 1—"not at all" to 5—"very much". Businesses' expectations of their customers' willingness to make a personal contribution through the five aforementioned support actions were evaluated in the third study (5-point willingness scale). The surveys concluded with profiling questions regarding business type in order to distinguish between food service businesses serving food and drinks (restaurant) and those only serving drinks (bar).

3.2. Data Collection

Data were collected in May 2020 (study 1), June 2020 (study 2) and November 2020 (study 3) through online surveys. The first and third study were conducted during the first and second wave of COVID-19 infections respectively, with food service businesses being mandatory closed. Data collection for the second study was performed in between waves of infections, when reopening was allowed. In order to facilitate data collection, questionnaires were integrated into Qualtrics software for both stakeholder groups. By using a convenience sampling procedure, the surveys were administered to the target groups, both food service businesses and food service customers. Stakeholders from sector organisations distributed the survey among their members, while social media channels were used to disseminate the survey to the general public. Regarding food service businesses, gourmet restaurants, bistros/brasseries, fast food restaurants, buffet restaurants and bars, whether or not serving food, were targeted. Food service customers had to eat or drink out at least once a year to be included. For both surveys, participation was further restricted to people with a minimum age of 18. After removing incomplete responses, the final sample consisted of 1083, 309 and 305 consumers and 306, 221 and 253 businesses for study 1, 2 and 3, respectively. The higher response rate for the first consumer study can be attributed to the strict lockdown in place at the time of survey administration.

Flanders (Belgium) was targeted as study location. Belgium reported its first confirmed case of COVID-19 on 4 February 2020 and first death on 10 March 2020. From 14 March 2020 onwards, restaurants and bars had to close their businesses, only to be allowed to reopen on 8 June 2020. Mandatory closure of dine-in services during lockdown periods caused restaurants to set up takeaway and delivery services. Reopening opportunities were accompanied with strict safety measures designed to prevent the spread of the virus and protect the health of both employees and customers. Belgium successfully flattened the epidemic curve in April 2020, yet the country experienced its second wave of COVID-19 in the final months of 2020. The government decided to reclose restaurants and bars, starting from 19 October 2020 onwards, initially for four weeks, but ultimately for over six months, as a response to the third wave. Meanwhile, several financial support measures and actions were introduced by the Belgian federal government and the public respectively to ensure the survival of these businesses. Further, similar to many countries, the consumption pattern of out-of-home eating to satisfy daily dietary needs has over time gained prominence in Belgium [57,58], illustrating the relevance of Flanders as a study location. Table 1 presents an overview of the survey development and data collection process.

Table 1. Overview of survey development and data collection for the three studies.

	Study 1	Study 2	Study 3
Survey Development			
	Consumers (n = 1083)	**Consumers** (n = 309)	**Consumers** (n = 305)
Part 1: Behavioural variables	Visit frequency before COVID-19 Revisit intention	Visit frequency before COVID-19 Revisit behaviour	Takeaway frequency before and since COVID-19
Part 2: Attitudes and perceptions	Attitudes towards 21 expected safety measures	Attitudes towards 14 imposed safety measures Perceived safety and compliance Perceived business transparency of safety measures	Attitudes towards government decisions Willingness to support through 5 actions Perceived government transparency of measures
Part 3: Profiling variables	Socio-demographic	Socio-demographic	Socio-demographic
	Food service sector (n = 306)	**Food service sector** (n = 221)	**Food service sector** (n = 253)
Part 1: Attitudes and perceptions	Expectations of attitudes towards 21 expected safety measures	Perceived impact on profitability of 14 imposed safety measures	Expectations of willingness to support through 5 actions
Part 2: Profiling variables	Business type	Business type	Business type
Data Collection			
Timing	May 2020	June 2020	November 2020
Stage of the pandemic	1st wave of infections	In between waves	2nd wave of infections
Situation for food service businesses (start date)	Mandatory closure (14 March 2020)	Reopening (8 June 2020)	Mandatory closure (19 October 2020)

3.3. Data Analysis

Statistical analyses were performed using IBM SPSS Statistics 27. Principal component analysis (PCA) with Varimax rotation was used to explore the underlying structure both of consumers' attitudes towards the expected to be imposed (study 1) and imposed (study 2) safety measures on food service businesses and of consumers' transparency perceptions regarding government communication (study 3). The Kaiser–Meyer–Olkin (KMO) measure of sampling adequacy and the Bartlett's test of sphericity provided acceptable values indicating the meaningfulness of performing a factor analysis on the chosen variables. The latent root criterion was applied so to only retain factors with an eigenvalue above one [59]. Factor loadings above 0.5 were considered practically significant, following the rule of thumb of Hair, et al. [59]; items with factor loadings below 0.5 were omitted. Since businesses evaluated the same set of safety measures, the factorial structure from the consumer oriented PCA was used to group businesses' attitudes, allowing for a comparison of food service businesses and consumers. Internal consistency of the factors was tested through McDonald's omega in order to justify the creation of composite variables based on the average score on the underlying items of each factor.

Hierarchical binary logistic regressions (enter method) were performed to estimate the role of the aforementioned factors, past frequency of out-of-home food consumption and socio-demographic variables (age, gender, education level) on consumers' revisit intention (0—postpone visit; 1—retake visit; study 1) as well as revisit behaviour (0—visit postponed; 1—visit retaken; study 2). Outliers, i.e., cases with standardized residuals above $|2|$ [59], were listed, subjected to visual inspection of DFBeta and stepwise removed (<3% of cases) until a suitable model was achieved. For all estimated models, no indications of major problems with multicollinearity were apparent. Although positive bivariate correlations were found between established factors, the coefficients were below the threshold (0.7), while collinearity diagnostics showed VIF (variance inflation factor) values well below 10 [59,60]. Different goodness-of-fit statistics were calculated to estimate model fit. The significance of the likelihood ratio chi-square tests indicated that the models containing the independent variables represented a significant improvement in fit relative to the model without variables ('null' model). Nagelkerke R^2 values, reflecting the amount of variation accounted for by the logistic model, showed an improvement from Model 2 relative to Model 1.

Other data analysis techniques used included descriptive, univariate, and bivariate analyses. Differences in socio-demographics between samples were tested with one-way ANOVA and chi-square tests. Differences in factor means (study 1, 2) were tested with paired samples *t*-tests, while independent samples *t*-tests were used for differences in composite variables (derived from the factors; study 1, 2) and (expectations of) willingness to support (study 3) between the target groups. Differences in takeaway consumption frequency (study 3) were tested with both paired samples *t*-tests, independent samples *t*-tests and bivariate correlations. Bivariate correlations were also used to examine associations between perceived transparency and other variables (study 2, 3).

4. Results

4.1. Sample Descriptives

The total sample consisted of 1697 consumers and 780 food service businesses. Table 2 presents the characteristics of the sample of each study. There were no differences between the consumer samples in terms of age (one-way ANOVA; $F = 2.47$; $p = 0.085$) and gender (chi-square test; $\chi^2 = 3.63$; $p = 0.163$). The average age of the sampled consumers was between 42 and 44. Both female and higher educated people are slightly overrepresented in the samples. At the level of food service businesses, business type was equally distributed between samples (chi-square test; $\chi^2 = 4.68$; $p = 0.096$). Roughly three out of four sampled businesses were restaurants serving both food and drinks, while bars that only serve drinks were less represented.

Table 2. Socio-demographic characteristics of the sample per study.

	Study 1	Study 2	Study 3
Consumers	(n = 1083) (%)	(n = 309) (%)	(n = 305) (%)
Age			
Mean (SD)	42.40 (13.73)	43.99 (14.66)	43.98 (15.10)
Gender			
Male	38.1	36.9	43.6
Female	61.9	63.1	56.4
Education			
Primary or secondary	29.8	16.5	21.6
Higher	70.2	83.5	78.4
Food service sector	(n = 306) (%)	(n = 221) (%)	(n = 253) (%)
Business type			
Restaurant (serving food and drinks)	81.0	78.7	73.5
Bar (only serving drinks)	19.0	21.3	26.5

4.2. Expected Safety Measures in Pandemic Times (Study 1)

4.2.1. Consumers' Attitudes and Businesses' Expectations of Their Customers' Attitudes

PCA was performed to explore the underlying structure of consumers' attitudes towards 21 expected safety measures and resulted in a factorial structure with three factors. Three items were stepwise excluded (loadings < 0.5), while 18 items were retained, all loading well on one of the three factors. The results of the final factor analysis with 18 items are summarized in Table 3.

Table 3. Factor loadings from principal component analysis for consumers' attitudes towards expected safety measures (study 1; n = 1083).

Items	Mean	S.D.	Factor 1	Factor 2	Factor 3
Disinfectants available on the table	3.42	1.20	**0.729**	−0.002	0.157
Staff disinfects toilet after each visit	3.92	1.13	**0.645**	0.293	0.111
Staff disinfects hands after clearing each table	4.21	0.98	**0.640**	0.324	0.114
Service is performed with mouth mask	3.62	1.22	**0.638**	0.257	0.300
Service is provided with gloves	3.11	1.34	**0.631**	0.101	0.203
Tables and chairs are disinfected after each visit	4.06	1.03	**0.598**	0.385	0.248
Mandatory disinfection of hands upon arrival	4.41	0.87	**0.500**	0.299	0.266
Newspapers and magazines are not provided	3.80	1.23	0.204	**0.799**	0.201
No possibility of self-service or buffet	3.99	1.14	0.174	**0.736**	0.247
Menus and drinks menus are not interchangeable between tables	3.97	1.04	0.312	**0.720**	0.225
Clients must hang their own coat in the checkroom	3.39	1.10	0.095	**0.538**	0.268
Only disposable consumables on the table	3.33	1.36	0.404	**0.536**	0.119
Mandatory reservation by clients	3.16	1.39	0.019	0.215	**0.700**
Customers received in shifts per time block	3.11	1.22	0.247	0.108	**0.681**
Seating only under guidance	3.69	1.26	0.196	0.244	**0.673**
Presence of walking paths	3.17	1.26	0.399	0.112	**0.609**
Clients can only consume while seated	3.58	1.23	0.206	0.395	**0.556**
Availability of waiting zones upon arrival	3.45	1.10	0.302	0.301	**0.536**
McDonald's omega			0.827	0.812	0.799
Mean (S.D.)			3.82 (0.78)	3.70 (0.88)	3.36 (0.88)

Note: KMO measure of sampling adequacy: 0.939; Bartlett's test of sphericity: 7445.125 ($p < 0.001$); bold indicates on which factor an item loads highest (loading > 0.5).

Factor 1 represents hygiene measures, which particularly focus on disinfecting hands and surfaces. Factor 2 deals with measures aimed at avoiding the sharing of objects between customers, such as menus and salt shakers. Factor 3 includes organisational measures that assure a well-organised flow of clients in the establishment. McDonald's omega values for the three factors indicated good internal consistency and allowed for the development of composite variables for factor 1 (seven items), factor 2 (five items) and factor 3 (six items). Table 3 also presents item and factor means. Paired samples *t*-tests indicated significant differences between the factor means. The mean of factor 1 (hygiene; $\overline{x} = 3.82$) is significantly higher compared to factor 2 (avoidance;$\overline{x} = 3.70$) ($t = 5.73$; $p < 0.001$), which in turn has a significantly higher mean than factor 3 (organisation;$\overline{x} = 3.36$) ($t = 14.58$; $p < 0.001$). While all items and factors are considered important when revisiting food service businesses, sanitary measures appear to be the priority for consumers.

To allow for comparison with restaurants' and bars' expectations of their customers' attitudes towards safety measures in food service businesses, the same factors were developed. McDonald's omega values above 0.7 justified the calculation of composite variables of hygiene and avoidance measures at the level of businesses. Internal consistency for the organisational measures was lower (McDonald's omega = 0.667), yet close to the threshold of 0.7 and still acceptable for exploratory research [59]. Surprisingly, businesses' expectations of their customers' attitudes towards safety measures in their businesses are significantly different from consumers' stated attitudes (Figure 1). Independent samples *t*-tests indicated significantly lower mean scores for hygiene ($t = 8.71$; $p < 0.001$) and organisational measures ($t = 3.91$; $p < 0.001$) from the businesses' perspective compared to the consumers' perspective.

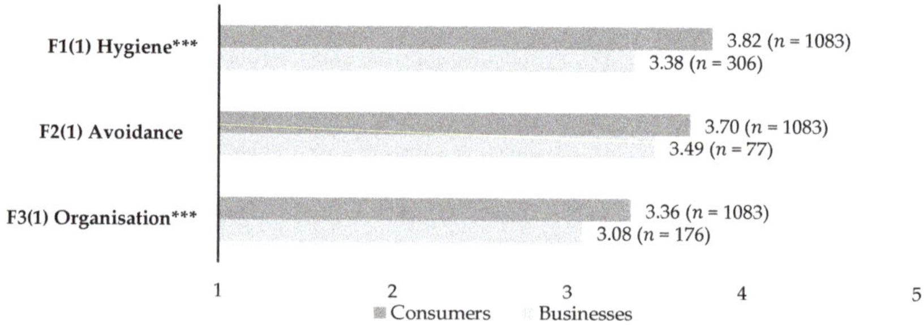

Figure 1. Expected safety measures: consumers' attitudes vs. businesses' expectations of their customers' attitudes. Note: not all statements were evaluated by all business owners due to irrelevance, explaining the differences in *n* between factors; *** $p < 0.001$; FX(Y) with X = number of factor, Y = number of study.

4.2.2. Determinants of Consumers' Revisit Intentions (Study 1)

The first study revealed that 58.2% of respondents had the intention to immediately revisit food service businesses once reopened. To understand which factors influenced consumers' intentions to either postpone or retake visits when restaurants and bars reopen, a binary logistic regression model was estimated. A hierarchical approach was used to assess the associations between attitudes (Model 2) and intention, controlling for socio-demographics and past behaviour (Model 1). Thirty outliers were omitted, yielding a total of 1053 valid responses. The results are summarized in Table 4. The significance of the likelihood ratio chi-square tests and the Nagelkerke R^2 values (0.125 and 0.279 for Models 1 and 2 respectively) indicated a moderate model fit.

Table 4. Coefficient estimates and diagnostics from hierarchical binary logistic regression explaining consumers' revisit intentions (study 1; n = 1053).

	Model 1: Consumer Profiling Variables					Model 2: Consumer Profiling and Attitudes				
Variable	**B**	**S.E.**	**Wald**	***p***	**Exp(B)**	**B**	**S.E.**	**Wald**	***p***	**Exp(B)**
Socio-demographic										
Age	−0.001	0.005	0.039	0.843	0.999	0.009	0.005	3.099	0.078	1.009
Gender (1 = male)	0.488	0.142	11.910	**0.001**	1.630	0.204	0.154	1.760	0.185	1.226
Education (1 = higher)	−0.141	0.151	0.872	0.350	0.868	−0.159	0.163	0.958	0.328	0.853
Past behaviour										
Visit frequency	0.335	0.049	47.305	**<0.001**	1.398	0.300	0.050	36.555	**<0.001**	1.349
Attitudes										
F1(1) Hygiene						−0.479	0.138	12.043	**0.001**	0.620
F2(1) Avoidance						−0.449	0.121	13.727	**<0.001**	0.638
F3(1) Organisation						−0.505	0.122	17.197	**<0.001**	0.604
Constant	−0.256	0.276	0.862	0.353	0.774	4.849	0.578	70.427	**<0.001**	127.613
Model										
Likelihood ratio	101.971			<0.001		244.007			<0.001	
Nagelkerke R^2	0.125					0.279				

Note: Predictive accuracy of 63.6% (Model 1) and 69.3% (Model 2) compared to 59.8% in the 'null' model; dependent variable (revisit intention) is a dummy variable: postpone visit (0), retake visit (1); bold indicates significant coefficients ($p < 0.05$); FX(Y) with X = number of factor, Y = number of study.

The first model (Model 1, one block) indicates that the probability of having the intention to revisit restaurants and bars was positively influenced by respondents' visit frequency before the lockdown. An increase in the frequency of out-of-home consumption by one visit per week increased the odds of intending to revisit by 40%. Furthermore, gender had an effect on revisit intention; being male increased the odds of intending to revisit (odds ratio: 1.63). Age and education did not significantly affect revisit intention. From the final model (Model 2, two blocks), attitudes towards safety measures related to food service businesses were identified as significant determinants of consumers' revisit intention; the importance of hygiene, avoidance and organisational measures had a significant negative influence. A one-unit increase in the attitude score decreased the odds of revisit intention by a factor in the range of 0.60 to 0.64. Respondents who valued the safety measures more were less likely to plan to retake their visits immediately and rather intended to postpone, with the risk of contamination as the major reason to do so (65%). This demonstrates the importance for businesses to strictly adhere to imposed measures in order to persuade customers to revisit their establishments during the COVID-19 pandemic. The effect of gender decreased in Model 2 and was no longer significant.

4.3. Imposed Safety Measures in Pandemic Times (Study 2)

4.3.1. Consumers' Attitudes and Perceived Impact on Businesses' Profitability

PCA was performed to explore the underlying structure of 14 actual imposed safety measures. A two-factor solution was recognized, with factors conceptually similar to the factors 'hygiene' and 'organisation' identified in study 1. Here, the factor 'avoidance' was not identified, and items related to avoid the sharing of objects between customers loaded on other factors. To compare both studies, all three items related to 'avoidance' were removed in this analysis. This yielded high factor loadings for the remaining 11 items. The results are summarized in Table 5.

Table 5. Factor loadings from principal component analysis for consumers' attitudes towards imposed safety measures (study 2; $n = 309$).

Items	Mean	S.D.	Factor 1	Factor 2
Tables and chairs are disinfected after each visit	4.26	0.96	**0.794**	0.227
Only paper towels and lockable bins in the toilets	4.47	0.76	**0.769**	0.176
Payment terminal is disinfected after each use or hand gels/cotton buds available	4.17	1.00	**0.756**	0.270
Disinfectants available for clients	4.38	0.79	**0.745**	0.208
Service is performed with mouth mask	4.02	1.10	**0.686**	0.462
Kitchen staff wears mouth mask or keeps distance	4.06	1.11	**0.668**	0.363
Glasses are washed with soap	4.50	0.74	**0.654**	0.240
Mandatory closure at 1 am	3.00	1.38	0.117	**0.801**
Clients can only consume while seated	3.69	1.22	0.332	**0.787**
Maximum of 10 clients per table	3.76	1.14	0.272	**0.748**
Distance of 1.5 m is maintained outside and inside	4.21	0.96	0.422	**0.587**
McDonald's omega			0.892	0.805
Mean (S.D.)			4.27 (0.72)	3.66 (0.93)

Note: KMO measure of sampling adequacy: 0.896; Bartlett's test of sphericity: 1778.693 ($p < 0.001$); bold indicates on which factor an item loads highest (loading > 0.5).

Similar to study 1, factor 1 deals with hygiene measures and factor 2 is clearly linked to organisational measures. McDonald's omega values for the two factors indicated good internal consistency and allowed calculations of composite variables for factor 1 (seven items) and factor 2 (four items). Table 5 also presents item and factor means, as derived from attitude scores on a 5-point importance scale. Paired samples t-test indicated significant differences between factor means ($t = 14.84$; $p < 0.001$), with hygiene measures ($\overline{x} = 4.27$) considered to be more important than organisational measures ($\overline{x} = 3.66$).

The factors grouping hygiene and organisational measures were also used to compare consumers' attitudes towards the measures with the perceived impact of the measures on the profitability of restaurants and bars. As for the latter, composite variables were calculated (McDonald's omega > 0.7). It becomes clear that both hygiene and organisational measures have a large perceived impact on businesses' profitability. Independent samples t-tests indicated significantly lower mean scores for organisational measures ($t = 7.17$; $p < 0.001$) from consumers' perspectives (importance) compared to businesses' perspectives (perceived impact on profitability) (Figure 2).

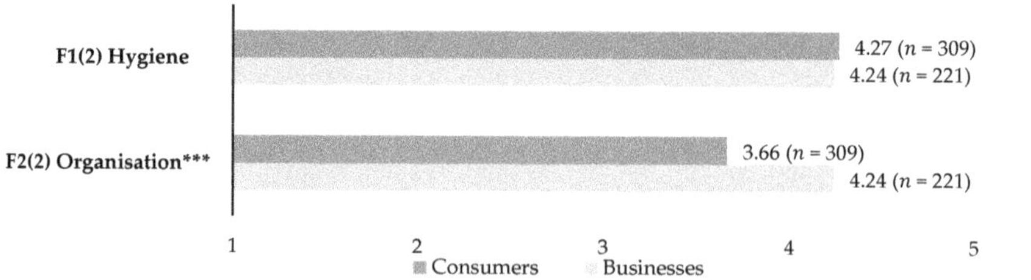

Figure 2. Imposed safety measures: consumers' attitudes vs. perceived impact on businesses' profitability. Note: *** $p < 0.001$; FX(Y) with X = number of factor, Y = number of study.

4.3.2. Determinants of Consumers' Revisit Behaviour (Study 2)

In the second study, 69.3% of respondents indicated that they immediately revisited food service businesses as soon as they were allowed. To understand which factors influence consumers' actual behaviour to either postpone or retake visits to restaurants and bars since reopening, another binary logistic regression model was estimated. A hierarchical approach was used to assess the associations between attitudes (Model 2) and current behaviour, controlling for socio-demographics and past behaviour (Model 1). Eight outliers were omitted, yielding a total of 301 valid responses. The results are summarized in Table 6.

Certain goodness-of-fit statistics (likelihood ratio, Nagelkerke R^2 of 0.289 and 0.348 for Models 1 and 2 respectively) were calculated and indicated moderate to good model fit.

Table 6. Coefficient estimates and diagnostics from hierarchical binary logistic regression explaining consumers' revisit behaviour (study 2; n = 301).

	Model 1: Consumer Profiling Variables					Model 2: Consumer Profiling and Attitudes				
Variable	B	S.E.	Wald	p	Exp(B)	B	S.E.	Wald	p	Exp(B)
Socio-demographic										
Age	−0.028	0.010	7.509	**0.006**	0.972	−0.022	0.011	4.097	**0.043**	0.978
Gender (1 = male)	0.608	0.313	3.771	0.052	1.837	0.465	0.324	2.057	0.151	1.591
Education (1 = higher)	−1.023	0.467	4.797	**0.029**	0.360	−1.034	0.482	4.602	**0.032**	0.356
Past behaviour										
Visit frequency	0.965	0.197	24.060	<0.001	2.626	0.968	0.210	21.245	**<0.001**	2.632
Attitudes										
F1(2) Hygiene						−0.830	0.329	6.371	**0.012**	0.436
F2(2) Organisation						−0.170	0.233	0.531	0.466	0.844
Constant	1.664	0.750	4.918	**0.027**	5.279	5.743	1.392	17.023	**<0.001**	311.952
Model										
Likelihood ratio	68.029			<0.001		84.051			<0.001	
Nagelkerke R^2	0.289					0.348				

Note: Predictive accuracy of 75.1% (Model 1) and 76.1% (Model 2) compared to 71.1% in the 'null' model; dependent variable (revisit behaviour) is a dummy variable: visit postponed (0), visit retaken (1); bold indicates significant coefficients (p < 0.05); FX(Y) with X = number of factor, Y = number of study.

The first model (Model 1, one block) indicates that the probability of revisiting restaurants and bars was positively influenced by respondents' visit frequency before the lockdown. Increasing the frequency of out-of-home consumption by one visit per week, increased the odds of revisiting by a factor of 2.63. Moreover, age and education had an impact on revisit behaviour. A 10-year increase in age was associated with an 28% decrease in the probability of revisiting; higher educated people were less likely to retake visits immediately (odds ratio: 0.36). When looking at the complete model (Model 2, two blocks), attitudes towards hygiene measures were identified as another significant determinant of consumers' revisit behaviour; a one-unit increase in the attitude score decreased the odds of revisiting by a factor of 0.44. The more respondents value the hygiene measures, the less likely they are to retake their visits immediately, hence more likely to postpone. The attitudinal variable related to organisational measures as well as gender did not significantly affect revisit behaviour.

4.4. Post-Pandemic Behaviour and Willingness to Support (Study 3)

Mandatory closure of dine-in services during lockdown periods led to a significant increase in consumers' ordering frequency of takeaway meals (paired samples t-test; $t = 9.35$; $p < 0.001$). Whereas before the COVID-19 pandemic people chose takeaway on average once per month (0.21 times/week, S.D. = 0.33), this doubled during lockdown periods (0.45 times/week, S.D. = 0.51). However, consumers expected to reinstate their pre-pandemic behaviour in terms of takeaway and out-of-home consumption once the pandemic was over. No significant differences in lockdown takeaway consumption were found for gender and education level, nor was there a correlation with age.

Figure 3 shows consumers' willingness to contribute versus businesses' expectations of their customers' willingness for five different support actions. While consumers and businesses ranked the options nearly identical, independent samples t-tests revealed significant differences between the two groups. Consumers' willingness to support exceeded businesses' expectations for all support actions evaluated.

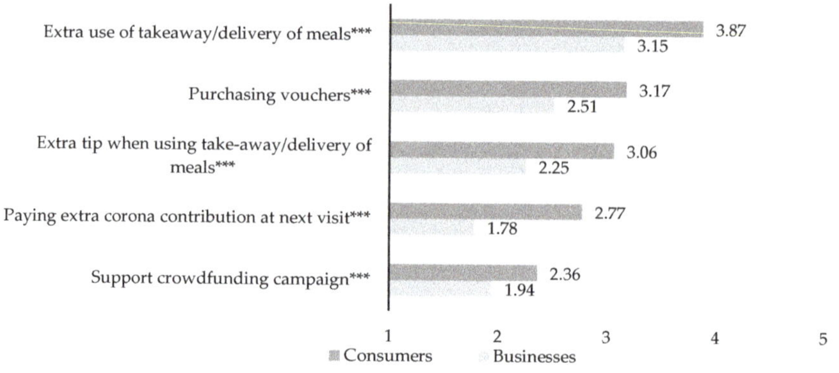

Figure 3. Consumers' willingness to personally contribute ($n = 305$) and businesses' expectations of their customers' willingness ($n = 253$). Note: *** $p < 0.001$.

4.5. Consumers' Transparency Perceptions of Safety Measures in Pandemic Times

4.5.1. Communication by Food Service Businesses (Study 2)

Transparency perceptions regarding businesses' communication of safety measures were calculated as a mean score of the eight items (McDonald's omega = 0.940). No significant differences in perceived transparency were found for gender and education level, neither was there a correlation with age. However, perceived transparency was highly positively correlated with perceived compliance and perceived safety (Table 7), which were also strongly, and positively correlated. The more transparently businesses communicate, the more customers believe that businesses adhere to the imposed safety measures and the more they felt safe during their visit.

Table 7. Bivariate correlations between consumers' perceived transparency, compliance, and safety (study 2; $n = 214$).

	Mean	S.D.	Perceived Transparency	Perceived Compliance
Perceived transparency	3.92	0.84	1	
Perceived compliance	4.05	1.04	0.596 ***	1
Perceived safety	4.18	0.95	0.602 ***	0.785 ***

Note: *** $p < 0.001$.

4.5.2. Communication by the Government (Study 3)

Consumers' transparency perceptions regarding government communications were measured for both lockdown periods. PCA was conducted on all 16 items and a factorial structure with two factors was recognized, i.e., eight items per lockdown. McDonald's omega values justified the creation of composite measures. Table 8 summarizes the results. A significant increase in perceived transparency was observed with the progression of the pandemic (paired samples *t*-test; $t = 5.79$; $p < 0.001$), although consumers considered the government's transparency for both lockdown periods to be fairly neutral.

Perceived transparency was also found to be positively correlated with support for the government's decision to close food service businesses. The correlation was stronger for perceived transparency related to the communication in the second lockdown ($r = 0.623$; $p < 0.001$) compared to the first lockdown ($r = 0.392$; $p < 0.001$).

Table 8. Factor loadings from principal component analysis for perceived transparency (study 3; $n = 305$).

Items	1st Lockdown	2nd Lockdown
	Factor 1	Factor 2
Information is timely	0.700	0.634
Information is relevant	0.763	0.793
Information is consistent	0.795	0.692
Information is complete	0.786	0.799
Information is easy to understand	0.806	0.774
Information is accurate	0.843	0.784
Information is reliable	0.769	0.829
Information explains the rationale	0.716	0.745
McDonald's omega	0.917	0.909
Mean (S.D.)	2.86 (0.88)	3.16 (0.88)

Note: KMO measure of adequacy: 0.910; Bartlett's test of sphericity: 3225.558 ($p < 0.001$); only factor loadings above 0.5 are presented.

5. Discussion

By using the COVID-19 pandemic as a case, this study addresses the need for research on consumers' changed behaviour regarding out-of-home food consumption during and following a pandemic. Based on three online surveys with 1697 consumers and 780 food service businesses, this study analysed (1) attitudes, intentions and behaviour regarding safety measures and dining out in pandemic times, and (2) transparency perceptions of safety measures. This research contributes to the current body of literature on out-of-home food consumption during the COVID-19 pandemic by integrating the perspectives of two stakeholder groups, i.e., food service consumers and businesses, and analysing their perceptions at different moments in time. Our findings provide important insights that will enable food service businesses to better understand consumers' perceptions so they can anticipate them and ensure their own survival. The discussion is structured according to the research questions posed.

5.1. Consumers' and Businesses' Attitudes towards Safety Measures (RQ1a, RQ1b)

Our results indicate that, although consumers were generally concerned about all safety measures, attitudes towards them can be categorized into three factors related to hygiene, avoidance of object sharing and organisation. For both expected and imposed safety measures, sanitary measures, which focus on disinfection of hands and surfaces, are prioritized. This highlights the importance consumers attach to disinfecting when it comes to preventing virus transmission while consuming food out-of-home. The priority given to hygiene measures is similar to the results of previous consumer studies, where availability of disinfectants, staff wearing masks, extensive cleaning of surfaces, strict handwashing and training employees about sanitary practices were considered the most important precautions to be taken by restaurants [31,61]. However, food service businesses themselves did not expect their customers to attribute that much importance to the safety measures in place at their establishment. In addition, the profitability of restaurants and bars was severely compromised by the safety measures imposed.

5.2. Determinants of Consumers' Revisit Intentions and Behaviour (RQ2)

Further, this study identifies different determinants of consumers' intention and behaviour related to visiting food service businesses post-lockdown. When comparing consumers' intentional and actual visiting behaviour, several differences can be recognized. While consumers' attitudes towards all measures (hygiene, avoidance, organisation) had significant effects on the intention to revisit, only their attitudes towards hygiene measures were a significant factor influencing the likelihood to actually revisit and attitudes towards organisational measures were not that influential. In sum, the higher the importance attributed to (hygiene) measures, the less likely consumers were to (intend to) revisit. As such, it indicates that good compliance with (hygiene) measures appears to be an important strategy for businesses to regain customers when reopening is allowed. These findings are in line with previous studies that indicate the importance of cleanliness and sanitation when selecting and visiting restaurants [38,39] and confirm past studies in this field that identified consumers' attitudes towards hygiene as a determinant of (re)visit intention [36,37]. Moreover, consumers' cleanliness concerns are even heightened in times of global health crises, such as the current pandemic [30,32]. Our results underpin the importance of cleanliness and safety measures to draw customers back in by meeting their expectations [30]. Similar, cleanliness was used as selling proposition to recover from SARS [40]. However, Wei, et al. [61] observed different results and suggested that perceived importance of preventive COVID-19 measures indirectly enhanced customers' intentions to dine out during the reopening period, through brand trust, i.e., customers' reliance on a certain business. Implementing safety measures helps restaurants to build brand trust, even more for those who perceive high risk of COVID-19, and more trustworthy restaurants attract more customers [61,62]. Customers with a low risk perception of COVID-19 are less willing to adapt their lifestyle to comply with safety measures, hence their trust in restaurants is less impacted by the adoption of preventive measures [62]. Despite the seemingly contradictory results, the findings are similar to ours: by implementing and strictly complying with safety measures, restaurants might convince customers to resume dining out during the pandemic.

In addition, our results indicate that men state that they are more likely to revisit immediately while women state that they are more likely to postpone. This might be explained by gender differences in health-protective behavioural response to a respiratory pandemic [63]. Women are more concerned about COVID-19 and therefore take more precautions to avoid contamination [64], even though the severity and mortality are higher for male COVID-19 patients [65]. However, when it comes to actual revisit behaviour, gender is eventually not a determinant in our study, while age and educational level are. Being older as well as having achieved a higher level of education significantly decreases the likelihood of revisiting immediately. The age effect might be linked to a higher probability of severe illness for older people [65]. In a study by Hakim, et al. [34], though age did not impact visit intention, older customers' visit intentions were less affected by denial of COVID-19 compared to younger people. Regarding educational status, Byrd, et al. [66] found that higher educated consumers have more concerns about the risk of contracting COVID-19 from restaurant food, which might explain their postponing behaviour, despite eating out-of-home more often [27]. Consumers' pre-pandemic frequency of out-of-home consumption had a significant positive effect on the likelihood to intend to revisit and actually revisit. The more often people went out to eat or drink before COVID-19, the more likely they are to (plan to) do so again. Current results are consistent with Lee, et al. [67], who showed that the frequency of past travel behaviour was positively associated with the intention to travel during the 2009 H1N1 outbreak, and Mehrolia, et al. [26], who indicated that the probability of ordering food though online food delivery services during the COVID-19 lockdown was higher for customers with a higher purchasing frequency before.

5.3. Post-Pandemic Behaviour and Willingness to Support (RQ3a, RQ3b)

Our results indicate that consumers' ordering frequency of takeaway meals doubled during mandatory closure periods of restaurants and bars, findings that are in line with Poelman, et al. [23]. Once the pandemic is over, consumers expect to return to their initial frequency of dining out and ordering takeaway.

Further, not only is the Belgian federal government financially helping food service businesses to overcome the current crisis, but consumers are also very willing to make personal contributions to support the sector, even more than was expected by restaurants and bars. Consumers indicate that they are highly willing to financially support to help food service businesses to survive, contrary to the expectations of the food service sector. Previous studies argued that high levels of solidarity with the food service sector have a positive effect on consumers' intention to visit restaurants and bars [34] and to continue using food delivery applications [54] during the COVID-19 pandemic; however, this situation-specific effect is expected to diminish over time. The notable discrepancy that was found between consumers' solidarity intentions and businesses' expectations provides knowledge and opportunities for businesses to facilitate their survival of the pandemic.

5.4. Consumers' Transparency Perceptions of Safety Measures in Pandemic Times (RQ4)

Perceived transparency of businesses' communications about the imposed measures is positively correlated with perceived compliance of businesses with those measures and perceived safety of customers during their visit. The correlations between these variables suggest the importance of transparent communication to appear well-compliant and to make customers feel safe during their visit. These findings are in line with previous research, which argues that food service businesses can restore customers' trust and encourage them to dine out during the pandemic by communicating in a transparent way [34,53].

Perceived transparency of government communication is positively correlated with the support for the mandatory closure decision. Similar results were observed by Scholz, et al. [47], who identified a positive correlation between the comprehensibility of a certain COVID-19-related decision, in particular of its underlying rationale, and its acceptance. Providing timely, clear, and consistent policy recommendations improves compliance [46,49]. As consumers' perceptions of transparency increased throughout the pandemic, this might have positive implications for the acceptance of more recent government decisions.

6. Conclusions

6.1. Implications

This study contributes to the growing body of COVID-19-related literature in the hospitality domain. It is one of the first studies to assess the role of safety measures in predicting consumers' revisit intention and behaviour. By demonstrating the importance consumers attribute to safety measures and hygiene when resuming visits to restaurants and bars, this study helps to better understand consumers' preferences regarding out-of-home food consumption during a pandemic. Further, while previous research has mostly focused on either consumers or food service businesses at one moment in time, this study extends the existing literature by integrating perspectives of two key stakeholder groups at multiple stages of the pandemic.

The COVID-19 pandemic has tremendously impacted food service business operations. The findings of this study can help the food service sector in developing adequate survival strategies. As both consumers' revisit intention and behaviour were determined by their attitudes towards sanitary measures, this study highlights how adoption of and adherence to safety measures may be an effective approach for food service businesses to attract customers in pandemic times. Further, it is suggested that by communicating transparently about these measures, business owners will make their customers feel safe during their visits. This study also revealed consumers' high willingness to contribute financially to the continued existence of restaurants and bars. The food service sector should benefit from these solidarity intentions as they are likely to decline over time. Understanding customers' expectations and willingness to provide support, both in terms of financial contributions during lockdown periods and physical visits during reopening periods, might help business owners face the challenges posed by this and future health crises.

6.2. Limitations and Future Research

The present study has some limitations. Our results are based on data from Flanders, Belgium, collected during the first and second wave of COVID-19. Perceptions and attitudes may differ from country to country, as the COVID-19 pandemic has affected countries in various ways and to various extents and has been tackled by various policy decisions. Moreover, although the pandemic is still ongoing, perceptions and attitudes may change over time as both consumers and food service businesses are gradually adjusting to the new normal. Future research could investigate the long-term effects of COVID-19 on out-of-home food consumption behaviour, both in later stages of the pandemic and when the pandemic would be over. Furthermore, caution is needed when interpreting consumers' views on visit intention and solidarity actions as they may deviate from actual behaviour, known as the intention–behaviour gap [68]. Finally, future research could further expand the variables used in this study. Besides attitudes towards safety measures and past behaviour, consumers' revisit intentions and behaviours could be affected by other factors, e.g., risk perception related to COVID-19 infection could also be relevant. To further explore the role of perceived transparency regarding communication in pandemic times, its effects on trust and compliance could be investigated. Future studies may also deepen the understanding of the discrepancies found between consumers' attitudes and businesses' expectations and elaborate on consumers' changed consumption behaviour.

Author Contributions: Conceptualization, H.V., X.G., H.D.S.; methodology, H.V., X.G., H.D.S.; formal analysis, H.V.; investigation, H.V.; writing—original draft preparation, H.V.; writing—review and editing, H.V., X.G., H.D.S.; supervision, X.G., H.D.S. All authors have read and agreed to the published version of the manuscript.

Funding: This research received no external funding.

Institutional Review Board Statement: This study has been conducted in accordance with the principles of the Helsinki Declaration, the EU General Data Protection Regulation 2016/679 and the 2018 EU guidelines on Ethics in Social Science and Humanities. Human participants to the study were adult volunteers who were informed about the overall purpose, methods, and implications of the research, who provided their informed consent for participation and who were free to withdraw their consent at any time during the research without any consequences. No deception has been used in this study. Participants were informed about the contact details of the principal researcher and of the institutional Data Protection Officer to be contacted in case of queries. No personal identifying data were recorded, and data records were anonymized for storage and further analysis.

Informed Consent Statement: Informed consent was obtained from all subjects involved in the study.

Data Availability Statement: Data will be made available on request.

Conflicts of Interest: The authors declare no conflict of interest.

References

1. WHO. WHO Coronavirus (COVID-19) Dashboard. Available online: https://covid19.who.int/ (accessed on 29 January 2022).
2. WHO. *COVID-19 Weekly Epidemiological Update18 January 2022*; WHO: Geneva, Switzerland, 2022.
3. CDC. How to Protect Yourself & Others: Stay 6 Feet Away from Others. Available online: https://www.cdc.gov/coronavirus/2019-ncov/prevent-getting-sick/social-distancing.html (accessed on 16 February 2022).
4. ECDC. Questions and Answers on COVID-19: Prevention. Available online: https://www.ecdc.europa.eu/en/covid-19/questions-answers/questions-answers-prevention (accessed on 16 February 2022).
5. Wilder-Smith, A.; Freedman, D.O. Isolation, quarantine, social distancing and community containment: Pivotal role for old-style public health measures in the novel coronavirus (2019-nCoV) outbreak. *J. Travel Med.* **2020**, *27*, 1–4. [CrossRef] [PubMed]
6. Lewnard, J.A.; Lo, N.C. Scientific and ethical basis for social-distancing interventions against COVID-19. *Lancet Infect. Dis.* **2020**, *20*, 631–633. [CrossRef]
7. FAO. *Agri-Food Markets and Trade in the Time of COVID-19*; FAO: Rome, Italy, 2020.
8. Laborde, D.; Martin, W.; Swinnen, J.; Vos, R. COVID-19 risks to global food security. *Science* **2020**, *369*, 500–502. [CrossRef]
9. Hobbs, J.E. Food supply chains during the COVID-19 pandemic. *Can. J. Agric. Econ.* **2020**, *68*, 171–176. [CrossRef]
10. Ben Hassen, T.; El Bilali, H.; Allahyari, M.S. Impact of COVID-19 on Food Behavior and Consumption in Qatar. *Sustainability* **2020**, *12*, 6973. [CrossRef]
11. Lachat, C.; Nago, E.; Verstraeten, R.; Roberfroid, D.; Van Camp, J.; Kolsteren, P. Eating out of home and its association with dietary intake: A systematic review of the evidence. *Obes. Rev.* **2012**, *13*, 329–346. [CrossRef] [PubMed]
12. Orfanos, P.; Naska, A.; Rodrigues, S.; Lopes, C.; Freisling, H.; Rohrmann, S.; Sieri, S.; Elmadfa, I.; Lachat, C.; Gedrich, K.; et al. Eating at restaurants, at work or at home. Is there a difference? A study among adults of 11 European countries in the context of the HECTOR* project. *Eur. J. Clin. Nutr.* **2017**, *71*, 407–419. [CrossRef]

13. Rodríguez-Pérez, C.; Molina-Montes, E.; Verardo, V.; Artacho, R.; García-Villanova, B.; Guerra-Hernández, E.J.; Ruíz-López, M.D. Changes in Dietary Behaviours during the COVID-19 Outbreak Confinement in the Spanish COVIDiet Study. *Nutrients* **2020**, *12*, 1730. [CrossRef] [PubMed]
14. Scarmozzino, F.; Visioli, F. COVID-19 and the Subsequent Lockdown Modified Dietary Habits of Almost Half the Population in an Italian Sample. *Foods* **2020**, *9*, 675. [CrossRef] [PubMed]
15. Marty, L.; de Lauzon-Guillain, B.; Labesse, M.; Nicklaus, S. Food choice motives and the nutritional quality of diet during the COVID-19 lockdown in France. *Appetite* **2021**, *157*, 105005. [CrossRef] [PubMed]
16. Ammar, A.; Brach, M.; Trabelsi, K.; Chtourou, H.; Boukhris, O.; Masmoudi, L.; Bouaziz, B.; Bentlage, E.; How, D.; Ahmed, M.; et al. Effects of COVID-19 Home Confinement on Eating Behaviour and Physical Activity: Results of the ECLB-COVID19 International Online Survey. *Nutrients* **2020**, *12*, 1583. [CrossRef] [PubMed]
17. Jayawardena, R.; Sooriyaarachchi, P.; Chourdakis, M.; Jeewandara, C.; Ranasinghe, P. Enhancing immunity in viral infections, with special emphasis on COVID-19: A review. *Diabetes Metab. Syndr. Clin. Res. Rev.* **2020**, *14*, 367–382. [CrossRef] [PubMed]
18. Mills, S.; Brown, H.; Wrieden, W.; White, M.; Adams, J. Frequency of eating home cooked meals and potential benefits for diet and health: Cross-sectional analysis of a population-based cohort study. *Int. J. Behav. Nutr. Phys. Act.* **2017**, *14*, 109. [CrossRef] [PubMed]
19. Wellard-Cole, L.; Davies, A.; Allman-Farinelli, M. Contribution of foods prepared away from home to intakes of energy and nutrients of public health concern in adults: A systematic review. *Crit. Rev. Food Sci. Nutr.* **2021**, 1–12. [CrossRef] [PubMed]
20. Cherikh, F.; Frey, S.; Bel, C.; Attanasi, G.; Alifano, M.; Iannelli, A. Behavioral Food Addiction During Lockdown: Time for Awareness, Time to Prepare the Aftermath. *Obes. Surg.* **2020**, *30*, 3585–3587. [CrossRef] [PubMed]
21. Di Renzo, L.; Gualtieri, P.; Pivari, F.; Soldati, L.; Attina, A.; Cinelli, G.; Leggeri, C.; Caparello, G.; Barrea, L.; Scerbo, F.; et al. Eating habits and lifestyle changes during COVID-19 lockdown: An Italian survey. *J. Transl. Med.* **2020**, *18*, 229. [CrossRef] [PubMed]
22. Murphy, B.; Benson, T.; McCloat, A.; Mooney, E.; Elliott, C.; Dean, M.; Lavelle, F. Changes in Consumers' Food Practices during the COVID-19 Lockdown, Implications for Diet Quality and the Food System: A Cross-Continental Comparison. *Nutrients* **2021**, *13*, 20. [CrossRef] [PubMed]
23. Poelman, M.P.; Gillebaart, M.; Schlinkert, C.; Dijkstra, S.C.; Derksen, E.; Mensink, F.; Hermans, R.C.J.; Aardening, P.; de Ridder, D.; de Vet, E. Eating behavior and food purchases during the COVID-19 lockdown: A cross-sectional study among adults in the Netherlands. *Appetite* **2021**, *157*, 105002. [CrossRef]
24. FDA. Food Safety and the Coronavirus Disease 2019 (COVID-19). Available online: https://www.fda.gov/food/food-safety-during-emergencies/food-safety-and-coronavirus-disease-2019-covid-19 (accessed on 1 December 2021).
25. Troise, C.; O'Driscoll, A.; Tani, M.; Prisco, A. Online food delivery services and behavioural intention—A test of an integrated TAM and TPB framework. *Brit. Food J.* **2021**, *123*, 664–683. [CrossRef]
26. Mehrolia, S.; Alagarsamy, S.; Solaikutty, V.M. Customers response to online food delivery services during COVID-19 outbreak using binary logistic regression. *Int. J. Consum. Stud.* **2021**, *45*, 396–408. [CrossRef]
27. Zang, J.J.; Luo, B.Z.; Wang, Y.P.; Zhu, Z.N.; Wang, Z.Y.; He, X.; Wang, W.J.; Guo, Y.; Chen, X.; Wang, C.F.; et al. Eating Out-of-Home in Adult Residents in Shanghai and the Nutritional Differences among Dining Places. *Nutrients* **2018**, *10*, 951. [CrossRef] [PubMed]
28. Zhong, Y.; Oh, S.; Moon, H.C. What Can Drive Consumers' Dining-Out Behavior in China and Korea during the COVID-19 Pandemic? *Sustainability* **2021**, *13*, 1724. [CrossRef]
29. Bilici, S.; Mortas, H.; Kose, S.; Varli, S.N.; Ayhan, B. Are restaurant menus vectors of bacterial cross-contamination? A pilot study in Turkey. *Brit. Food J.* **2017**, *119*, 401–410. [CrossRef]
30. Kim, T.J.; Almanza, B.; Ma, J.; Park, H.; Kline, S.F. The cleanliness of restaurants: ATP tests (reality) vs. consumers' perception. *Int. J. Contemp. Hosp. Manag.* **2021**, *33*, 893–911. [CrossRef]
31. Gursoy, D.; Chi, C.G. Effects of COVID-19 pandemic on hospitality industry: Review of the current situations and a research agenda. *J. Hosp. Market. Manag.* **2020**, *29*, 527–529. [CrossRef]
32. Taylor, S. The socially distant servicescape: An investigation of consumer preference's during the re-opening phase. *Int. J. Hosp. Manag.* **2020**, *91*, 9. [CrossRef] [PubMed]
33. Kim, J.; Lee, J.C. Effects of COVID-19 on preferences for private dining facilities in restaurants. *J. Hosp. Tour. Manag.* **2020**, *45*, 67–70. [CrossRef]
34. Hakim, M.P.; Zanetta, L.D.; da Cunha, D.T. Should I stay, or should I go? Consumers' perceived risk and intention to visit restaurants during the COVID-19 pandemic in Brazil. *Food Res. Int.* **2021**, *141*, 110152. [CrossRef]
35. Zanetta, L.D.A.; Hakim, M.P.; da Cunha, D.T. COVID-19 policies and recommendations for foodservice reopening: An integrative review. *J. Foodserv. Bus. Res.* **2021**, 1–25. [CrossRef]
36. Liu, P.; Lee, Y.M. An investigation of consumers' perception of food safety in the restaurants. *Int. J. Hosp. Manag.* **2018**, *73*, 29–35. [CrossRef]
37. Truong, N.; Nisar, T.; Knox, D.; Prabhakar, G. The influences of cleanliness and employee attributes on perceived service quality in restaurants in a developing country. *Int. J. Cult. Tour. Hosp. Res.* **2017**, *11*, 608–627. [CrossRef]
38. Barber, N.; Scarcelli, J.M. Clean restrooms: How important are they to restaurant consumers? *J. Foodserv.* **2009**, *20*, 309–320. [CrossRef]
39. Cullen, F. Factors Influencing Restaurant Selection in Dublin. *J. Foodserv. Bus. Res.* **2005**, *7*, 53–85. [CrossRef]

40. Tse, A.C.B.; So, S.; Sin, L. Crisis management and recovery: How restaurants in Hong Kong responded to SARS. *Int. J. Hosp. Manag.* **2006**, *25*, 3–11. [CrossRef] [PubMed]
41. Kim, J.; Kim, J.; Lee, S.K.; Tang, L.L. Effects of epidemic disease outbreaks on financial performance of restaurants: Event study method approach. *J. Hosp. Tour. Manag.* **2020**, *43*, 32–41. [CrossRef]
42. Clark, C.; Davila, A.; Regis, M.; Kraus, S. Predictors of COVID-19 voluntary compliance behaviors: An international investigation. *Glob. Transit.* **2020**, *2*, 76–82. [CrossRef]
43. Han, Q.; Zheng, B.; Cristea, M.; Agostini, M.; Belanger, J.J.; Gutzkow, B.; Kreienkamp, J.; Leander, N.P.; PsyCorona, C. Trust in government regarding COVID-19 and its associations with preventive health behaviour and prosocial behaviour during the pandemic: A cross-sectional and longitudinal study. *Psychol. Med.* **2021**, 1–11. [CrossRef]
44. DiGiovanni, C.; Conley, J.; Chiu, D.; Zaborski, J. Factors influencing compliance with quarantine in Toronto during the 2003 SARS outbreak. *Biosecurity Bioterrorism-Biodefense Strategy Pract. Sci.* **2004**, *2*, 265–272. [CrossRef]
45. Nivette, A.; Ribeaud, D.; Murray, A.; Steinhoff, A.; Bechtiger, L.; Hepp, U.; Shanahan, L.; Eisner, M. Non-compliance with COVID-19-related public health measures among young adults in Switzerland: Insights from a longitudinal cohort study. *Soc. Sci. Med.* **2021**, *268*, 113370. [CrossRef]
46. Rubin, G.J.; Amlot, R.; Page, L.; Wessely, S. Public perceptions, anxiety, and behaviour change in relation to the swine flu outbreak: Cross sectional telephone survey. *Br. Med. J.* **2009**, *339*, b2651. [CrossRef]
47. Scholz, J.; Wetzker, W.; Licht, A.; Heintzmann, R.; Scherag, A.; Weis, S.; Pletz, M.; Betsch, C.; Bauer, M.; Dickmann, P.; et al. The role of risk communication in public health interventions. An analysis of risk communication for a community quarantine in Germany to curb the SARS-CoV-2 pandemic. *PLoS ONE* **2021**, *16*, e0256113. [CrossRef] [PubMed]
48. Van Bavel, J.J.; Baicker, K.; Boggio, P.S.; Capraro, V.; Cichocka, A.; Cikara, M.; Crockett, M.J.; Crum, A.J.; Douglas, K.M.; Druckman, J.N.; et al. Using social and behavioural science to support COVID-19 pandemic response. *Nat. Hum. Behav.* **2020**, *4*, 460–471. [CrossRef] [PubMed]
49. Webster, R.K.; Brooks, S.K.; Smith, L.E.; Woodland, L.; Wessely, S.; Rubin, G.J. How to improve adherence with quarantine: Rapid review of the evidence. *Public Health* **2020**, *182*, 163–169. [CrossRef] [PubMed]
50. Moon, M.J. Fighting COVID-19 with Agility, Transparency, and Participation: Wicked Policy Problems and New Governance Challenges. *Public Adm. Rev.* **2020**, *80*, 651–656. [CrossRef]
51. Song, C.; Lee, J. Citizens' use of social media in government, perceived transparency, and trust in government. *Public Perform. Manag. Rev.* **2015**, *39*, 430–453. [CrossRef]
52. Wilson, A.M.; Withall, E.; Coveney, J.; Meyer, S.B.; Henderson, J.; McCullum, D.; Webb, T.; Ward, P.R. A model for (re)building consumer trust in the food system. *Health Promot. Int.* **2017**, *32*, 988–1000. [CrossRef]
53. Yost, E.; Cheng, Y. Customers' risk perception and dine-out motivation during a pandemic: Insight for the restaurant industry. *Int. J. Hosp. Manag.* **2021**, *95*, 102889. [CrossRef]
54. Zanetta, L.D.; Hakim, M.P.; Gastaldi, G.B.; Seabra, L.M.J.; Rolim, P.M.; Nascimento, L.G.P.; Medeiros, C.O.; da Cunha, D.T. The use of food delivery apps during the COVID-19 pandemic in Brazil: The role of solidarity, perceived risk, and regional aspects. *Food Res. Int.* **2021**, *149*, 110671. [CrossRef]
55. FOD Economie, KMO, Middenstand en Energie. *Guide for a Safe Restart of Hospitality*; FOD Economie: Brussels, Belgium, 2020.
56. Rawlins, B. Measuring the relationship between organizational transparency and employee trust. *Public Relat. J.* **2008**, *2*, 1–21.
57. Vandevijvere, S.; Lachat, C.; Kolsteren, P.; Van Oyen, H. Eating out of home in Belgium: Current situation and policy implications. *Br. J. Nutr.* **2009**, *102*, 921–928. [CrossRef]
58. Deliens, T.; Clarys, P.; De Bourdeaudhuij, I.; Deforche, B. Determinants of eating behaviour in university students: A qualitative study using focus group discussions. *BMC Public Health* **2014**, *14*, 53. [CrossRef] [PubMed]
59. Hair, J.F.; Black, W.C.; Babin, B.J.; Anderson, R.E. *Multivariate Data Analysis*, 7th ed.; Pearson Education Limited: Essex, UK, 2014.
60. Field, A. *Discovering Statistics Using SPSS*, 3rd ed.; SAGE: London, UK, 2009.
61. Wei, C.H.; Chen, H.; Lee, Y.M. Factors influencing customers' dine out intention during COVID-19 reopening period: The moderating role of country-of-origin effect. *Int. J. Hosp. Manag.* **2021**, *95*, 102894. [CrossRef]
62. Wei, C.; Chen, H.; Lee, Y.M. COVID-19 preventive measures and restaurant customers' intention to dine out: The role of brand trust and perceived risk. *Serv. Bus.* **2021**. [CrossRef]
63. Moran, K.R.; Del Valle, S.Y. A Meta-Analysis of the Association between Gender and Protective Behaviors in Response to Respiratory Epidemics and Pandemics. *PLoS ONE* **2016**, *11*, e0164541. [CrossRef] [PubMed]
64. Hesham, F.; Riadh, H.; Sihem, N.K. What Have We Learned about the Effects of the COVID-19 Pandemic on Consumer Behavior? *Sustainability* **2021**, *13*, 4304. [CrossRef]
65. Jin, J.-M.; Bai, P.; He, W.; Wu, F.; Liu, X.-F.; Han, D.-M.; Liu, S.; Yang, J.-K. Gender Differences in Patients With COVID-19: Focus on Severity and Mortality. *Front. Public Health* **2020**, *8*, 152. [CrossRef] [PubMed]
66. Byrd, K.; Her, E.; Fan, A.; Almanza, B.; Liu, Y.R.; Leitch, S. Restaurants and COVID-19: What are consumers' risk perceptions about restaurant food and its packaging during the pandemic? *Int. J. Hosp. Manag.* **2021**, *94*, 102821. [CrossRef]
67. Lee, C.K.; Song, H.J.; Bendle, L.J.; Kim, M.J.; Han, H. The impact of non-pharmaceutical interventions for 2009 H1N1 influenza on travel intentions: A model of goal-directed behavior. *Tour. Manag.* **2012**, *33*, 89–99. [CrossRef] [PubMed]
68. Webb, T.L.; Sheeran, P. Does changing behavioral intentions engender bahaviour change? A meta-analysis of the experimental evidence. *Psychol. Bull.* **2006**, *132*, 249–268. [CrossRef] [PubMed]

foods

MDPI

Article

Facing Food Risk Perception: Influences of Confinement by SARS-CoV-2 Pandemic in Young Population

Fernando Cantalapiedra, Cristina Juan * and Ana Juan-García

Laboratory of Food Chemistry and Toxicology, Faculty of Pharmacy, University of Valencia, Av. Vicent Andrés Estellés s/n, Burjassot, 46100 València, Spain; fernando.cantalapiedra@uv.es (F.C.); ana.juan@uv.es (A.J.-G.)
* Correspondence: cristina.juan@uv.es

Abstract: A new food safety level of trust in food risk perception has been noticed, as a consequence of the SARS-CoV-2 pandemic. The pandemic made-up to review nutritional recommendations for the population, mainly for the young population. Here, the results of a designed survey for the young population, from the University of Valencia, Spain, belonging to grades in the health branch of knowledge, and in charge of carrying out the shopping task for their household, are reported. The study reports three different scenarios and years, as defined by the SARS-CoV-2 pandemic: before the pandemic (period January–December 2019), during the pandemic lockdown (period March 2020–August 2020), and after the pandemic lockdown (September 2020–June 2021). The survey was designed with questions, profiling responses using the best–worst elicitation (BWE) format. Results reported that trust and evaluation of information differed in all three scenarios. In the SARS-CoV-2 pandemic, there was (i) a high increase in trust in the information provided *inside* (by) the shopping place, while there were no changes for the *outside* (kept in medium score); (ii) trust in cooperative stakeholders went from a medium-low to medium-high score, while, for individual stakeholders, it was maintained as a medium score, and (iii) trust in information on food products was kept in high score. Regarding the evaluation of the information provided by stakeholders, a tendency in medium score was maintained, while that from the channels of distribution went from medium-low to medium-high for buying on-site. A uniform tendency was observed for online/other distribution channels for all three years and descriptors studied: "Internet", "Farmer on-demand", and "Cooperative consumers" (<50%). This research provides findings of implications that contribute to changing the perception of food risk, due to the COVID-19 pandemic, i.e., the adaptation of the young population, trust in safety and quality, and importance of coordination from all communication points to avoid negative or strongest consequences, in case of future lockdowns or health crisis.

Keywords: young population; food perception; risk perception; food safety; COVID-19

Citation: Cantalapiedra, F.; Juan, C.; Juan-García, A. Facing Food Risk Perception: Influences of Confinement by SARS-CoV-2 Pandemic in Young Population. *Foods* **2022**, *11*, 662. https://doi.org/10.3390/foods11050662

Academic Editors: Derek V. Byrne and Arun K. Bhunia

Received: 17 January 2022
Accepted: 21 February 2022
Published: 24 February 2022

1. Introduction

Food safety, in a globalized world, is a major concern in the food supply chain for consumers. The concept of food safety seems to be completed, when there is no food risk or perceived, while the evaluation of risk perception implies the trust in purchasing a product and behavior [1,2]. In this sense, two points can be considered of great importance in food safety: (i) trust in food products and (ii) the levels of communications, which vary depending on the country or cultural contexts. Both points influence food perception across countries [3]. There are plenty of stages, indirectly implicit in the acquisition, for consumers to choose and consume a food product, which makes it necessary to give a wider trust in food, for the population, that guarantees food safety.

Food has several factors and connotations, when spreading risk messages, related to culture, symbolism, family, and even religion; in fact, there is evidence suggesting differences between food and non-food risks [4]. There are risks and benefits associated with food, such as the presence of contaminants and nutritional contribution. From both perspectives, when there is an intention of spreading this information, it is necessary for all to be considered as communication strategies [5].

Literature reports that the response of consumers to food safety or confidence is different, depending on the risk origin [6,7], so that, if a different behavior of potential risks is subsequently adopted, if it is referred to as technological or natural risks [8,9]. Similarly, this happens in the type of exposure to such risk: chronic or acute. In the acute context, the natural risk might increase the risk perception; while, in chronic context, the information provided in the risk assessment process by authorities is available for specific groups of vulnerable populations, so that the relationship among different circumstances causes different behavior and perception of risks [10].

During COVID-19, adequate nutrition was correlated with several indicators that influenced nutritional education (family member at home, watching TV during mealtime, country, maternal education, etc.) [11]. It pointed out that it was necessary to review and reshape nutritional recommendations for the young population, due to the different nutrition behavior reported during the COVID-19 lockdown [11].

Besides the aspect commented, it is of great importance to highlight the globalization that the market has developed in the last years. Warranties of good functioning in globalization can be supported, if there is trust in the food supply chain at different levels: general (understood as an interpersonal trust), stakeholders on the food chain, food authorities, and food products [12]. Nowadays, there has been a new level of trust, as a consequence of the SARS-CoV-2 pandemic, which is reflected in the study, presented here, for the young population. In this sense, the main focus of this study was to evaluate how the young population was facing food risk perception before, during, and after the lockdown pandemic, as well as what the most influential source/guide that provided safety and quality was.

National institutes of statistics have had to face the new scenario with COVID-19 by working on guidelines to obtain new methodologies of generating statistics, but most importantly to continue providing help in obtaining data across the globe [12–16]. One of the first steps in communicating food risk is to understand how consumers perceive that risk, and this is supported by involving science, evidence, and data regarding risk analysis; nevertheless, decisions in risk are also sustained by instinctual and/or emotional factors. These aspects have been indicated in previous studies; however, in the last year (referred to 2021), after the SARS-CoV-2 pandemic situation (after strict lockdowns), it could affect this perception and denote an increase in the demands of safety and quality on the part of the consumer and consumer agencies. Hence, the results of a designed survey for the young population from the University of Valencia, Spain, belonging to grades in the branch of health sciences and in charge of carrying out the shopping task for their household, are here reported (Table 1). The study collects the trust, regarding different sources of information, that gives warranties in the food chain, as well as how it is perceived for the young population. The study reports three different scenarios, defined by the SARS-CoV-2 pandemic: before the pandemic (period January–December 2019 "*normal life*"), during the pandemic lockdown (period March 2020–August 2020 "*during lockdown*"), and after the pandemic lockdown (September 2020–June 2021 "*after lockdown*").

Table 1. Characteristics of respondents, corresponding to the young population from the University of Valencia.

Students	%
Gender	
Male	16
Female	84
Age	
18–25 years old	66
26–35 years old	24
>35 years old	10
Education Level Degree in	
Pharmacy	37
Food Science	20
Dietetics and Human Nutrition	25
Master of Quality and Food Safety	18

2. Material and Methods

2.1. Respondents

Data were collected in Spain from students of the University of Valencia, through a validated survey, from 2019 to 2021 (Ethics Committee at University of Valencia—UV-INV-1942475). The sample, consisting of 600 individuals, represents the young population, who are in charge of shopping baskets for the household (Table 1). The information collects responses of students from different health branch degrees: human nutrition and dietetics, pharmacy, food science, and masters' degrees in quality and food safety. Students were contacted in a class by leaving the survey design open, with an age range from 18 to >35 years old. Characteristics of the population studied are collected in Table 1 and Figure 1. The survey was open for answering, for the groups described above; 84% were female respondents, and 16% were male. The order of respondents, per educational level, was pharmacy degrees, followed by dietetics and human nutrition, food science, and quality and food safety master's degrees (more details of enlisted respondents are in Section 3).

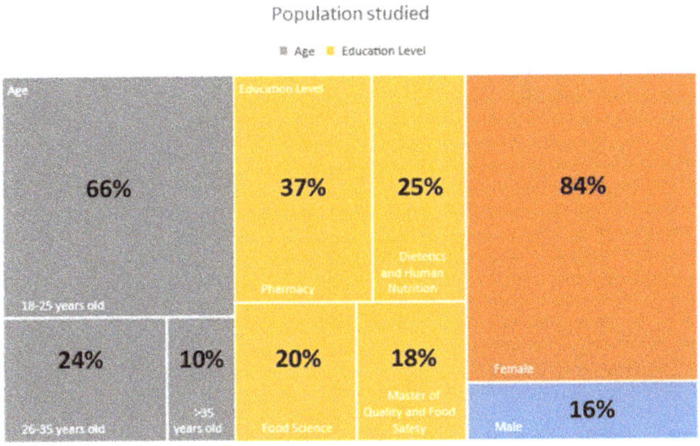

Figure 1. Characteristics of the young population studied, according to age, education level, and gender.

2.2. Survey Design and Recruitment

A survey was designed, with questions profiling responses using the best–worst elicitation (BWE) format (described in Petrolia et al., 2016 [17]). The BWE format refers to getting answers that indicate only the first-best choice, and it has been used in several studies, as reported in the literature [18–23]. It describes, somehow, an order of what is the best alternative, followed by the worst and those remaining, again, the "best" and worst, until all options are ranked. The methodology results are easier to obtain than the standard question format, and it also permits our group to understand the best–worst ranked answers for the risk perception of the population studied. The structure of the questionnaire corresponded to three sections: trust in those providing information of food (source and stakeholders of the food chain) (Figures 2 and 3), frequency of consumers in checking the information present in food (Figure 4), and evaluation of the information provided (Figures 5 and 6).

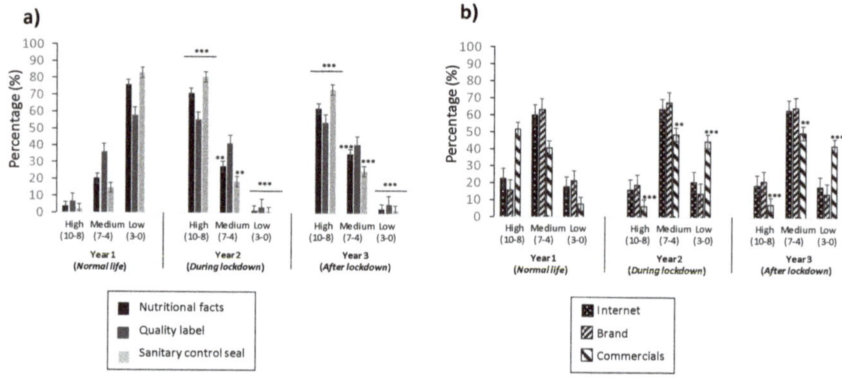

Figure 2. Trust of young population in information source of food products inside the shopping place (**a**) ("nutritional facts", "quality label", and "sanitary control seal") and outside the shopping place (**b**) ("internet", "brand", and "commercials"). Values correspond to the mean ± SD of population responding to the questionnaire. $^{**}p \leq 0.01$ and $^{***}p \leq 0.001$, with respect to the control (Year 1).

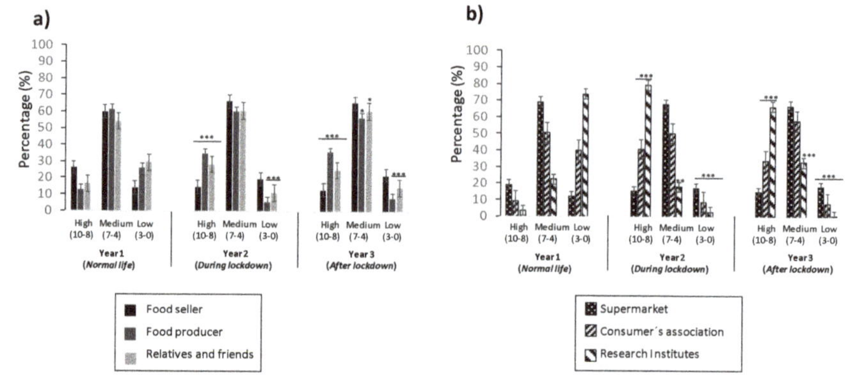

Figure 3. Percentage of trust of young population in stakeholders involved in the food chain production: individual ("Food seller", "Food producer", and "Relatives and friends") (**a**) and cooperative ("Supermarket", "Consumer's association", and "Research institutes") (**b**). Values correspond to the mean ± SD of the population responding the questionnaire. $^{*}p \leq 0.05$, $^{**}p \leq 0.01$, and $^{***}p \leq 0.001$, with respect to the control (Year 1).

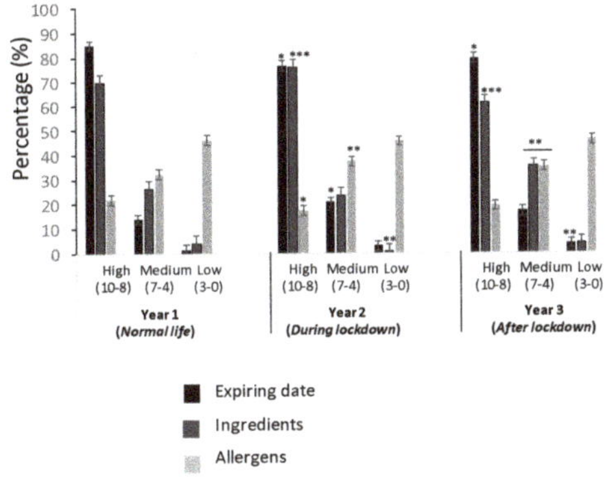

Figure 4. Percentage of young population checking information in food products, referred to: "Expiring date", "Ingredients", and "Allergens". Values correspond to the mean ± SD of the population responding to the questionnaire. * $p \leq 0.05$, ** $p \leq 0.01$, and *** $p \leq 0.001$, with respect to the control (Year 1).

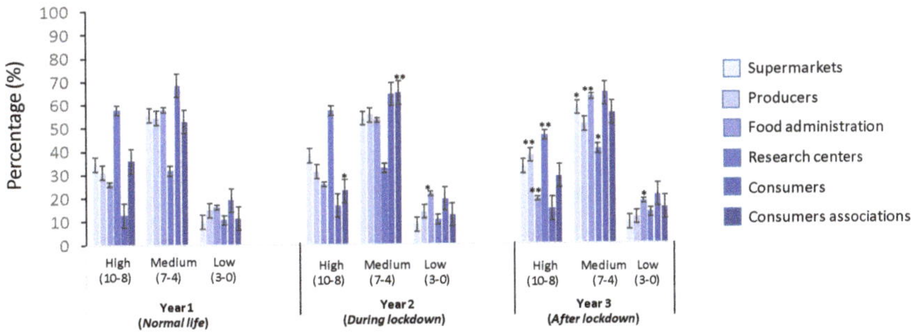

Figure 5. Evaluation of stakeholder in providing information of food products for the young population ("Supermarkets", "Producers", "Food administration", "Research centers", "Consumers", and "Consumers associations"). Values correspond to the mean ± SD of the population responding to the questionnaire. * $p \leq 0.05$, ** $p \leq 0.01$, with respect to the control (Year 1).

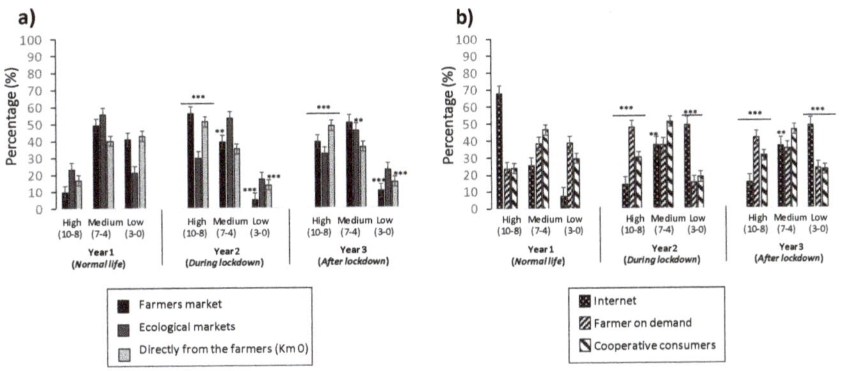

Figure 6. Evaluation of different channels for the young population to buy food: (**a**) on-site ("farmers market", "ecological markets", and "directly from the market"); (**b**) online/others ("internet", "farmer on demand", and "cooperative consumers"). Values correspond to the mean ± SD of the population responding to the questionnaire. ** $p \leq 0.01$, and *** $p \leq 0.001$, with respect to the control (Year 1).

2.3. *Questionnaire Used for the Survey*

The questionnaire was developed based on "best-worst elicitation" (BWE), in order to analyze whether the pandemic, associated with SARS-CoV-2 attributes, was associated with food safety and risk perception. The complete initial questionnaire consisted of 17 questions, previously tested through a pilot study, for its validation by the Organization of Consumers and Users—OCU organism in 2017 [24]. The questions selected for the survey were based on the existing data in the literature, carried out as a preceded task, while the discussed and reported results are based on those that gave a greater relevance. A scale from 0 to 10 points was used to describe the profile of trust or information from stakeholders, and the following score-levels were defined: (i) low: from 0 to 3 points; (ii) medium: from 4 to 7 points, and (iii) high: from 8 to 10 points. The questionnaire was answered in three different periods, as defined by the SARS-CoV-2 pandemic: before pandemic (*normal life*—Year 1), during pandemic lockdown (Year 2), and after pandemic lockdown (*end of lockdown*—Year 3).

2.4. *Statistical Analysis*

BWE allowed us to know the number of times an attribute of COVID-19 was selected as the most (best) or least influential (worst), as well as the average score for each attribute, for the entire sample, which allowed us to build the different figures described in the Results sections. Statistical analysis of data was carried out using the IBM SPSS Statistic version 23.0 (SPSS, Chicago, IL, USA) statistical software package and GraphPad Prism 8.0 (GraphPad Software, Inc., San Diego, CA, USA). Data were expressed as mean ± SD of three independent experiments. The statistical analysis of the results was performed by student's *t*-test for paired samples. Differences between groups were analyzed statistically with ANOVA, followed by the Tukey HDS post hoc test for multiple comparisons. The level of $p \leq 0.05$ was considered statistically significant.

3. Results

3.1. *Best–Worst Scores and Respondents*

The best–worst methodology allowed us to identify the most influential COVID-19 attributes, as considered by food companies during the pandemic. It was contemplated to have the different informative risk perceptions from each respondent that would influence their choices. Respondents were asked to provide a numerical rating for what they perceived to be the level of food safety for food, in general, without specifications but with a background on their studies (following BWE, described in Section 2.2). Results could change in populations with non-informative risk perceptions. Scores were divided into low, medium, and high categories, according to the description in Section 2.3.

The respondents' profiles are reported in Table 1 and Figure 1. The total number of respondents was 600 students, corresponding to 16% males and 84% females; ages were between 18 to >35 years old, while the distribution by educational level was as follows: 37% studying for a pharmacy degree, 20% studying for a food science degree, and 25% studying for a dietetics and human nutrition degree. Finally, the population studying for the master's in quality and food safety corresponded to 18%.

3.2. *Trust in the Information Source*

The information source in the young population revealed a markedly different score of trust, according to the survey, which is reported in Figure 2. It was divided by the information perceived (i) in the market, referred to as before or during purchasing (*inside* the shopping place) and contained on the packaging label (nutritional facts, quality seal, and sanitary control seal) (Figure 2a), and (ii) external (*outside*) to the shopping place, referred to as information not printed on the label but perceived from media (internet, brand, and commercials) (Figure 2b).

Figure 2a reports that, before COVID-19 "*normal live*" (Year 1), the trust of information contained on the label (*inside* the shopping place) was low for all three descriptors (nutritional facts, quality seal, and sanitary control seal), ranging from 58% to 83%, corresponding to "Quality seal" and "Sanitary control seal", respectively. This tendency changes drastically during and after COVID-19's lockdown (Year 2 and 3), as it went to high trust for all the three descriptors, above mentioned, with percentages of trust similar to those reported in low trust, from 54% to 73%, for the same descriptors as in Year 1 (Figure 2a). The medium trust for all three years studied revealed the following order of descriptors: "Quality seal" (from 36% to 41%) > "Nutritional facts" (from 20% to 35%) > "Sanitary control seal" (from 15% to 25%) (Figure 2a).

The results regarding the information perceived from "outside" of the shopping place, and mostly through media, is reported in Figure 2b. It reveals that the trust for the young population is in the medium score for the three descriptors (internet, brand, and commercials) and three years studied. The order that followed was: "Brand" (from 63% to 68%), "Internet" (from 60% to 62%), and "Commercials" (from 41% to 50%) (Figure 2b). When observing the "Commercial" factor, an interesting behavior, in all three years studied, for the high score of trust, was observed, as it went from 50% to 7% for Year 1 to Years 2 and 3, respectively; while, inversely, behavior was observed for the low score of trust, as it went from 8% to 45% for Year 1 to Years 2 and 3, respectively (Figure 2b). The "Internet" and "Brand" factors were maintained very similarly in all three years, for both high and low trust, in the ranges from 15% to 23% and 14% to 21%, respectively.

3.3. *Trust in Stakeholders in Food Chain Production*

Figure 3 reports the trust of the young population in the stakeholders involved in the food chain production, divided into individuals (Figure 3a) and cooperatives (Figure 3b). Trust by "Individuals" was maintained in the medium score for all three descriptors (sellers, producers, and relatives and friends), and the three years studied in the highest values, ranging from 54% to 61%. When observing the high score, increases of trust were observed for the (i) "Producer" descriptor, ranging from 13% to 36% for Years 1 to 3, respectively, and (ii) "Family and Friends" descriptor, ranging from 16% to 28% for Years 1 to 2, respectively, while decreases were observed for the "Seller" descriptor, ranging from 26% to 13% for Years 1 to 3 (Figure 3a). Results opposite to this were observed in the low score, as a decrease of trust values was observed for the (i) "Producer" descriptor, ranging from 26% to 8% for Years 1 to 3, respectively, and (ii) "Relatives and Friends" descriptor, ranging from 30% to 11% for Years 1 to 2, respectively, while an increase was observed for the "Seller" descriptor, ranging from 14% to 22%, from Years 1 to 3 (Figure 3a).

Trust by "Cooperatives" includes three descriptors: supermarket, consumer´s association, and research institutes. Results revealed that "Supermarkets" had similar values in all three years, from 12% to 69%, 15% to 68%, and 15% to 67% for Years 1, 2, and 3, respectively (Figure 3b). Notice that the upper bond values corresponded to the medium score. "Consumer´s association" provided a change in the values of trust from Years 1 to 2, as it went from a low to high score of trust (40% in both cases), although the highest values were for the medium score (50%). In Year 3, the percentage of trust was maintained, as well as for Year 2, although an increase of 8% was reported for the medium score, with a decrease of 8% for high score. Regarding the factor of "Research Center", it was the factor that suffered the major variation, as it went from low score of trust (with 74%) in Year 1 to a decrease of 3% and 1% for Years 2 and 3, respectively; in consequence, the high score of trust went from 4% in Year 1 to 79% in Year 2, reaching 66% in Year 3 (Figure 3b).

3.4. *Checking Information in Food Products*

Figure 4 reports the results related to checking the information in food products at the moment of buying, referred to as the "Expiring date", "Ingredients", and "Allergens". The highest values of trust were reported for the high score for "Expiring date", ranging from 77% to 85%, followed by "Ingredients" (from 61% to 76%). Notice that this behavior was maintained equally for all three years, with the following order of trust: high > medium > low. For "Allergens", the highest values were reported for a low score, with 46% in all three years, followed by a medium score, ranging from 32% to 35%, and, finally, a high score from 17% to 22%. In summary, "Ingredients" and "Expiring date" are highly checked and were maintained practically equally in all three of the years studied, while "Allergens" maintained low values, without changing in all three years.

3.5. Evaluating Information from Stakeholders

The results of evaluating the information of the food products provided from stakeholders, such as "Supermarkets", Producers", "Food Administration", "Research Centers", "Consumers", and "Consumers Associations", are reported in Figure 5.

In general terms, it is noticed that the evaluation is maintained in the medium score for all three years, followed by high and low score in the last position. In detail, the medium score ranged from 32% to 68%, 32% to 65%, and 40% to 64% for Years 1, 2, and 3, respectively; the "Consumers" factor had the highest score, and the "Research Centers" factor had the lowest score. The high score of the evaluation ranged from 13% to 58%, 17% to 57%, and 15% to 46% for Years 1, 2, and 3, respectively; the "Research Centers" factor had the highest score, and the "Consumers" factor had the lowest score. Finally, the low score was scored with the lowest percentages, as follows: from 10% to 19%, 8% to 21%, and from 9% to 21% for Years 1, 2, and 3, respectively; the "Food Administration" factor had the highest score, and the "Supermarkets" factor had the lowest score. In summary, the medium score was reported for "Supermarket" and "Food administrations", while high score was reached for "Research centers".

3.6. Evaluating Channels for Purchasing Food Products

A predisposition for purchasing food products by using different channels of distribution, such as "Farmer's market", "Ecological markets", "Directly from the market", "Internet", "Farmers on-demand", and "Cooperative consumers", was evaluated. Figure 6 is divided by channels that require us to move to a specific marketplace "on-site" (Figure 6a) and channels that allow for purchasing food products online (Figure 6b).

The tendency observed in Figure 6a was very similar for all three years, with evaluations fitting the medium scores from 40% to 56%, 35% to 53%, and 36% to 51% for Years 1, 2, and 3, respectively. In Year 2, the evaluation was very close to the medium score, with percentages ranging from 30% to 56%, due to the "Farmer's market" and "Directly from the market" factors; however, in Year 3, there was an increase in the evaluation of the low score for all three factors, studied from 10% to 22%.

The results reported in Figure 6b reveal that "Internet" was the factor experiencing the strongest changes among all three of the years studied. It went from a high score evaluation of 68% in Year 1 to 15% in Year 3, subsequently reaching a low score in Year 3, with 48%. "Farmers on-demand" had a maximum medium score in Year 1 (38%), which was similar in Years 2 (37%) and 3 (35%). Similarly, this happened for "Cooperative consumers", with medium score values of 46%, 51%, and 46% for Years 1, 2, and 3, respectively.

4. Discussion

Information sources in risk communication have a principal role in spreading the voice when facing and engaging food safety. According to several studies [25,26], the inclusion of new technologies (mainly through apps) has become the main source checked by consumers, with a special focus on mass media, as the main contributor, but also providing education to the population, allegedly due to the accurate reports [27]. However, the uncontrollable impact of such sources can be considered a negative contribution that emphasizes risks [28]. It has been demonstrated that behavior and decisions in food products are shaped by the consumer´s perception, contained in the information source [29]. Results, reported in Figure 2b, confirm the relevance of this fact, showing that trust in the information perceived by studied young population from a source *"outside"* the shopping place is highly distributed, rather than that from *"inside"* the shopping place, referring to that which is printed on the label (Figure 2a). The influence of this information source changed during and after the pandemic lockdown (Years 2 and 3 of study) by trusting more of the information contained on labels (Figure 2a). This observed behavior could be associated with an uncertain situation at the beginning of the pandemic, as news (and mass media) in the first lockdown reported the possibility of the virus spreading in food: "food could contribute and contain the virus SARS-CoV-2", another fact that could justify the changes observed in the population target. While sources evaluated in Figure 2a could be perceived as closer to our target population (students with a background in food safety), the ones in Figure 2b are closer to a broader audience, no matter the studies or background. These results coincide with an extensive study carried out in the UK population [30], Netherlands [31], and USA (Texas) [32]. It also puts in manifestation that the target audience is a factor to consider in food risk communication and, indeed, food risk perception.

In a study carried out for the Italian population, the quality of a food product and reason for purchasing such products have been associated [33,34]. Similarly, this happens with the typical products or PDO-certified products related to a brand [33,34]. Quality brands and certifications of origin are indicators that make it easier for consumers to judge and strengthen the perception of quality; in fact, not looking for information on the label can increase uncertainty, regarding the ability to appraise the quality, and encourage a tendency to rely on certification [33]. Sometimes this quality is reinforced with seals of quality or sanitary control or details in the nutritional facts, all contained in the label of the food and/or food product. All three indicators are reported in Figure 2a, which had been highly trusted during and after the lockdown pandemic, Years 2 and 3, respectively.

When adding value to a food product, it brings an increase of quality in consumers´ perception. That value can be provided in different ways, as reported by Mascarello et al. (2015) [33], with a coordinated flow within the food product, based on creating, maintaining, and enhancing characteristics in the food product [33]. When providing these advantages, the information contained on labels is crucial to communicate them to the consumer; the broader the audience, the greater the role of labels, quality seals, and communication. Communication is the tool used by institutions that look into scientific evidence and focus on specific groups that define perception and target actions that help to promote healthy behaviors [35].

Additionally, the lifestyle, household composition, age of residents, and employment of a determined local area or population group affect the determination of quality and food risk perception [33]. However, when a scandal involves the food industry, only brands are associated with a guarantee of food safety, which sometimes can also fall into distrust [33,36,37]. This point is important, as it has been demonstrated that COVID-19 has a human-to-human transmission, which causes a complete, indirect effect on the food industry [38].

When focusing outside the shopping place, in a study carried out in Chinese population (aged 40 years old), regarding the information source reported by different channels, it was observed that television (TV) was the most-used channel, followed by the internet and "other sources" [39]. In Turkey, similar results were obtained, with TV and mass media as the main information sources of trust; however, government publications were highly trusted, which reinforced the point of helping to educate consumers by food authorities [40]. All this can vary, when focused on a specific type of food, as demonstrated in a study in population from South Korea, regarding genetically modified food (GMO); when exploring risk perception, in general, journalists and science journalists' were the latest that had more trust [41].

Facing food safety during COVID-19, by the food industry, sparked special attention, and several studies started to come up describing or reporting the issue that had to be strengthened [42]. In a survey study, carried out in 16 countries and by more than 800 food companies, it was revealed that the most important attributes faced were the staff awareness and the implementation of restrictive hygiene procedures, following the two main documents that WHO had developed for the food supply chain [38,43]. The industry was not compromised at any moment, regardless of food safety, despite not having any protocol or emergency plan for a previous pandemic [42]. Since then, protocols of the food chain have been reinforced, emphasizing the hygiene of hands, disinfecting packages, use of adequate equipment of protection, and preparation of food [44].

The stakeholders implicated in food chain production, in providing information, supposes a factor that can alter the risk perception in the trust that the population puts in the information source, so that there are perceptions of the participants playing a determinant role, according to stakeholders in the food chain. In general, it is expected to have greater trust in those perceived as more knowledgeable, honest, or closer to the information, as demonstrated in a study carried out in different populations [35,39,45]; nevertheless, there are country cultural factors that influence such responses. Similar to the results obtained here, the punctuation obtained in this study can be related to the stakeholders involved in the food chain production, either individually or from cooperatives. In the study of Liu et al., 2014 [39], the factors "Food producer" and "Relatives and friends" were perceived with honesty and concern, respectively, for citizen´s health perception. In our study, both were perceived with the highest values, especially during the lockdown pandemic (Year 2), and classified as individual stakeholders involved in food chain production (Figure 3a). Among that, the factors "Consumer´s association" and "Research institutes" were perceived as honesty in providing accurate information [39]; in our study, the percentage of trust varied indistinctly for each year studied but were classified as cooperative stakeholders involved in food chain production (Figure 3b).

Consumers´ behaviors and attitudes toward safe food should be taken into account, in regard to perceived food safety, i.e., checking the package information of the food before purchasing, for example, the content of allergens, ingredients, expiring date, origin, calories, nutrition facts, and brand were measured in this study (Figure 4). One of the observations before purchasing a food product is the expiration date, referred to as the last date that a food should be eaten or used, i.e., understanding that, after that date, the characteristics of the food product are altered, and risk might occur. In a study carried out with the Turkish population, it was revealed that there is a rejection behavior to expiring date information [40]. In our study, this information was highly checked before pandemic (Year 1) and decreased during lockdown (Year 2) and after it (Year 3) (Figure 4). There was not further investigation carried out here to explain this, but it could be hypothesized that there was a high trust in food products, due to the no association of infection through them, despite the initial message from some news sources. EFSA declared that there was no scientific evidence that food was a risk or transmission route of the virus [46]. Publicity on TV and media were providing security in all food products, related to SARS-CoV-2, after a few months of the pandemic lockdown. Another issue reported in this study, as well as in Figure 4, is checking the information from the ingredients, which reported a similar behavior as that for expiring date; for allergens, this inversely correlated behavior, as compared to the expiring date and ingredients. The allergens content is information usually checked by consumers.

In evaluating the information provided by several stakeholders of the food production, in a study carried out with the Turkish population and referred to the information provided to them, it was evaluated as unreliable from that of scientists and specialists, while the government was evaluated with the highest value and responsibility to ensure food safety [40]. This does not coincide with our results, due to the different profile of population studied, not only in age but also in the country and studies background, which was the base of our population recruitment (Figure 5). Food manufacturers, scientists, and media were the greatest valuable stakeholders in providing product information, according to Rohr et al. (2005) [47], and even more trustworthy by consumers or environmental organizations [40]. This coincides with our results, as consumers were highly evaluated, and the highest values during the pandemic lockdown (Year 2). On the other side, it has been reported that nutritionists, which constitutes the group of the population studied here, have a high value in spreading information of food safety [40]; jointly, consumers´ association was extremely reliable. This fact was also observed here.

Finally, the preference for purchasing food products during the pandemic lockdown was also asked in our young population of study, as there was an increase of news in TV and media related to online (internet) shopping, coinciding with Brugarolas et al. (2020) [48], who also noticed a stockpiled food tendency during COVID-19, due to buying more often. That fact brought producers and distribution companies to develop strategies to decrease this effect or stock non-perishable foods. However, according to the group of the population studied here, there were no difference during the three years of study, in relation to the SARS-CoV-2 pandemic (Figure 6). The population was asked about their disposition to buying food products from different channels; besides the marked young age of the population studied, there was a high tendency of buying in markets of proximity, but also on the internet (online). Results are very hopeful, considering the new tendencies around food products and market introduced for new styles of life. In the study of [48], the population studied was broader (from 18 to more than 65) than the one reported here (Table 1).

It is important to mention that this study was focused on the risk perception for a specific group of the population, which coincides with the pandemic lockdown, due to SARS-CoV-19, without paying attention to any specific product type. The interest lays in analyzing the topic of risk perception, associated with food safety, as a co-complex field, as well as defining the behavior of young consumers. Additionally, a very unique and specific circumstance is reported, as several factors shifted the behavior, while, for others, this was maintained; it would be interesting to analyze the approach for a specific type of food product.

5. Conclusions

Consumers' education starts with a strong trust in the basis of the information that is provided by the different stakeholders involved in food production/manufacturers. The alterations in food risk perception are produced when a pandemic sprouted/arrived and changed trust and confidence in several aspects. This situation has shown a high reliance and trust with nutritional facts, quality labels, and sanitary control seals, after and during the pandemic lockdown, as well as an increase in the trust of the farmers' market, farmers of demand, and internet shopping. Results of this study put, in evidence, the importance of trust, regarding the information spread for food risk perception for the young population, with background studies in the field of food safety, as well as the implications of the legislation for some labels and stakeholders, which could be more influential in some aspects. Questionnaires of food risk perception, per group of population, help to give us a better idea of the perception of food safety, as well as to make comparisons between groups of population and focus the campaigns of education in food production. Nevertheless, further studies of collaboration would be necessary to have a broader picture of more countries for this group of the population.

Author Contributions: Data curation: F.C., A.J.-G. and C.J.; formal analysis: F.C. and C.J.; visualization: A.J.-G., C.J. and F.C.; investigation: F.C., C.J. and A.J.-G.; methodology: F.C. and C.J.; supervision: A.J.-G. and C.J.; writing—original draft: F.C. and C.J.; writing—review and editing: C.J. and A.J.-G. All authors have read and agreed to the published version of the manuscript.

Funding: This research received no external funding.

Institutional Review Board Statement: The study was conducted according to the guidelines of the Declaration of Helsinki and approved by the Institutional Ethics Committee of University of Valencia (UV-INV-1942475).

Informed Consent Statement: Ethics Committee at University of Valencia—UV-INV-1942475.

Data Availability Statement: Not applicable.

Conflicts of Interest: The authors declare no conflict of interest.

References

1. Hong, I.B.; Cho, H. The impact of consumer trust on attitudinal loyalty and purchase intentions in B2C e-marketplaces: Intermediary trust vs. seller trust. *Int. J. Inf. Manag.* **2011**, *31*, 469–479. [CrossRef]
2. Hong, I.B.; Cha, H.S. The mediating role of consumer trust in an online merchant in predicting purchase intention. *Int. J. Inf. Manag.* **2013**, *33*, 927–939. [CrossRef]
3. Murphy, B.; Benson, T.; McCloat, A.; Mooney, E.; Elliott, C.; Dean, M.; Lavelle, F. Changes in consumers' food practices during the COVID-19 lockdown, implications for diet quality and the food system: A cross-continental comparison. *Nutrients* **2021**, *13*, 20. [CrossRef] [PubMed]
4. Food and Agriculture Organization of the United Nations/World Health Organization (FAO/WHO). Forthcoming. In *A Handbook on Risk Communication Applied to Food Safety*; Food and Agriculture Organization: Rome, Italy, 2015. Available online: http://www.auv-ks.net/repository/docs/Final_version_Handbook_28-11-2014.pdf (accessed on 16 January 2022).
5. Van Dijk, H.; Fischer, A.R.H.; de Jonge, J.; Rowe, G.; Frewer, L.J. The Impact of Balanced Risk-benefit Information and Prior Attitudes on Post-information Attitudes. *J. Appl. Soc. Psychol.* **2012**, *42*, 1958–1983. [CrossRef]
6. Nauta, M.J.; Fischer, A.R.H.; van Asselt, E.D.; de Jong, A.E.I.; Frewer, L.J.; de Jonge, R. Food Safety in the Domestic Environment: The Effect of Consumer Risk Information on Human Disease Risks. *Risk Anal.* **2008**, *28*, 179–192. [CrossRef]
7. Frewer, L.J.; van der Lans, I.A.; Fischer, A.R.H.; Reinders, M.J.; Menozzi, D.; Zhang, X.; van den Berg, I.; Zimmermann, K.L. Public Perceptions of Agri-food Applications of Genetic Modification–A Systematic Review and Meta-analysis. *Trends Food Sci. Technol.* **2013**, *30*, 142–152. [CrossRef]
8. Haleem, A.; Javaid, M.; Vaishya, R. Effects of COVID-19 pandemic in daily life. *Curr. Med. Res. Pract.* **2020**, *10*, 78–79. [CrossRef]
9. Fanelli, R.M. Changes in the Food-Related Behaviour of Italian Consumers during the COVID-19 Pandemic. *Foods* **2021**, *10*, 169. [CrossRef]
10. Kaptan, G.; Fischer, A.R.H.; Frewer, L.J. Extrapolating understanding of food risk perceptions to emerging food safety cases. *J. Risk Res.* **2018**, *21*, 996–1018. [CrossRef]
11. Ruíz-Roso, M.B.; de Carvalho Padilha, P.; Matilla-Escalante, D.C.; Brun, P.; Ulloa, N.; Acevedo-Correa, D.; Ferreira Peres, W.A.; Martorell, M.; Bousquet Carrilho, T.R.; de Oliveira Cardoso, L.; et al. Changes of Physical Activity and Ultra-Processed Food Consumption in Adolescents from Different Countries during COVID-19 Pandemic: An Observational Study. *Nutrients* **2020**, *12*, 2289. [CrossRef]
12. Murphy, B.; Benson, T.; Lavelle, F.; Elliott, C.; Dean, M. Assessing differences in levels of food trust between European countries. *Food Control* **2021**, *120*, 107561. [CrossRef]
13. EUROSTAT–European Statistics Office. 2021. Available online: https://ec.europa.eu/eurostat/web/COVID-19/population-health (accessed on 15 January 2022).
14. Martinez-Ferran, M.; de la Guía-Galipienso, F.; Sanchis-Gomar, F.; Pareja-Galeano, H. Metabolic Impacts of Confinement during the COVID-19 Pandemic Due to Modified Diet and Physical Activity Habits. *Nutrients* **2020**, *12*, 1549. [CrossRef]
15. Górnicka, M.; Drywień, M.E.; Zielinska, M.A.; Hamułka, J. Dietary and Lifestyle Changes During COVID-19 and the Subsequent Lockdowns among Polish Adults: A Cross-Sectional Online Survey PLifeCOVID-19 Study. *Nutrients* **2020**, *12*, 2324. [CrossRef]
16. INE–National Institute of Statistics from Spain. 2021. Available online: https://www.ine.es/COVID/COVID_salud.htm (accessed on 15 January 2022).
17. Petrolia, D.R. Risk preferences, risk perceptions, and risky food. *Food Policy* **2016**, *64*, 37–48. [CrossRef]
18. Marley, A.A.J.; Louviere, J.J. Some probabilistic models of best, worst, and best-worst choices. *J. Math. Psychol.* **2005**, *49*, 464–480. [CrossRef]
19. Flynn, T.N.; Louviere, J.J.; Peters, T.J.; Coast, J. Best-worst scaling: What it can do for health care research and how to do it. *J. Health Econ.* **2007**, *26*, 171–189. [CrossRef] [PubMed]
20. Potoglou, D.; Burge, P.; Flynn, T.; Netten, A.; Malley, J.; Forder, J.; Brazier, J.E. Best-worst scaling vs. discrete choice experiments: An empirical comparison using social care data. *Soc. Sci. Med.* **2011**, *72*, 1717–1727. [CrossRef]
21. Scarpa, R.; Notaro, S.; Louviere, J.; Raffaelli, R. Exploring scale effects of best/worst rank ordered choice data to estimate benefits of tourism in alpine grazing commons. *Am. J. Agric. Econ.* **2011**, *93*, 813–828. [CrossRef]
22. Flynn, T.; Marley, A.J. Best worst scaling: Theory and methods. In *Handbook of Choice Modelling*; Hess, S., Daly, A., Eds.; Edward Elgar Publishing: Cheltenham, UK, 2014; pp. 178–201.
23. Rigby, D.; Burton, M.; Lusk, J.L. Journals, preferences, and publishing in agricultural and environmental economics. *Am. J. Agric. Econ.* **2015**, *97*, 490–509. [CrossRef]
24. OCU. 2017. Available online: https://www.ocu.org/alimentacion/seguridad-alimentaria/noticias/encuesta-seguridad-alimentaria (accessed on 10 January 2019).
25. Kuttschreuter, M.; Rutsaert, P.; Hilverda, F.; Regan, Á.; Barnett, J.; Verbeke, W. Seeking Information about Food-Related Risks: The Contribution of Social Media. *Food Qual. Prefer.* **2014**, *37*, 10–18. [CrossRef]
26. Rutsaert, P.; Barnett, J.; Gaspar, R.; Marcu, A.; Pieniak, Z.; Seibt, B.; Lima, M.L.; Fletcher, D.; Verbeke, W. Beyond Information Seeking: Consumers' Online Deliberation about the Risks and Benefits of Red Meat. *Food Qual. Prefer.* **2015**, *39*, 191–201. [CrossRef]
27. Tiozzo, B.; Mari, S.; Ruzza, M.; Crovato, S.; Ravarotto, L. Consumers' perceptions of food risks: A snapshot of the Italian Triveneto area. *Appetite* **2017**, *111*, 105–115. [CrossRef] [PubMed]

28. Laestadius, L.I.; Lagasse, L.P.; Smith, K.C.; Neff, R.A. Print News Coverage of the 2010 Iowa Egg Recall: Addressing Bad Eggs and Poor Oversight. *Food Policy* **2012**, *37*, 751–759. [CrossRef]
29. Zhang, Y.; Yalin, S.; Yeolib, K. The Influence of Individual Differences on Consumer's Selection of Online Sources for Health Information. *Comput. Hum. Behav.* **2017**, *67*, 303–312. [CrossRef]
30. Redmond, E.C.; Griffith, C.J. Consumer Perceptions of Food Safety Risk, Control and Responsibility. *Appetite* **2004**, *43*, 309–313. [CrossRef]
31. Kornelis, M.; de Jonge, J.; Frewer, L.; Dagevos, H. Consumer selection of food-safety information sources. *Risk Anal.* **2007**, *27*, 327–335. [CrossRef]
32. Whatley, K.W.; Doerfert, D.L.; Kistler, M.; Thompson, L. An examination of the food safety information sources and channels utilized and trusted by residents of Lubbock, Texas. *J. Agric. Educ.* **2005**, *46*, 70–81. [CrossRef]
33. Mascarello, G.; Pinto, A.; Parise, N.; Crovato, S.; Ravarotto, L. The perception of food quality. Profiling Italian consumers. *Appetite* **2015**, *89*, 175–182. [CrossRef]
34. Keningham, T.; Aksoy, L.; Perkins-Munn, T.; Vavra, T. The brand customer connection. *Mark. Manag.* **2005**, *14*, 33–37.
35. Verbeke, W.; Frewer, L.J.; Scholderer, J.; De Brabanderd, H.F. Why consumers behave as they do with respect to food safety and risk information. *Anal. Chim. Acta.* **2007**, *586*, 2–7. [CrossRef]
36. Van Rijswijk, W.; Frewer, L.J. Consumer perceptions of food quality and safety and their relation to traceability. *Br. Food J.* **2008**, *110*, 1034–1046. [CrossRef]
37. Yeung, R.; Yee, W.M.S. Food safety concern. Incorporating marketing strategies into consumer risk coping framework. *Br. Food J.* **2010**, *114*, 40–53. [CrossRef]
38. World Health Organization & Food and Agriculture Organization of the United Nations. COVID-19 and Food Safety: Guidance for Food Businesses: INTERIM Guidance, 7 April 2020. 2020. Available online: https://apps.who.int/iris/handle/10665/331705 (accessed on 15 January 2022).
39. Liu, R.; Pieniak, Z.; Verbeke, W. Food-related hazards in China: Consumers' perceptions of risk and trust in information sources. *Food Control* **2014**, *46*, 291–298. [CrossRef]
40. Ergönül, B. Consumer awareness and perception to food safety: A consumer analysis. *Food Control* **2013**, *32*, 461–471. [CrossRef]
41. Kim, S.-H.; Kim, H.; Oh, S.-H. Talking About Genetically Modified (GM) Foods in South Korea: The Role of the Internet in the Spiral of Silence Process. *Mass Commun. Soc.* **2014**, *17*, 713–732. [CrossRef]
42. Djekic, I.; Nikolić, A.; Uzunović, M.; Marijke, A.; Liu, A.; Han, J.; Brnčić, M.; Knežević, N.; Papademas, P.; Lemoniati, K.; et al. COVID-19 pandemic effects on food safety-Multi-country survey study. *Food Control* **2021**, *122*, 107800. [CrossRef]
43. World Health Organization & Food and Agriculture Organization of the United Nations. COVID-19 and Food Safety: Guidance for Competent Authorities Responsible for National Food Safety Control Systems: Interim Guidance, 22 April 2020. 2020. Available online: https://apps.who.int/iris/handle/10665/331842 (accessed on 15 January 2022).
44. Maragoni-Santos, C.; de Souza, T.S.P.; Matheus, J.R.V.; Nogueira, T.B.B.; Xavier-Santos, D.; Miyahira, R.F.; Antunes, A.E.C.; Fai, A.E.C. COVID-19 pandemic sheds light on the importance of food safety practices: Risks, global recommendations, and perspectives. *Crit. Rev. Food Sci. Nutr.* **2021**, *16*, 1–13. [CrossRef]
45. Verbeke, W.; Vermeir, I.; Brunsø, K. Consumer evaluation of fish quality as basis for fish market segmentation. *Food Qual. Prefer.* **2007**, *18*, 651–661. [CrossRef]
46. EFSA. Coronavirus: No Evidence That Food Is a Source or Transmission Route. 2020. Available online: https://www.efsa.europa.eu/es/news/coronavirus-no-evidence-food-source-or-transmission-route (accessed on 15 January 2022).
47. Röhr, A.; Lüddecke, K.; Drusch, S.; Müller, M.J.; Alvensleben, R.V. Food quality and safety—Consumer perception and public health concern. *Food Control* **2005**, *16*, 649–655. [CrossRef]
48. Brugarolas, M.; Martinez-Carrasco, L.; Rabadan, A.; Bernabeu, R. Innovation strategies of the Spanish agri-food sector in response to the black swan COVID-19 pandemic. *Foods* **2020**, *9*, 1821. [CrossRef]

foods

MDPI

Article

Factors Influencing Customer Decisions to Use Online Food Delivery Service during the COVID-19 Pandemic

Kyungyul Jun [1], Borham Yoon [2,*], Seungsuk Lee [3] and Dong-Soo Lee [3]

1 Department of Food and Nutrition, Kosin University, Busan 49104, Korea; kjun@kosin.ac.kr
2 Department of Food and Nutrition, Sunchon National University, Suncheon 57922, Korea
3 Department of Parks, Recreation and Hospitality Administration, Arkansas Tech University, Russellville, AR 72801, USA; slee17@atu.edu (S.L.); dlee13@atu.edu (D.-S.L.)
* Correspondence: byoon@scnu.ac.kr; Tel.: +82-61-750-3658

Abstract: Despite the popularity of online food delivery systems in the foodservice industry, there have been few studies into customers' decision-making process to use online food delivery services during the Coronavirus disease (COVID-19) pandemic. This study applied the technology acceptance model (TAM) to examine the factors affecting customers' intention to use online food delivery services. Results showed (a) the perceived usefulness affects customer's online food delivery usage directly and indirectly through customer attitude; (b) enjoyment and trust are also key factors determining behavior intention toward customer attitude using online food delivery services; (c) positive relationship between social influence and customer attitude; and (d) a positive relationship between customer attitude and behavior intention in the online food delivery service context. These findings provide theoretical and managerial implications that contribute to the online food delivery service industry.

Keywords: online food delivery service; COVID-19 pandemic; technology acceptance; trust; enjoyment; social influence

Citation: Jun, K.; Yoon, B.; Lee, S.; Lee, D.-S. Factors Influencing Customer Decisions to Use Online Food Delivery Service during the COVID-19 Pandemic. *Foods* **2022**, *11*, 64. https://doi.org/10.3390/foods11010064

Academic Editor: Derek V. Byrne

Received: 3 November 2021
Accepted: 22 December 2021
Published: 28 December 2021

1. Introduction

According to the World Health Organization (WHO), the 2019 coronavirus disease (COVID-19) erupted in China in December 2019 and expanded as a global pandemic on 11 March 2020 [1]. Because COVID-19 has a high risk of death and human-to-human transmission, self-quarantine, wearing a mask in public, social distancing, and restriction of people's movement have been strongly recommended by WHO [1]. Consequently, most of the United States required residents to stay at home and forced foodservice operations to be closed or restricted [2].

With the restriction of dine-in service due to COVID-19, many restaurants adapted and heavily relied on contactless and online food delivery systems to survive. The number of foodservice and users using online food delivery systems has surged during COVID-19 [3]. About 67 percent of residents preferred using online delivery services to purchase food during the COVID-19 pandemic in the US [3]. Online food delivery service refers to internet-based food ordering delivery services that connect customers with partner foodservice operations via their websites or mobile applications [4]. Online food delivery services provide a wide range of restaurant lists, allowing customers to compare menus, prices, and even reviews from other users by restaurant types. Furthermore, the distribution of mobile devices has provided customers with a new platform—food delivery apps—that is available when they order food online. Moreover, it is expected that more customers and restaurants utilized online food delivery services during and after the COVID-19 pandemic.

Existing studies provide an understanding of customers' motivations to use online food delivery systems [5,6] and factors affecting online food delivery service usage [4,7,8].

Additionally, previous studies have examined the factors based on the technology acceptance model (TAM) [9] that determine whether or not an individual adapts to innovation. Although TAM is a robust and powerful theoretical framework of users' acceptance and usage of technology, testing and extending TAM by integrating it with other factors (social influence, trust, and enjoyment) may provide insight for food service industry management to develop the strategies of online food delivery services. Moreover, few studies have examined the factors influencing customers' decision making toward the use of online food delivery services, especially under pandemic conditions [10]. As the COVID-19 pandemic has changed customers' dining and consumption behaviors, it is necessary to consider the COVID-19 pandemic as a context factor affecting customers' online food delivery service [11]. Therefore, the purpose of the study is to examine the factors affecting customers' online food delivery services usage by applying TAM and other factors (e.g., enjoyment, trust, and social influence) to provide a comprehensive model during the COVID-19 pandemic. This research provides a theoretical foundation by using the TAM in the food delivery context and practical implications for online food delivery and the foodservice industry during the COVID-19 pandemic.

2. Literature Review

2.1. Online Food Delivery Service

The progress in informational technology has introduced a new business model into the food service industry. Along with the advent of internet technology, some big fast-food chains, especially pizza franchises, have been the pioneers to embrace online food ordering with their websites. Restaurants have adopted online food ordering because it has met or exceeded expectations in several ways for restaurant operations [12]. Online food ordering has grown in popularity among customers and restaurants because of its benefits [13].

Online food ordering through websites was introduced with several different concepts. Aside from the websites operated by restaurant chains, as mentioned above, the predecessors of online food ordering services only aggregated and listed restaurants' names with their basic information, such as phone numbers or addresses, on their website platforms. Those platforms have begun to provide more information, including menus or prices. Subsequently, online food ordering websites have taken food orders from allied restaurants. In this stage, the food ordering platforms have grabbed the food orders solely. Restaurants took care of the delivery by themselves if delivery was available. The latest approach in the food ordering systems has been for the platform to take care of the delivery. Conclusively, when restaurants utilize online food ordering, they may operate their websites or receive the orders through multiple-restaurant platforms. In addition, the food delivery may be carried out directly by the restaurants to the customer (e.g., Domino's), or the platform picks up the meals at the restaurant and delivers them to customers (e.g., Uber Eats). Some platforms (e.g., GrubHub) provide both services [13]. The online food delivery services began with online food ordering; the online food delivery service is separately a significant business model. Recently, online food delivery was defined as the process that food ordered online is prepared and delivered to the customers by connecting customers with partner foodservice operations via their websites or mobile applications [4].

The demand for online food delivery services has dramatically increased over the last few years and is expected to grow. The global online food delivery platform market already amounts to US $31 billion [13]. As COVID-19 has changed, customers prefer a contactless and online-to-delivery system to face-to-face and dine-in service [11,14]. The online food delivery market continues to attract new customers. Therefore, factors motivating customers to use online food delivery services under the COVID-19 pandemic are needed to understand customers' decision-making process and therefore help the foodservice business survive in this era.

2.2. Technology Acceptance Model

The TAM is a widely adopted conceptual framework to explain the acceptance of new technology [9,15]. The underlying foundation of TAM is a series of concepts that explains and predicts a certain human behavior with beliefs, attitudes, and behavioral intention (BI). The relationship among belief, attitude, intention, and behavior initially provides the theoretical base to the famous and robust theories in social psychology, including the theory of reasoned action (TRA) [16] and the theory of planned behavior [17]. An individual's attitude toward conduct is considered proportional to a weighted sum of their evaluations of relevant beliefs about the predicted consequences of that behavior in these early theories [16]. However, in TAM, general beliefs (e.g., perceived ease of use (EOU) and perceived usefulness (PU)) rather than salient beliefs are considered to play an essential role in shaping attitudes toward utilizing a particular technology [9].

One of the most significant differences between TRA and TAM is the mediator role of attitude. Attitude fully mediates the effect of beliefs on behavioral intention in TRA, so the causal relations between the beliefs (EOU and PU) are not assumed. TRA posits that a person's attitude toward a behavior is directly proportional to the sum of the beliefs about the behavior [17,18]. Therefore, an attitude captured separately is found as a formative construct and a composite of beliefs. However, after TAM without attitude was proposed [19], many empirical studies have suggested the direct impact of beliefs on behavioral intention and the relations among beliefs (e.g., PU, EOU, enjoyment, and/or trust). In a meta-analysis review with articles in the e-commerce context, Ingham et al. [20] confirms that TAM including attitude as a mediator is a better explanatory model than TAM without attitude.

Hypothesis 1 (H1). *Attitude significantly influences behavioral intention.*

2.3. Perceived Usefulness (PU) and Ease of Use (EOU)

Along with attitude, TAM frequently discusses the relationships between beliefs. Attitude is explained as a partial mediator of beliefs, and EOU is claimed to be a direct causal predecessor of PU [9]. The causal relations between beliefs are proposed in many empirical studies. However, there are many inconsistencies in the relationships between beliefs. Moreover, as an example, the relations between beliefs are suggested in the reverse direction with equally persuasive logical arguments by respected scholars [21,22]. Whereas Pavlou [21] argues that trust affects EOU, Gefen et al. [22] argue that EOU affects trust. Nevertheless, Ingham et al. [20] reason the introduction of causal relationships between beliefs is based on the common usage of structural equations rather than on well-established theoretical grounds. This is a notable explanation for Davis's [15] initial justification of the causal relationship between EOU and PU, which is more circumstantial and data-dependent than theoretical.

PU is often proposed to affect the behavioral intention directly as well as indirectly through attitude. The mediator role of PU between EOU and BI is confirmed in several reviews of TAM [23–25]. The direct effects of beliefs on behavioral intention were introduced in the early visions of TAM.

Hypothesis 2a (H2a). *PU significantly influences attitude.*

Hypothesis 2b (H2b). *PU significantly influences behavioral intention.*

Hypothesis 3 (H3). *EOU significantly influences attitude.*

2.4. Enjoyment (EJM)

Since Davis et al. [9] introduced the concept of enjoyment into TAM, this concept is a significant factor that drives users to use a new technology [20]. Davis et al. [9] adopted enjoyment as the extrinsic motivation to test its direct effect on the behavioral intention and the indirect effect through usefulness. Davis et al. [9] stated that whatever positive or negative feelings may be brought to mind toward a specific behavior have a causal link to intention. The direct impacts of perceived risk and trust on intention were tested by many academic studies, including Pavlou [21]. Enjoyment was driven by the motivation theory [19]. The motivation to perform an activity is broadly classified into two categories: extrinsic (instrumental) motivation and intrinsic (hedonic) motivation. Whereas perceived usefulness is an example of extrinsic motivation, enjoyment is an example of intrinsic motivation. Hederson et al. [26] argue that enjoyment is the most important predictor of intention in the study that uses TAM as the reference model for the electronic supermarket. Childers et al. [27] find that enjoyment is vital in predicting customers' attitudes toward target behaviors. In several e-commerce studies, enjoyment is a meaningful direct predictor of the intention to use e-shopping [28–32] or an indirect predictor through a positive attitude toward using it [33,34]. Although many studies introduce playfulness instead of enjoyment, both enjoyment and playfulness are used similarly in empirical studies [19,31,35–37].

Hypothesis 4a (H4a). *Enjoyment significantly influences attitude.*

Hypothesis 4b (H4b). *Enjoyment significantly influences behavioral intention.*

2.5. Trust (TR)

Safety is one of the main reasons why many customers hesitate to purchase online [38]. Trust is the customers' beliefs about the retailers' safety and internet technology. In previous studies, perceived risk is treated as a distinctive variable from trust, and even the casual relationship of these two variables was studied [21]. However, they are very similar in their conceptualization except in opposite directions. Customers' perceptions of risk are often characterized as their expectations of probable losses or other unfavorable outcomes from a transaction. Both the vendors and the transaction itself are associated with negative views [39,40]. Otherwise, trust is defined as a collection of precise ideas about the vendor's trustworthiness [21,33,38], a sense of confidence and security about online transactions [41,42], or a combination of trustworthiness in the vendor and trustworthiness in the transaction [30,43]. Therefore, with some modifications, trust and perceived risk can be treated as alternative variables. In this study, the authors use trust as a comprehensive concept, including perceived risk.

Previous research has shown that trust directly affects intended use [38,44–46]. Furthermore, risk perception is a direct negative predictor of intention [21,43,47–51]. Trust plays an essential role in developing a good attitude towards e-shopping [35,47,49,52–54]. Risk perception is a direct negative predictor of attitude [47–49,55].

Hypothesis 5a (H5a). *Trust significantly influences attitude.*

Hypothesis 5b (H5b). *Trust significantly influences behavioral intention.*

2.6. Social Influence (SI)

According to Fishbein and Ajzen [16], subjective norms reflect how the customer is affected by the perception of some significant references to one's behavior. According to Venkatesh et al. [56], social influence is a broad notion that encompasses the concepts of subjective norm, social factors, and image. In TAM research, social influence, including subjective norms, does not effectively predict, especially in a voluntary setting [20,53–57]. However, in the case of e-shopping acceptance, social influence is a direct positive antecedent of intention [26,31,42,49,58,59].

According to two research studies, social influence has a favorable effect on attitude [31,60]. Barkhi and Wallace [60] conceptualize peer influence as a salient belief to shape customer attitude toward purchasing decisions in virtual stores. Kim et al. [31] argue that subjective norms significantly affect attitude toward the use and perceived usefulness and intention to reuse in the context of customer acceptance of airline B2C e-commerce websites.

Hypothesis 6a (H6a). *Social influence significantly influences attitude.*

Hypothesis 6b (H6b). *Social influence significantly influences behavioral intention.*

Based on the previous discussion, the causal relations among beliefs are set aside, and the beliefs, including PU, EOU, trust, and enjoyment, are treated as exogenous variables and modeled as attitude components toward the behavioral intention. Consequently, attitude is employed as a mediator between beliefs and intention. Also, the direct effects of beliefs on behavioral intention are allowed (Figure 1).

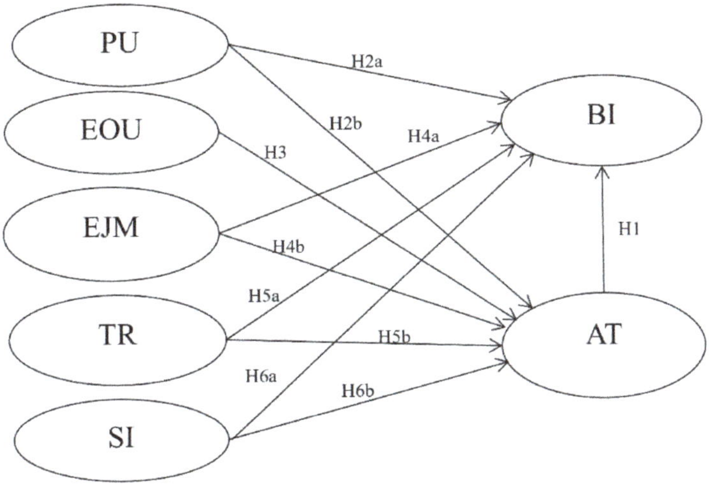

Figure 1. Research model. Note: PU = perceived usefulness; EOU = perceived ease of use; EJM = enjoyment; TR = trust; SI = social influence; AT= attitude; BI=behavior intention.

3. Methodology

3.1. Measurement

A self-administered questionnaire was developed based on a comprehensive literature review of online food delivery and technology-oriented quality attributes. A total of 23 items were adopted from existing literature related to technology acceptance and online food delivery [9,17,27,29,56,61]. At the beginning of the survey, a definition of an online food delivery service was presented. The questionnaire consisted of four parts measuring the constructs including (a) TAM variables of perceived usefulness, perceived ease of use, attitude toward online food delivery service, and intention to use food delivery service; (b) trust, enjoyment, and social factors for online food delivery services; (c) experiences with online food delivery services; and (d) demographic data, including gender, ethnicity, education, and household income. Specifically, items of usefulness and ease of use were adapted from Davis [9]. Items of trust and enjoyment were utilized from Pavlou [21] and Childers et al. [27], respectively. Social influence items were taken from Ajzen [17]. Attitude items were adapted from Suh and Han [61]. Finally, behavior intention items were adapted from Suh and Han [61] and Venkatesh et al. [56]. All constructs, except for socio-demographic information, were measured using a 7-point Likert scale anchored by "strongly disagree (1)" and "strongly agree (7)."

3.2. Sample and Data Collection

An online questionnaire was developed using Qualtrics and distributed via Amazon's Mechanical Turk system (MTurk). The current study targeted the general U.S. customers over 18 years who have used online food delivery ordering during the ongoing COVID-19 pandemic. At the beginning of the survey, each participant was screened to confirm they live in the U.S. and they had at least one online food delivery ordering experience within the past three months. The data were collected through MTurk over two weeks, from 6 July 2020 to 19 July 2020.

Among 450 responses collected, 20 respondents did not fully complete the survey. After reviewing their submissions, four samples were deleted because they failed to answer the attention check question correctly (i.e., "For this question, please select "Strongly disagree" to demonstrate your attention"). As a result of this data screening process, a total of 426 responses were used for data analysis.

3.3. Data Analysis

The data were analyzed using IBM SPSS 22 and AMOS 22 software. The data analysis followed the two-step approach by Anderson and Gerbing [62]. The first step assessed the reliability and validity of the measurement model. A confirmatory factor analysis (CFA) was conducted to assess the reliability and validity of the measurement model in the first step. To evaluate reliability and validity, Cronbach's alpha values and factor loadings were estimated. Reliability and convergence of the factors were also examined by composite reliability (CR) and average variance extracted (AVE). Discriminant validity was determined by comparing AVEs with the squared multiple correlations between constructs.

The second step tested the research model and the proposed hypotheses by the structural equation model (SEM). Seven common model-fit measures were used to assess the model's overall goodness of fit: the ratio of $\chi 2$ to degrees of freedom (df), comparative fit index (CFI), the goodness of fit index (GFI), adjusted goodness of fit index (AGFI), normalized fit index (NFI), Tucker-Lewis index (TLI), and root mean square error of approximation (RMSEA).

4. Results

4.1. Descriptive Analysis

As shown in Table 1, the percentage of male respondents (54.4%) was slightly higher than the percentage of female respondents (45.5%). About 74.6% of respondents were Caucasian, followed by African American (10.1%), Asian American (7.5%), Hispanic (5.4%), Native American (0.7%), and Other (1.6%). About half of the respondents had Bachelor's and Graduate degrees (57.0%), and nearly half of the respondents lived in suburban areas (51.9%). Approximately 77.1% of respondents used online food delivery services more than once per month.

4.2. Validity and Reliability of Measurements

A confirmatory factor analysis (CFA) was conducted to validate the internal and external consistency of the constructs used in the study. As shown in Table 2, the CFA results found satisfactory goodness of fit indices ($\chi 2$ = 525.962, df = 67, CMIN/df = 2.517, RMR = 0.065, GFI = 0.903, AGFI = 0.872, NFI = 0.945, IFI = 0.966, CFI = 0.966, RMSEA = 0.060) [63]. The adequacy of the measurement model was evaluated based on the criteria of reliability and convergent validity. Reliability was examined based on the composite reliability (CR) value. In Table 2, all of the values are above 0.7, indicating adequate composite reliability [63]. The average variance extracted (AVE) of all seven latent variables was higher than the suggested threshold value of 0.5, suggesting the convergent validity of the scale [63]. Thus, the reliability and convergent validity of the constructs applied in the study were supported.

Table 1. Profile of the respondents.

Demographic Characteristics		Frequency	Percentage
Gender	Male	232	54.5
	Female	194	45.5
Ethnicity	White/Caucasian	318	74.6
	African American	43	10.1
	Hispanic or Latino	23	5.4
	Asian American	32	7.5
	Native American or American Indian	3	0.7
	Others	7	1.6
Educational level	High school diploma and under	120	28.2
	Associate degree	63	14.8
	Bachelor's degree	180	42.4
	Graduate degree (Master or Doctoral)	62	14.6
Annual household income	Less than $20,000	47	11.0
	$20,000~$39,999	94	22.1
	$40,000~$59,999	76	17.8
	$60,000~$79,999	85	20.0
	$80,000~$99,999	50	11.7
	$100,000 or more	74	17.4
Living area	Urban	158	37.1
	Suburban	221	51.9
	Rural	47	11.0
Frequency of use	Several times a day	4	0.9
	Once a day	4	0.9
	Several times a week	68	16.0
	Once a week	105	24.6
	At least once a month	148	34.7
	At least once every two months	40	9.4
	At least once every three months	37	8.7
	Only used once	20	4.7

Table 2. Results of confirmatory factory analysis.

Constructs and Measurement Items	Standardized Loading	CR	AVE
Perceived Usefulness (PU, Cronbach's Alpha = 0.871)			
Online food delivery platform makes my food ordering efficient	0.853	0.874	0.698
Online food delivery platform enhances my effectiveness in food ordering	0.814		
Online food delivery platform is useful in food ordering	0.840		
Perceived ease of use (EOU, Cronbach's Alpha = 0.894)			
Learning to operate the online food delivery platform is easy for me	0.844	0.896	0.743
The online food delivery platform is clear and understandable	0.834		
The online food delivery platform is easy to use	0.906		
Enjoyment (EJM, Cronbach's Alpha = 0.896)			
I have fun using the online food delivery platform	0.823	0.896	0.683
Using the online food delivery platform is exciting	0.813		
Using the online food delivery platform is enjoyable	0.854		
Using the online food delivery platform is interesting	0.816		
Trust (TR, Cronbach's Alpha = 0.899)			
The online food delivery platform is trustworthy	0.889	0.900	0.751
The online food delivery platform keeps promises and commitments	0.833		
I trust in the online food delivery platform	0.876		

Table 2. *Cont.*

Constructs and Measurement Items	Standardized Loading	CR	AVE
Social influence (SI, Cronbach's Alpha = 0.902)			
People who influence my behavior think that I should use the online food delivery platform	0.870	0.904	0.759
People who are important to me think that I should use the online food delivery platform	0.944		
My friends want me to use the online food delivery platform	0.793		
Attitude (AT, Cronbach's Alpha = 0.921)			
Using the online food delivery platform is a pleasant idea	0.889	0.921	0.795
Using the online food delivery platform is a positive idea	0.905		
Using the online food delivery platform is an appealing idea	0.881		
Behavior Intention (BI, Cronbach's Alpha = 0.967)			
I intend to continue using the online food delivery platform in the future	0.959	0.967	0.880
I predict I would use the online food delivery platform in the future	0.932		
I plan to use the online food delivery platform in the future	0.934		
I expect my use of the online food delivery platform to continue in the future	0.927		

χ^2/df = 2.517 ($p < 0.001$), CFI = 0.966, GFI = 0.903, AGFI = 0.872, NFI = 0.945, TLI = 0.959, RMSEA = 0.060. Note. CR = composite reliability; AVE = average variance extracted.

To examine discriminant validity, we compared the squared root of the AVE of each construct and its correlation coefficients with other constructs [64]. The result in Table 3 shows that all square roots of the AVEs ranging from 0.827 to 0.938 were larger than those corresponding correlation coefficients among the constructs. Thus, the discriminant validity of the constructs was supported [64]. In summary, the measurement model demonstrated adequate reliability, convergent validity, and discriminant validity.

Table 3. Correlations and discriminant validity.

Variable	Perceived Usefulness	Perceived Ease of Use	Enjoyment	Trust	Social Influence	Attitude	Behavior Intention
Perceived usefulness	0.836						
Perceived ease of use	0.783	0.862					
Enjoyment	0.637	0.475	0.827				
Trust	0.799	0.724	0.675	0.866			
Social influence	0.357	0.230	0.464	0.476	0.871		
Attitude	0.809	0.642	0.728	0.801	0.466	0.892	
Behavior intention	0.763	0.615	0.499	0.725	0.406	0.763	0.938

Note: Diagonal elements show square root of the average variance extracted (AVE). Below the diagonal is the correlation coefficient.

4.3. Hypotheses Testing

To use the path coefficients supplied by SEM to test the hypotheses, it is necessary to assess the model's goodness-of-fit for the variables. The goodness-of-fit tests are used to determine how well a model fits the data. The goodness-of-fit measures (χ^2 = 526.048, df = 66, CMIN/df = 2.505, RMR = 0.065, GFI = 0.903, NFI = 0.945, IFI = 0.966, CFI = 0.966, RMSEA = 0.060) were found to largely satisfy the evaluation criteria. Figure 2 and Table 4 show the results of structural model analysis and results of testing hypotheses, respectively. Perceived usefulness (β = 0.461, $p \leq 0.000$) was found to have a significantly positive effect on attitude (β = 0.461, $p \leq 0.001$) and BI (β = 0.475, $p \leq 0.001$), which means H2a and H2b were supported by the model. EOU (β = −0.031, p = 0.675) was not found to have a significant effect on attitude, thus not supporting H3. EJM was found to have a significant positive effect on attitude (β = 0.238, $p \leq 0.001$) thus supporting H4a. EJM was found to have a significant effect on BI (β = 0.241, $p \leq 0.001$), thus supporting H4b. Trust was found to positively affect attitude (β = 0.302, $p \leq 0.001$) and BI (β = 0.240, p = 0.009), thus supporting H5a and H5b. Social influence was not found to affect attitude significantly, thus not supporting H6a. Social influence was found to positively affect BI (β = 0.084, p =0.032), thus supporting H6b. Finally, attitude was found to significantly influence BI (β = 0.499, $p \leq 0.001$), thus supporting H1. As shown in Table 4 and Figure 2, all the hypotheses are supported except for hypotheses H3 and H6a.

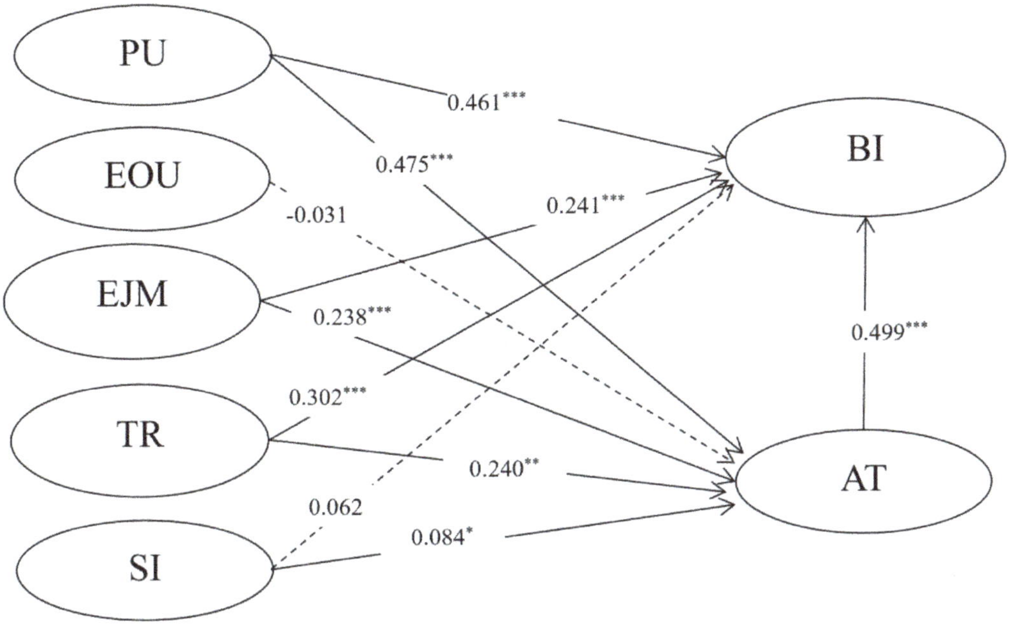

Figure 2. Structural equation model with parameter estimates. * $p < 0.05$, ** $p < 0.01$, *** $p < 0.001$. Non-significant paths are shown in dotted lines. Note. PU = perceived usefulness; EOU = perceived ease of use; EJM = enjoyment; TR = trust; SI = social influence; AT= attitude; BI=behavior intention.

Table 4. Result of structural model analysis.

Hypotheses		Beta	S.E.	Critical Ratio	p-Value	Decision
H1	AT -> BI	0.499 ***	0.092	5.413	0.000	Supported
H2a	PU -> AT	0.461 ***	0.086	5.366	0.000	Supported
H2b	PU -> BI	0.475 ***	0.098	4.864	0.000	Supported
H3	EOU -> AT	−0.031	0.075	−0.419	0.675	Not supported
H4a	EJM -> AT	0.238 ***	0.047	5.052	0.000	Supported
H4b	EJM -> BI	0.241 ***	0.061	−3.968	0.000	Supported
H5a	TR -> AT	0.302 ***	0.077	3.929	0.000	Supported
H5b	TR -> BI	0.240 **	0.092	2.602	0.009	Supported
H6a	SI -> AT	0.062	0.032	1.942	0.052	Not supported
H6b	SI -> BI	0.084 *	0.039	2.143	0.032	Supported

* $p < 0.05$, ** $p < 0.01$, *** $p < 0.001$. Note. PU = perceived usefulness; EOU = perceived ease of use; EJM = enjoyment; TR = trust; SI = social influence; AT = attitude; BI = behavior intention.

5. Discussion

This study examined the factors affecting customers' intentions to use online food delivery services by using the extended approach of the TAM. The findings of this study confirmed the significant effect of enjoyment (EJM), trust (TR), and social influence (SI) on customers' acceptance of online food delivery services. Data analysis results demonstrated that perceived usefulness (PU), EJM, and TR were determinants that positively influenced

BI directly. Furthermore, PU, EJM, TR, and SI were found to have an impact on BI, with attitude (AT) serving as a mediating variable. As a result, the PU, EJM, TR, SI, and AT influence the intention to use online food delivery services. According to the path analysis, it can be concluded that the proposed model in the current study fits to explain the antecedents of online food delivery service usage intention during the COVID-19 pandemic. Four of the five proposed variables (PU, EJM, TR, SI) were found as statistically significant factors influencing customers' intention to adopt online food delivery services.

In comparing the path coefficients of antecedents of AT, PU was the most powerful predictor of AT toward online food delivery service relative to the other belief factors. Among the factors resulting in BI, PU was found to be the most influential factor affecting customers' online food delivery service used. This result confirms previous studies related to adopting new technologies and services in the online shopping context [20,27,34]. Moreover, these results are consistent with prior research [5,10], showing that customers are more likely to use the online food delivery service if they perceive it as useful.

The two main beliefs in the construction of BI toward adopting new technology in the traditional form of TAM are PU and EOU. Furthermore, EOU is often used as an antecedent of PU in structural equations in recent TAM research. However, the causal relationships between beliefs are not based on well-established theoretical ground [15]. Especially in the context of e-commerce, the causal relationships among beliefs are questionable [20]. This is consistent with a finding that as attitude captured more beliefs, the more the influence of EOU declined on the model in a meta-analytic study [9].

Additionally, the current study found that EOU is not a significant factor in AT. These results are counter to previous findings [23–25] but consistent with results from Yuan et al. [65] and Zhao and Bacao [7]. Because customers gained enough experience from their previous usage of online food delivery services, the ease of use of online food delivery services will no longer determine customers' attitudes after their initial adoption of online food delivery services. Furthermore, during the COVID-19 pandemic, other factors, such as safety, efficiency, and trust, are more important determinants and can provide more benefits for customers.

The second most significant factor of customers' online food delivery service usage intention was TR. This finding is similar to previous studies [35,38,44–46] showing that TR has a significant effect on customer technology adoption intention in the online shopping context. Customers may be unsure whether the restaurant accurately receives orders or whether the quality of food delivered is as excellent as the quality of food served at the restaurant, which underlines the necessity of TR in the online food delivery service context [10]. During the pandemic, when contactless delivery was essential and required, trust significantly formulated customers' intention to use online food service under the COVID-19 pandemic situation.

Interestingly, enjoyment was also found as a strongly significant determinant of customer attitude and intention toward using online food delivery services. This result is consistent with previous studies [35–37], which discovered a positive relationship between enjoyment/playfulness and attitude toward new technology in the online shopping context. This study revealed that the more customers thought that utilizing online food delivery services was exciting, fun, and enjoyable, the more they positively used those services. In the hospitality industry, hedonic motivation was the critical factor affecting customers' service quality evaluation [66]. Likewise, customers seek excitement, pleasure, and fun when purchasing food via an online delivery service. Therefore, the current study clarified that enjoyment mattered when using the online food delivery service. Furthermore, it is a notable suggestion that these studies could replace EOU as a determinant of AT or BI in a TAM. Eventually, different beliefs influence attitude and intention to use a platform differently because different environments for the platform differentiate the importance of each belief [27].

However, the influence of SI was a statistically significant factor influencing AT, but it does not have a strong impact on BI towards online food delivery usage. These results differ from the previous findings [31] in the fashion context but are similar to previous studies [20,56,57]. Compared to the fashion products that customers tend to interact with others to reduce their uncertainty or anxiety about adopting innovation in fashion, customers' perceived pressure is relatively low from the social community in terms of adopting or using online food delivery services.

6. Implications and Future Research

6.1. Theoretical Implications

This study contributes to the current literature with several theoretical implications. First, the present study demonstrated and provided empirical evidence that the TAM model worked in online food delivery services during the COVID-19 pandemic. The study enriched the literature on technology usage in an emergency, especially during a pandemic. With intensifying competition across the foodservice industry and increasing difficulty for companies to reach their customers during the pandemic, this study provides a theoretical framework for online food delivery literature.

Second, this study employed the TAM to evaluate customer behavior intentions to utilize online food delivery services theoretically. The comprehensive approach of TAM with the other factors (i.e., enjoyment, trust, and social influence) is considered to have better explanatory power than the standard TAM model. This study is notable in that this study attempted to apply the TAM model to measure customer behavior, which has rarely been used in the realm of online food delivery services. Furthermore, this study confirmed that utilitarian value (i.e., perceived usefulness) is more critical than hedonic value (i.e., enjoyment) in online food delivery services used. The finding of this study makes a substantial academic contribution to e-commerce and online food services. Therefore, this study attempted to provide a comprehensive model to understand customers' online food delivery services usage.

6.2. Practical Implications

The global pandemic has limited and even obliterated in-person experiences that the food services industry depends on for survival. Many food and catering service businesses are turning to virtual events to keep their customers and generate new revenue streams [67]. Due to the restrictions on in-person service in the food service industry, customer demand for online food delivery services has skyrocketed. The COVID-19 pandemic provided numerous opportunities for customers and food service operators to adopt new technologies and develop new platforms for online food delivery. The study's findings can help food service operators formulate effective strategies for running profitable food service businesses that use online food delivery. During the pandemic, online food delivery services improved their marketing to encourage people to use more online food services following the Centers for Disease Control and Prevention's (CDC) social distancing guideline [1]. As a result, more jobs were created to cover the online food delivery markets. This research has managerial implications for online food delivery service providers as well. Food service operators should emphasize social influence, accurate product information, convenience, and online food service quality.

First, the study's findings suggest that having excellent perceived usefulness is essential to promote customers' intention to use online food delivery services. The perceived usefulness of online food delivery services could be improved by providing high quality information [8]. It is suggested that online food delivery service providers offer up-to-date information by periodically updating product information according to consumption trends (e.g., restaurant list, price, or menu information). The service providers also need to ensure that they provide accurate and reliable information, such as business hours, delivery areas, and time, which increase customers' food ordering effectiveness.

Second, the online food delivery platform should consider providing detailed product information, which elicits enjoyment of the online service [37]. For example, service providers can offer information about the reviews of the restaurant and food, healthy menu options with calorie information, organic and locally sourced ingredients, and detailed descriptions and pictures of food, which lead customers to feel enjoyment in the food ordering process. Furthermore, the visual features and graphic design of the online food delivery service platform could provide fun for customers while using the service.

Third, one of the most effective ways to increase customer satisfaction with online food delivery is to build customer trust. Trust is crucial in influencing positive attitudes and behaviors toward online food delivery services. Given that a firm privacy policy can increase trust in an online service [40], online food delivery service providers should commit to protecting customer privacy (e.g., personal and credit card information). Service providers should post a clearly stated privacy policy to build trust and reduce risk.

Finally, online food delivery services managers should understand what they want and need to meet customers' expectations during the COVID-19 pandemic. In the present pandemic, it is critical to emphasize the benefits of online food delivery services (e.g., food safety, hygiene, and contactless delivery) in reducing the annoyance caused by COVID-19 in people's daily lives. Because of the ban on social gatherings and dine-in service, it is believed that individuals will be ready to eat at home and protect themselves during the pandemic. Therefore, food service managers must figure out how to positively reach out to their customers while persuading them to continue using online food delivery services during the pandemic. This will assist customers in perceiving highly personalized treatment from online food delivery service providers, resulting in improved business performance and increased customer satisfaction.

6.3. Limitations and Future Study

Despite its implications, some limitations and suggestions for future research should be discussed. First, the sample of this study is the online food delivery service users in the United States, which may limit the generalization of the findings to other countries. Results may vary across countries due to cultural differences, technology acceptance, and other factors. As a result, the research model employed in this study should be replicated and tested in other countries to confirm its validity and usefulness. In the future, longitudinal research will be needed to fine-tune the findings of this study and compare the COVID-19 pandemic to post-COVID-19 conditions. Second, other variables such as the frequency with which an online food delivery service is used and demographic traits (such as age and gender) could be regarded as moderating factors. The relationship between variables related to TAM and customers' intention to use online food delivery services could be different depending on the level of food delivery user experience. For example, the relationship between perceived ease of use (EOU) and attitude could be supported for customers who use online services infrequently. Due to a lack of experience, they may be unable to understand and interpret the information easily; thus, EOU could be a key factor influencing customers' intentions to purchase food through an online food delivery platform. In future studies, it is recommended that an additional variable should be included to gain a better understanding of customer behaviors in the context of an online food delivery service.

Author Contributions: Conceptualization, K.J. and B.Y.; Data curation, S.L. and D.-S.L.; Formal analysis, K.J.; Investigation, B.Y.; Methodology, K.J.; Project administration, K.J.; Writing—original draft, K.J. and B.Y.; Writing—review & editing, S.L. and D.-S.L. All authors have read and agreed to the published version of the manuscript.

Funding: This research received no external funding.

Institutional Review Board Statement: The study was conducted in accordance with the Declaration of Helsinki, and approved by the Institutional Review Board of The University of Tennessee, Knoxville (UTK IRB-20-05928-XM, 2 July 2020).

Informed Consent Statement: Informed consent was obtained from all subjects involved in the study.

Data Availability Statement: Not applicable.

Conflicts of Interest: The authors declare no conflict of interest.

References

1. World Health Organization. Coronavirus. 2020. Available online: https://www.who.int/health-topics/coronavirus#tab=tab_1 (accessed on 27 November 2021).
2. Restaurant Law Center. Official Orders Were Closing or Restricting Foodservice Establishments in Response to COVID-19. 2020. Available online: https://restaurant.org/downloads/pdfs/business/covid19-official-orders-closing-or-restricting.pdf (accessed on 21 November 2021).
3. Statista. 2020. Available online: https://www.statista.com (accessed on 27 November 2021).
4. Ray, A.; Dhir, A.; Bala, P.K.; Kaur, P. Why do people use food delivery apps (FDA)? A uses and gratification theory perspective. *J. Retail. Consum. Serv.* **2019**, *51*, 221–230. [CrossRef]
5. Yeo, V.C.S.; Goh, S.-K.; Rezaei, S. Customer experiences, attitude and behavioral intention toward online food delivery (OFD) services. *J. Retail. Consum. Serv.* **2017**, *35*, 150–162. [CrossRef]
6. Nguyen, T.T.H.; Nguyen, N.; Nguyen, T.B.L.; Phan, T.T.H.; Bui, L.P.B.; Moon, H.C. Investigating customer attitude and intention towards online food purchasing in an emerging economy: An extended tam approach. *Foods* **2019**, *8*, 576. [CrossRef]
7. Zhao, Y.; Bacao, F. What factors determining customer continuingly using food delivery apps during 2019 novel coronavirus pandemic period? *Int. J. Hosp. Manag.* **2020**, *91*, 102683. [CrossRef]
8. Kang, J.W.; Numkung, Y. The information quality and source credibility matter in customers' evaluation toward food O2O commerce. *Int. J. Hosp. Manag.* **2019**, *78*, 189–198. [CrossRef]
9. Davis, F.D.; Bagozzi, R.P.; Warshaw, P.R. User acceptance of computer technology: A comparison of two theoretical models. *Manag. Sci.* **1989**, *35*, 982–1003. [CrossRef]
10. Hong, C.; Choi, H.; Choi, E.; Joung, H. Factors affecting customer intention to use online food delivery services before and during the COVID-19 pandemic. *J. Hosp. Tour. Manag.* **2021**, *48*, 509–518. [CrossRef]
11. Gu, S.; Slusarczyk, B.; Hajizada, S.; Kovalyova, I.; Sakhbieva, A. Impact of the COVID-19 Pandemic on online customer purchasing behavior. *J. Theor. Appl. Electron. Commer. Res.* **2021**, *16*, 125. [CrossRef]
12. Kimes, S.E. The current state of online food ordering in the US restaurant industry. *Cornell Hosp. Rep.* **2011**, *11*, 6–18.
13. Statista. eServices Report 2019–Online Food Delivery. Available online: https://www.statista.com/study/40457/food-delivery/ (accessed on 15 July 2020).
14. Hwang, J.; Kim, D.; Kim, J.J. How to Form Behavioral Intentions in the Field of Drone Food Delivery Services: The Moderating Role of the COVID-19 Outbreak. *Int. J. Environ. Res. Public Health* **2020**, *17*, 9117. [CrossRef]
15. Davis, F.D. Perceived usefulness, perceived ease of use, and user acceptance of information technology. *MIS Q.* **1989**, *13*, 319–340. [CrossRef]
16. Fishbein, M.; Ajzen, I. Belief, Attitude, Intention, and Behavior: An Introduction to Theory and Research. *Philos. Rhetor.* **1977**, *10*, 177–188.
17. Ajzen, I. The theory of planned behavior. *Organ. Behav. Hum. Decis. Processes* **1991**, *50*, 179–211. [CrossRef]
18. Fishbein, M. An investigation of the relationships between beliefs about an object and the attitude toward that object. *Hum. Relat.* **1963**, *16*, 233–239. [CrossRef]
19. Davis, F.D.; Bagozzi, R.P.; Warshaw, P.R. Extrinsic and intrinsic motivation to use computers in the workplace. *J. Appl. Soc. Psychol.* **1992**, *22*, 1111–1132. [CrossRef]
20. Ingham, J.; Cadieux, J.; Berrada, A.M. e-Shopping acceptance: A qualitative and meta-analytic review. *Inf. Manag.* **2015**, *52*, 44–60. [CrossRef]
21. Pavlou, P.A. Customer acceptance of electronic commerce: Integrating Trust and risk with the technology acceptance model. *Int. J. Electron. Commer.* **2003**, *7*, 101–134.
22. Gefen, D.; Karahanna, E.; Straub, D.W. Trust and TAM in online shopping: An integrated model. *MIS Q.* **2003**, *27*, 51–90. [CrossRef]
23. King, W.R.; He, J. A meta-analysis of the technology acceptance model. *Inf. Manag.* **2006**, *43*, 740–755. [CrossRef]
24. Legris, P.; Ingham, J.; Collerette, P. Why do people use information technology? A critical review of the technology acceptance model. *Inf. Manag.* **2003**, *40*, 191–204. [CrossRef]
25. Yousafzai, S.Y.; Foxall, G.R.; Pallister, J.G. Technology acceptance: A meta-analysis of the TAM: Part 2. *J. Model. Manag.* **2007**, *2*, 281–304. [CrossRef]
26. Henderson, R.; Rickwood, D.; Roberts, P. The beta test of an electronic supermarket. *Interact. Comput.* **1998**, *10*, 385–399. [CrossRef]
27. Childers, T.L.; Carr, C.L.; Peck, J.; Carson, S. Hedonic and utilitarian motivations for online retail shopping behavior. *J. Retail.* **2001**, *77*, 511–535. [CrossRef]
28. Koufaris, M. Applying the technology acceptance model and flow theory to online customer behavior. *Inf. Syst. Res.* **2002**, *13*, 205–223. [CrossRef]
29. Cyr, D.; Hassanein, K.; Head, M.; Ivanov, A. The role of social presence in establishing loyalty in e-service environments. *Interact. Comput.* **2007**, *19*, 43–56. [CrossRef]
30. Wen, C.; Prybutok, V.R.; Xu, C. An integrated model for customer online repurchase intention. *J. Comput. Inf. Syst.* **2011**, *52*, 14–23.

31. Kim, J.; Ma, Y.J.; Park, J. Are US customers ready to adopt mobile technology for fashion goods? An integrated theoretical approach. *J. Fash. Mark. Manag.* **2009**, *13*, 215–230.
32. Shen, J. Social comparison, social presence, and enjoyment in the acceptance of social shopping websites. *J. Electron. Commer. Res.* **2012**, *13*, 198–212.
33. Hassanein, K.; Head, M. Manipulating perceived social presence through the web interface and its impact on attitude towards online shopping. *Int. J. Hum. Comput. Stud.* **2007**, *65*, 689–708. [CrossRef]
34. Ha, S.; Stoel, L. Customer e-shopping acceptance: Antecedents in a technology acceptance model. *J. Bus. Res.* **2009**, *62*, 565–571. [CrossRef]
35. Oh, S.H.; Kim, Y.M.; Lee, C.W.; Shim, G.Y.; Park, M.S.; Jung, H.S. Customer adoption of virtual stores in Korea: Focusing on the role of Trust and playfulness. *Psychol. Mark.* **2009**, *26*, 652–668. [CrossRef]
36. Van der Heijden, H. User acceptance of hedonic information systems. *MIS Q.* **2004**, *28*, 695–704. [CrossRef]
37. Ahn, T.; Ryu, S.; Han, I. The impact of web quality and playfulness on user acceptance of online retailing. *Inf. Manag.* **2007**, *44*, 263–275. [CrossRef]
38. Gefen, D.; Straub, D. Managing user trust in B2C e-services. *e-Serv J.* **2003**, *2*, 7–24. [CrossRef]
39. Jarvenpaa, S.L.; Tractinsky, N.; Vitale, M. Customer trust in an Internet store. *Inf. Technol. Manag.* **2000**, *1*, 45–71. [CrossRef]
40. Kim, D.J.; Ferrin, D.L.; Rao, H.R. A trust-based customer decision-making model in electronic commerce: The role of Trust, perceived risk, and their antecedents. *Decis. Support. Syst.* **2008**, *44*, 544–564. [CrossRef]
41. McCloskey, D.W.; Leppel, K. The impact of age on electronic commerce participation: An exploratory model. *J. Electron. Commer. Organ.* **2010**, *8*, 41–60. [CrossRef]
42. Wei, T.T.; Marthandan, G.; Chong, A.Y.; Ooi, K.; Arumugam, S. What drives Malaysian m-commerce adoption? An empirical analysis. *Ind. Manag. Data Syst.* **2009**, *109*, 370–388.
43. Dash, S.; Saji, K.B. The role of customer self-efficacy and website social-presence in customers' adoption of B2C online shopping: An empirical study in the Indian context. *J. Int. Consum. Mark.* **2008**, *20*, 33–48. [CrossRef]
44. Chiu, C.M.; Lin, H.Y.; Sun, S.Y.; Hsu, M.H. Understanding customers' loyalty intentions towards online shopping: An integration of technology acceptance model and fairness theory. *Behav. Inf. Technol.* **2009**, *28*, 347–360. [CrossRef]
45. Chai-Har, L.; Eze, U.C.; Ndubisi, N.O. Analyzing key determinants of online repurchase intentions. *Asia Pac. J. Mark. Logist.* **2011**, *23*, 200–221.
46. Van Slyke, C.; Lou, H.; Belanger, F.; Sridhar, V. The influence of culture on customer-oriented electronic commerce adoption. *J. Electron. Commer. Res.* **2010**, *11*, 30–40.
47. Lee, M.C. Predicting and explaining the adoption of online trading: An empirical study in Taiwan. *Decis. Support Syst.* **2009**, *47*, 133–142. [CrossRef]
48. Izquierdo-Yusta, A.; Calderon-Monge, E. Internet as a distribution channel: Empirical evidence from the service sector and managerial opportunities. *J. Internet Commer.* **2011**, *10*, 106–127. [CrossRef]
49. Zhu, D.S.; Lee, Z.C.; O'Neal, G.S.; Chen, Y.H. Mr. Risk! Please trust me: Trust antecedents that increase online customer purchase intention. *J. Internet Bank Commer.* **2011**, *16*, 1–23.
50. Featherman, M.S.; Pavlou, P.A. Predicting e-services adoption: A perceived risk facets perspective. *Int. J. Hum. Comput. Stud.* **2003**, *59*, 451–474. [CrossRef]
51. Chang, H.H. Intelligent agent's technology characteristics applied to online auctions' task: A combined model of TTF and TAM. *Technovation* **2008**, *28*, 564–577. [CrossRef]
52. Pavlou, P.A.; Fygenson, M. Understanding and predicting electronic commerce adoption: An extension of the theory of planned behavior. *MIS Q.* **2006**, *30*, 115–143. [CrossRef]
53. Kim, H.B.; Kim, T.T.; Shin, S.W. Modeling roles of subjective norms and eTrust in customers' acceptance of airline B2C eCommerce websites. *Tour. Manag.* **2009**, *30*, 266–277. [CrossRef]
54. Benamati, J.; Fuller, M.A.; Serva, M.A.; Baroudi, J. Clarifying the integration of Trust and TAM in e-commerce environments: Implications for systems design and management. *IEEE Trans. Eng. Manag.* **2009**, *57*, 380–393. [CrossRef]
55. Yang, S.; Park, J.; Park, J. Customers' channel choice for university-licensed products: Exploring factors of customer acceptance with social identification. *J. Retail. Consum. Serv.* **2007**, *14*, 165–174. [CrossRef]
56. Venkatesh, V.; Morris, M.G.; Davis, G.B.; Davis, F.D. User acceptance of information technology: Toward a unified view. *MIS Q.* **2003**, *27*, 425–478. [CrossRef]
57. Venkatesh, V.; Davis, F.D. A theoretical extension of the technology acceptance model: Four longitudinal field studies. *Manag. Sci.* **2000**, *46*, 186–204. [CrossRef]
58. Vijayasarathy, L.R. Predicting customer intentions to use online shopping: The case for an augmented technology acceptance model. *Inf. Manag.* **2004**, *41*, 747–762. [CrossRef]
59. Bosque, I.R.D.; Crespo, A.H. How do internet surfers become online buyers? An integrative model of e-commerce acceptance. *Behav. Inf. Technol.* **2011**, *30*, 161–180. [CrossRef]
60. Barkhi, R.; Wallace, L. The impact of personality type on purchasing decisions in virtual stores. *Inf. Technol. Manag.* **2007**, *8*, 313–330. [CrossRef]
61. Suh, B.; Han, I. Effect of Trust on customer acceptance of Internet banking. *Electron. Commer. Res. Appl.* **2002**, *1*, 247–263. [CrossRef]

62. Anderson, J.C.; Gerbing, D.W. Structural equation modeling in practice: A review and recommended two-step approach. *Psychol. Bull.* **1988**, *103*, 411–423. [CrossRef]
63. Hair, J.F., Jr.; Hult, G.T.M.; Ringle, C.M.; Sarstedt, M. *A Primer on Partial Least Squares Structural Equation Modeling (PLS-SEM)*; Sage Publications: Thousand Oaks, CA, USA, 2016.
64. Fornell, C.; Larcker, D.F. Evaluating structural equation models with unobservable variables and measurement error. *J. Mark. Res.* **1981**, *18*, 39–50. [CrossRef]
65. Yuan, S.; Liu, Y.; Yao, R.; Liu, J. An investigation of users' continuance intention towards mobile banking in China. *Inf. Dev.* **2016**, *32*, 20–34. [CrossRef]
66. Kim, M.J.; Hall, C.M. A hedonic motivation model in virtual reality tourism: Comparing visitors and non-visitors. *Int. J. Inf. Manag.* **2019**, *46*, 236–249. [CrossRef]
67. Heritage Radio Network the World's Pioneer Found Radio Station. How Do You Produce Successful Virtual Events? 2021, Volume 34. Available online: https://heritageradionetwork.org/episode/how-do-you-produce-successful-virtual-events (accessed on 27 November 2021).

foods

MDPI

Article

Trends in Food Preferences and Sustainable Behavior during the COVID-19 Lockdown: Evidence from Spanish Consumers

Shanshan Li [1,*], Zein Kallas [2,3,*], Djamel Rahmani [2] and José Maria Gil [2,3]

1 Institute for Research in Sustainability Science and Technology (IS-UPC), Polytechnic University of Catalonia, 08034 Barcelona, Spain
2 CREDA-UPC-IRTA (Centre for Agro-food Economy and Development), 08860 Castelldefels, Spain; djamel.rahmani@upc.edu (D.R.); Chema.Gil@upc.edu (J.M.G.)
3 DEAB (Department of Agrifood Engineering and Biotechnology), Polytechnic University of Catalonia, 08860 Barcelona, Spain
* Correspondence: shanshan.li@upc.edu (S.L.); zein.kallas@upc.edu (Z.K.)

Abstract: The COVID-19 pandemic poses a threat to global food security, and it changes consumers' food buying and consumption behavior. This research not only investigates trends in Spanish consumers' general food shopping and consumption habits during the lockdown, but also investigates these trends from the perspective of sustainable purchasing. Specifically, total food consumption (C), food expenditure (E), and purchase of food with sustainable attributes (S) were measured. Data were collected from a semi-structured questionnaire which was distributed online among 1203 participants. The logit models showed that gender, age, employment status, and consumers' experiences were associated with total food consumption and expenditure during the lockdown. In addition, consumers' risk perceptions, shopping places, trust level in information sources, and risk preference were highly essential factors influencing consumers' preferences and sustainable behavior. Consumers' objective knowledge regarding COVID-19 was related to expenditure. Furthermore, family structure only affected expenditure, while income and place of residence influenced food consumption. Mood was associated with expenditure and the purchase of sustainable food. Household size affected purchasing behavior towards food with sustainable attributes. This research provides references for stakeholders that help them to adapt to the new COVID-19 situation.

Keywords: COVID-19 lockdown; food preferences; risk preference; risk perceptions; food purchasing behavior; food consumption behavior; sustainable behavior

Citation: Li, S.; Kallas, Z.; Rahmani, D.; Gil, J.M. Trends in Food Preferences and Sustainable Behavior during the COVID-19 Lockdown: Evidence from Spanish Consumers. *Foods* **2021**, *10*, 1898. https://doi.org/10.3390/foods10081898

Academic Editor: Derek V. Byrne

Received: 30 June 2021
Accepted: 12 August 2021
Published: 16 August 2021

1. Introduction

Novel coronavirus disease, named "COVID-19" by the World Health Organization (WHO), was initially reported in the city of Wuhan, China in December 2019 [1]. Subsequently, it began rapidly spreading around the world, resulting in a global pandemic. Spain took many preventive measures, including lockdowns, stay-at-home orders, mass quarantine, and transportation halts when COVID-19 started to spread in Spain. The Spanish government declared a state of emergency on 14 March 2020 and increased the severity of the state of alarm from 30 March to 14 April 2020, which was a strict lockdown period. People could only leave home when they were working in essential services (health, security, social, and economic wellbeing of citizens) or when they needed to buy necessary products (groceries and medicine) during the lockdown [2]. The COVID-19 pandemic situation not only affected human health, but also caused several economic and social changes. On the one hand, the rate of unemployment increased and financial strain became more severe [3], which led to an increase in depression risk, stress, and feelings of helplessness [4]. On the other hand, the COVID-19 pandemic created new working and family situations (e.g., teleworking, e-learning, homes with narrow space, and living space without direct access to sunlight), which also induced stress and depression [5].

In this context, a significant share of consumers increased their food consumption due to higher anxiety levels [6]. A previous study showed that consumers in ten European countries consumed more food as a result of the COVID-19 lockdowns across Europe and an increase in homeworking that led people to spend more time at home, influencing their consumption behavior and food choices [7]. Another study reported that almost half of the respondents stated that they increased food consumption during the lockdown in Italy, with twenty percent of them gaining weight [6]. On the contrary, compared to the period before the COVID-19 outbreak, Polish youth had a better dietary intake during the outbreak, as the pandemic changed the determinants of food choices, reinforcing the importance of health and weight control [8]. The Italian lockdown allowed consumers to make positive habits towards food consumption [9]. In addition, the COVID-19 outbreak led Spanish consumers to adopt a healthier eating habit/behavior, as evidenced by a higher level of adherence to the Mediterranean diet (MedDiet) [10].

Additionally, the COVID-19 lockdown also changed consumers' shopping behavior. Individuals focused on buying food items as a behavioral reaction to feelings of stress and uncertainty [11]. Negative feelings (e.g., fear, stress, and uncertainty) could cause a panic buying situation [12,13]. Panic buying behavior exacerbates stock-out situations and often leads to a price increase in food products [12]. Spanish consumers were shown to be stockpiling non-perishable food and other supplies during the COVID-19 lockdown [14]. Some people stockpiled food items and bought more on each trip to minimize store visits, aiming to reduce the risk of infection [15]. According to previous research, 64% of consumers experienced product shortages at stores from which they were attempting to purchase, and 50% of consumers stocked up on products to avoid deficiencies in the future during the COVID-19 outbreak in India [16]. Additionally, consumers' food spending increased dramatically during the COVID-19 outbreak [17,18], and another report indicated that grocery spending increased in Spain due to COVID-19 [19]. Furthermore, the COVID-19 pandemic enabled people to turn to purchasing food products online in an attempt to limit their perceived risk of exposure to infection [20]. Moreover, a previous study indicated that consumers turned to purchasing organic food or buying food products directly from farmers [21].

In addition, consumers' shifts to more sustainable behavior can dramatically reduce their carbon impact [22], which contributes to the achievement of sustainable development in Spain. There is considerable literature that has explored consumers' attitudes, purchasing, and consumption behavior towards food products with sustainable attributes (e.g., organic food, animal welfare food, fair-trade food, environmentally friendly food, and local food) before the COVID-19 lockdown [23–26]. However, little research attempted to measure them during the lockdown, and it is of great importance and necessity to conduct such a study that ensures the availability of sustainable food in the market during the pandemic. To date, few studies have explored how COVID-19 affected Spanish consumers' purchasing or consumption behavior [2,27], and these studies focused on the evolution of people's information searches or only on food consumption/dietary behavior. Evidence on trends in expenditure and purchases of food with sustainable attributes during the Spanish lockdown and their related determinants is insufficient. This research includes more comprehensive potential impact factors and, to our knowledge, is the first study that not only investigates trends in Spanish consumers' general food buying and consumption behavior during the lockdown, but also investigates these trends from the perspective of sustainable purchasing. In this context, the main objective of this study is to analyze trends in consumers' food preferences and sustainable behavior during the COVID-19 lockdown. To reach the main objective, three secondary objectives were proposed as intermediate steps. Firstly, this study aimed to measure changes in consumers' total food consumption, which reflects a preference for consumer behavior during the lockdown, and to identify its determinant factors. Secondly, this study aimed to explore how consumers' food expenditure (representing a behavioral preference) changes and to identify its impact factors.

Thirdly, this study aimed to examine trends in shopping behaviors toward food attributes with sustainable behaviors.

2. Literature Review

A large number of studies have been conducted on the determinants of consumers' food purchases and consumption. In summary, these can be divided into four dimensions: (1) food-related characteristics (e.g., appearance, packaging, label, and price); (2) individual socio-demographic variables (e.g., age, gender, household size, family structure, and income); (3) psychological factors (e.g., mood); and (4) cognitive factors (e.g., attitude or preference, beliefs, trust, perception, and knowledge) [28]. In this research, we focused on consumers' socio-demographic characteristics, as well as psychological and cognitive factors.

2.1. Socio-Demographic Factors

Numerous studies have shown that socio-demographic characteristics have a significant impact on food purchasing and consumption behavior [23,29,30]. For example, many studies have suggested that females purchase sustainable food more frequently [31,32]. This may be because, on the one hand, women are more health-conscious, and they consider sustainable food (e.g., organic food) to be healthier than conventional food [29]. On the other hand, women are often responsible for household food purchases and are therefore more aware of sustainable food [32]. However, another study showed that although women have positive attitudes towards organic vegetables, there is no significant impact on the actual consumption of organic vegetables [33]. These different findings may be related to the attitude-behavior gap.

In some previous empirical studies carried out in Europe, income is identified as a factor influencing the purchase of organic food, for example, consumers with higher incomes are more likely to purchase higher levels of organic food [23,34]. Conversely, a study conducted in the United States did not find an association between income and organic food purchasing behavior [35]. These different outcomes may be related to regional differences. Moreover, another study suggested that income has no influence on the regularity of organic food consumption, but it affects individual spending on organic food [30].

With regard to age, previous research has shown that young consumers (18–32 years old) frequently buy organic food [32]. However, another study revealed that older consumers (over 55 years) purchase sustainable food more often [31]. Additionally, older people spend less on food, which may be related to the fact that calorie requirements decrease with age; therefore, older consumers buy a lower amount of food [36]. In addition, previous literature suggested that income and age are indicators of food spending behavior [37]. In terms of family structure, previous research has demonstrated that consumers who have children tend to buy sustainable products [38]. Another study also indicated that the presence of children in the household is positively associated with the likelihood of consuming organic food [33] and food expenditure [39]. Moreover, another study also showed that household income, size, and composition (with children) positively affect food expenditures [39]. Therefore, according to these previous findings and the ongoing global novel coronavirus pandemic, this research explored the influence of consumers' profiles on purchasing and consumption behavior during the lockdown.

2.2. Psychological Factors (Mood)

Mood is one motive which may drive consumers' food choices [40,41]. Some studies have explored the relationship between mood and food, and they have found that negative moods positively influence food intake [42,43]. In addition, an early study revealed that people are more likely to consume healthy foods in positive moods and are more prone to eat unhealthy foods (e.g., snacks high in sugar and salt) in negative moods [44]. This may be because foods high in sugar or fat can reduce the effects of negative emotions through the neurotransmission of dopamine, making people happier [45].

In contrast, Mehrabian and Riccioni pointed out that a positive mood is related to high appetite levels [46]. Another review study also indicated that positive mood is a neglected trigger for eating more food due to the close correlation between socialization and food consumption [47,48]. In particular, in terms of positive emotions, research has shown that consumers may eat more pleasantly and extend time duration of the meal, and therefore consume more food, when eating with familiar and friendly people [48].

Moreover, changes in food consumption behavior due to fluctuations in emotional states may be triggered by situations or events outside of a person's daily routine, such as adapting to certain environments or motivating themselves [49]. Therefore, given the current global pandemic, consumers' emotional states fluctuate and may change consumption behavior; therefore, we explored whether emotional states have an impact on consumers' food preferences and sustainable behavior defined in this research.

2.3. *Cognitive Factors*

2.3.1. Trust in Information Sources

Trust is regarded as an important predictor of consumers' attitudes and food behavior [50]. Previous research has revealed that information from highly trusted sources is more likely to evoke changes in attitudes and behaviors [51]. Trust in information sources influences consumers' attitudes and purchase intentions during a food security crisis [52]. Consumers' level of trust and source of information during a health crisis may influence whether they adopt certain recommended food safety behaviors [53]. In addition, a previous study found a relationship between trust in information sources and risk perceptions [54]. Trust in authorities' sources of information is vital to reduce unnecessary fear and inappropriate risk perceptions [55], while trust in the information given by the media increases risk perception [54]. Moreover, panic buying (increased purchasing) has occurred in public health emergencies since ancient times [56]. For example, the COVID-19 pandemic generated fear of scarcity among consumers, which increased risk perception and ultimately lead to impulsive and panic buying behavior [57]. Very little research has studied the relationship between trust in information sources regarding COVID-19 and buying behavior during the lockdown. Therefore, this research filled this gap by including trust in information sources regarding COVID-19 to measure food preferences and sustainable behavior.

2.3.2. Risk Perceptions and Risk Preference

In social science research, "risk" is defined as the likelihood of physical, social, or financial harm/loss due to a hazard within a specified time frame [58]. A "hazard" is a situation, event, or substance that may be harmful to people, nature, or man-made facilities, whereas a risk is not; it is an inference about the impact of a hazard on people (or nature or assets) [59].

Risk perception plays an essential role in consumers' purchase intentions and behavior [60], and it is defined as people's subjective judgments about the frequency and severity of a particular risk [61]. Typically, risk perception is measured by asking participants about specific risk scenarios [61]. Several studies have shown that risk perception is an indicator of food consumption. For instance, increased risk perception of fish-eating negatively affects total consumer fish consumption [62].

Risk preference includes three types: risk-loving, risk-neutral, and risk-averse. For risk-averse consumers, sustainable attributes of food (e.g., food being organic) are risky and uncertain. Therefore, they prefer to buy and eat conventional products rather than purchase sustainable food [63]. Risk perception and risk preference (elicited through the lottery game) are key determinants of the acceptance of risky foods [64].

2.3.3. Knowledge

Knowledge is a crucial strategy for consumers to make purchase decisions [65]. There are three types of knowledge: subjective knowledge (self-perceived knowledge), objective knowledge (the content of knowledge), and usage experience [66]. Several studies showed that higher levels of COVID-19 knowledge are related to changes in dietary habits and depression [67]. Lower knowledge of COVID-19 is associated with COVID-19-related behavioral changes, such as purchasing more goods and stockpiling [68,69]. In addition, knowledge may potentially affect personal perceptions and purchasing decisions, especially when health issues like COVID-19 arise [70].

Based on existing literature, we introduced these variables as potential predictors influencing consumers' changes in food preferences and sustainable behavior, as defined in this research.

3. Materials and Methods

3.1. Data Collection and Questionnaire Design

A semi-structured questionnaire in an online survey (Qualtrics consumers' panels) among 1203 participants was conducted during the lockdown situation in Spain in May 2020. The questionnaire for this study was divided into seven sections: (1) changes in consumers' behavior during the lockdown, including food consumption, food expenditure, purchase of food with sustainable attributes, and shopping places; (2) consumers' trust level in information sources; (3) risk preference; (4) risk perceptions; (5) knowledge level; (6) mental status (mood states and concerns regarding COVID-19); and (7) socio-demographic variables. The questionnaire was reviewed and validated by a group of experts from different universities and countries. The Cronbach's alpha coefficients of the scales were tested, and all coefficients were above 0.68, indicating acceptable internal consistency. Factor analysis also confirmed the validity of the constructs. In order to have a representative sample, quota sampling stratified by age and gender was used. Consumers who were fully or partially responsible for purchasing food (over the age of 18) were recruited to participate in the present study. The questionnaire was available in Spanish. On average, each respondent spent 25 minutes filling out the questionnaire. Respondents participated in our survey voluntarily, and we explained to them the purpose of the study and that their information would not be disclosed. The questionnaire was approved by the Ethics Committee of the Centre for Agro-food Economy and Development (CREDA) and was carried out in accordance with the ethical norms of social science research.

3.2. Independent Variables Included in this Research

3.2.1. Risk Preference

Risk preference was a highly important factor in consumers' behavioral intention [71]. The MPL (Multiple Price List) has been widely used in psychology and economics research because of its easy and effective procedure, which was based on expected utility theory (EUT) [72,73]. Therefore, MPL was employed to measure consumers' risk preference in this research. In this MPL experiment, respondents were asked to choose between lottery A and lottery B twenty times. In the first task, they had a 100% chance of receiving €200 under lottery A; under lottery B they had a 50% chance of receiving €200 and a 50% chance of receiving nothing. By that analogy, 20 tasks, until lottery A with a 100% chance of receiving €10, and lottery B with the same, were conducted to measure consumers' risk preference. The payoff of lottery A decreased in turn, while the payoff of lottery B remained unchanged (€100). Lottery A is the "safe" choice whose payoff is more than the potential payoff in the "risky" lottery B among the top ten choices. In the 11th task, the payoff of lottery A is the same as that of lottery B. Starting from the 12th task, lottery A has less payoff than lottery B.

The number of "safe choices" (choosing lottery A) has often been used to describe risk preference [64]. In our research, the number of risk-loving individuals' "safe choices" should be less than or equal to 9, while the number of risk-neutral people's "safe choices" should be equal to 10, and the number of risk-averse people's "safe choices" should be more than or equal to 11.

3.2.2. Risk Perceptions

As concluded in the literature review, risk perception played an essential role in consumers' purchase intentions and behavior [60]. In this research, risk perception consisted of three aspects: risk of COVID-19, food security risk, and financial risk perceptions.

As for risk perception of COVID-19, previous studies indicated that risk perception was designated as a mix of perceived vulnerability (how likely a person thinks he/she will contract the disease) and perceived severity (how serious people think contracting the disease will be for him/her) [74], which was applied in a recent study to measure perceived risk regarding COVID-19 [75]. According to the previous research measuring SARS-related risk perceptions during the 2003 SARS outbreak [76] and another study during the 2009 H1N1 pandemic [74], we measured consumers' risk perception of COVID-19 by two items: (1) perceived risk of vulnerability, employing a 5-point Likert scale that ranges from 1 (very unlikely) to 5 (very likely) (How likely do you think you are to contract coronavirus in the next six months?); and (2) perceived risk of severity, using an 11-point Likert scale from 0 (not serious at all) to 10 (very serious) (How serious do you think your health will be if you contract the coronavirus in the next six months?). If consumers perceived a higher severity or a higher likelihood of contracting the virus (get a higher score on the 11-point or 5-point Likert scale), they had a higher risk perception of COVID-19. The 11-point Likert scale provided respondents with a wider range of options and yielded better predictive analysis. Additionally, previous research indicated that the 11-point Likert scale from 0 to 10 was popular due to its high composite reliability [77].

According to the recognized definition, food security was defined as "access to adequate food for all people at all times to have an active and healthy life" [78]. In this research, consumers' perceived food security risk was elicited using a 7-point Likert scale that ranges from 1 (very unlikely) to 7 (very likely), and they needed to answer how likely they thought it was that food shortages and food prices would rise in the next six months (How likely do you think it is that the following scenarios will occur in the next 6 months?—food shortages; food prices will go up). Regarding financial risk, a 5-point Likert scale ranging from 1 (not at all) to 5 (a great deal) (How threatened do you feel about your current financial situation?) was used.

3.2.3. Mood States, Experiences, Concerns, and Shopping Places

As introduced in the literature review, negative and positive moods influenced food choices [79]. COVID-19 put consumers under great stress and caused them to exhibit different moods, which may have influenced their purchasing and consumption behavior during the pandemic. Therefore, respondents were asked about their mood status (including positive moods and negative moods) via a 5-point Likert scale ranging from 0 (none of this feeling) to 4 (a great deal of this feeling) (Considering the COVID-19 situation where you currently live, do you feel . . . ?—irritated; confident; angry; reassured; annoyed; and aggravated). Anger as a negative mood was measured using four items that had been validated in previous studies: irritated, angry, annoyed, and aggravated [80]. Cronbach's alpha was $\alpha = 0.91$ in this research. The positive emotions included feeling reassured and confident, which were selected from previous research [81]. Cronbach's alpha was $\alpha = 0.80$.

In addition, the COVID-19 pandemic brought stress and uncertainty for people, which could result in panic buying, thus threatening global food security. For people who experienced food shortages or higher food prices during the COVID-19 outbreak, their purchasing behavior may have changed [82]. As a consequence, in this study, we measured food security experiences (food shortages, higher food prices, and neither) (During the outbreak, did you experience the following scenarios?—food shortages; higher food prices; and neither of them). In addition, we measured COVID-19 experiences, similar to a recent study [83], by asking respondents if they contracted COVID-19 or not (Have you contracted the COVID-19 virus?) (1 = Yes, I tested positive for the COVID-19 virus; 2 = No. I had the symptoms, but the test result came back negative; 3 = No. I did not have the symptoms, so I did not opt for a test; and 4 = I do not know. I had the symptoms but did not have access to a test) and asking if they knew someone who had been diagnosed or died due to COVID-19 (Do you know someone who has been diagnosed or died due to the COVID-19 virus?—members of my family; friends; neighbors; friends of my friends; colleagues; and no, I don't know any person) and examined whether experiences played an important role in consumers' behavior during the lockdown.

Additionally, previous work indicated that consumers' concerns were related to buying behavior [84]. Concerns regarding COVID-19 were wide-ranging, encompassing both health and financial issues [85]. Hence, we adopted a 7-point Likert scale ranging from 1 (not concerned at all) to 7 (extremely concerned) to evaluate consumers' health concerns about COVID-19 and ultimately to explore its impact on consumers' behavior during the lockdown (Please indicate your level of health concern about COVID-19). In addition, a previous study showed that there was a significant increase in food shopping online, with 45% of consumers in ten European countries making more online purchases during the lockdown [7]. Another study conducted in South Korea indicated that during the 2015 MERS outbreak in South Korea, consumers decreased their spending on food at department stores and outside the home, while they increased their spending on food purchased online, suggesting that changes in shopping location influenced changes in consumers' food expenditure [86]. Moreover, a shopping place, such as a large store, can stimulate consumers' emotions, which can further influence purchase decisions [87]. Therefore, change in shopping place as a potential indicator was included in this research. Respondents were asked to answer two questions to assess the shopping place variable before and during the lockdown (Where do you usually buy food products? (Before restrictions due to COVID-19) and Where do you usually buy food products? (During the lockdown)) (1 = hypermarkets, supermarkets; 2 = specialized food stores; 3 = malls; 4 = farmer's market/open markets; 5 = retailers' websites; 6 = organic food stores; and 7 = others).

3.2.4. Trust in Information Sources and Knowledge

Consumers look for health information from a wide cluster of sources and channels [88]. Trust in health organizations and government health agencies has been identified as an important correlate of health-related decision-making and behavior [89]. In public health emergencies (e.g., a flu flare-up), people with high trust in government health agencies react more rapidly and are more likely to comply with the health recommendations given by the agencies [90]. As mentioned in the literature review, trust in information sources influenced consumers' attitudes and purchase intentions during a food security crisis [52]. In this context, consumers' trust in information sources was elicited by using a 5-point Likert scale ranging from 1 (not trustworthy at all) to 5 (extremely trustworthy) (Consider the following sources of information regarding COVID-19. How trustworthy do you feel these sources are?—government; social media such as Twitter, Facebook; health professionals such as doctors; family, friends, and colleagues; scientists; and news such as papers, TV, and radio).

In addition, we assessed consumers' levels of subjective and objective knowledge regarding COVID-19 to determine if they have an impact on their shopping and consumption behavior. To be specific, a 7-point Likert scale ranging from 1 (not knowledgeable at all) to 7 (very knowledgeable) was employed to measure respondents' subjective knowledge (How well do you think you know COVID-19), with the results expressed as a percentage, i.e., from 0 (not knowledgeable at all) to 100 (very knowledgeable). In addition, the level of objective knowledge was displayed as the percentage of correct answers, and respondents were asked to judge whether symptoms of COVID-19 were correct or incorrect; symptoms presented included existing and non-existing symptoms (The following are 17 symptoms of COVID-19. Please judge whether they are true or false).

3.3. Measuring Consumers' Food Preferences and Sustainable Behavior

In this research, three dependent variables, including changes in total food consumption (C), food expenditure (E), and purchasing behavior towards food with sustainable attributes (S), were measured to determine trends in food preferences and sustainable behavior during the lockdown. Changes in food consumption and expenditure were measures of food behavioral preferences during the lockdown. Respondents were asked to answer a question (How has COVID-19 impacted your total consumption of food), reflecting consumers' consumption behavior during the lockdown. Individual scores ranged from "−3" (greatly decreased) to "+3" (greatly increased) regarding total food consumption (C). In addition, respondents were asked to respond to a question (How has COVID-19 impacted your food shopping behavior?—spending money on food purchases), with scores ranging from "−3" (greatly decreased) to "+3" (greatly increased) regarding food expenditure to measure consumers' purchasing behavior (E). Consumers' sustainable purchasing behavior (S) was assessed by their purchases of sustainable food (organic, local, animal welfare, and fair-trade food), with scores ranging from "−3" (greatly decreased) to "+3" (greatly increased) (During the COVID-19 lockdown, how did your purchases of the following foods change?—organic; local; animal welfare; and fair-trade). Cronbach's alpha was $\alpha = 0.68$.

Figure 1 illustrates the framework of this study. The independent variables included in this study are the factors mentioned earlier that may be associated with consumers' food shopping and consumption behavior. Table 1 presents the details of the sample profile. The Kolmogorov–Smirnov test was used to measure the normality of the variables, and the mean and standard deviation (SD) were computed.

As can be seen, among the 1203 respondents, 51.0% were females, and 57.0% stated that they were healthy. 56.1% of respondents (before the lockdown) and 53.6% (during the lockdown) had a monthly household income of 1000–3000 euros, and the majority were aged 40–59 years (36.9%). In addition, 36.3% of samples had a household size of 2 people, and 61.2% of households had no children aged 0–12 or adults aged over 70 years. 71.8% of participants lived in urban places, and 24.4% had a full-time job (without variation). According to the gender and age distribution, the sample reflected the population of Spain.

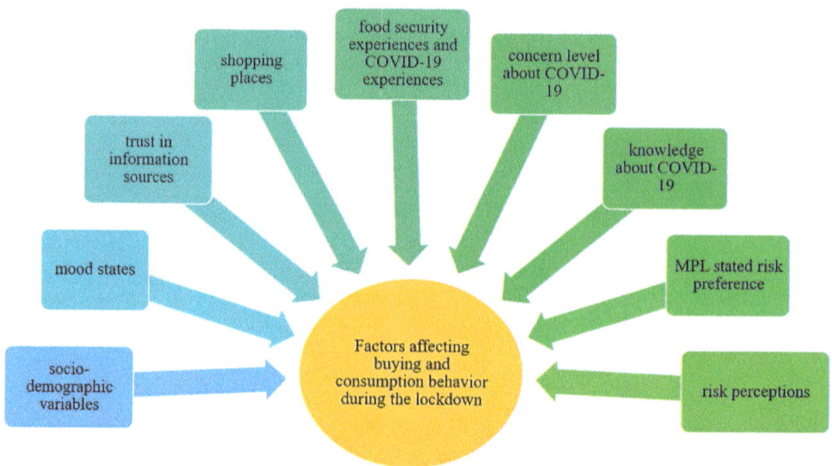

Figure 1. The framework of factors affecting consumers' food buying and consumption behavior during the lockdown.

Table 1. Socio-demographic variables in this research (n = 1203).

Socio-Demographic Variables		Percentage (%)
Gender	Male	49.0
	Female	51.0
Age	18–39 years	28.1
	40–59 years	36.9
	More than 60 years	35.0
Monthly household income before the lockdown	<999 euros	10.5
	1000–3000 euros	56.1
	>3001 euros	22.0
	"I prefer not to answer"	11.4
Monthly household income during the lockdown	<999 euros	19.0
	1000–3000 euros	53.6
	>3001 euros	15.9
	"I prefer not to answer"	11.5
Stated health status	Unhealthy	43.0
	Healthy	57.0
Household size	1 person	10.7
	2 persons	36.3
	3 persons	27.0
	4 persons	20.3
	5 persons	4.0
	6 persons or more	1.7
Family structure	There are children aged 0–6 years	Yes (13.5), No (86.5)
	There are children aged 7–12 years	Yes (15.5), No (84.5)
	There are adults over 70 years	Yes (14.1), No (85.9)
	None of the above	Yes (61.2), No (38.8)
Place of residence	Urban place	71.8
	Suburban place	14.8
	Rural place	13.4
Employment status	Student	2.3
	Full time (without variation)	24.4
	Full time (telecommuting)	16.5
	ERTE [a] (partial or total)	10.8
	A homemaker	5.2
	Sick leave	2.1
	Unemployed	15.0
	Retired	21.5
	Unable to work	2.2

[a] refers to a File of Temporary Regulation of Employment (ERTE). It consists of a temporary collective dismissal, in which the company temporarily suspends employment contracts, for reasons including the temporary stoppage of activities or insufficient income.

3.4. Data Analysis

In this study, consumer behavior change (the dependent variable) was a dichotomous variable with two categories: increase and no increase. The logit regression has often been used to analyze discrete dependent variables; therefore, the binary logistic regression model was deemed appropriate for this study. SPSS version 24.0 (IBM, Chicago, IL, USA) software was used. A descriptive analysis was also employed.

4. Results and Discussion

4.1. Results of the Independent Variables Included in the Model

Table 2 presents the results of the independent variables included in the model. Results revealed that Spanish consumers' subjective and objective knowledge level regarding COVID-19 was above average (77.26% > 50.00% and 67.44% > 50.00%). This may be due to the fact that the Spanish government, health experts, and the media have conveyed a considerable amount of information about COVID-19 to society. The results also showed that 65.7% of respondents were risk-averse, 13.6% were risk-neutral, and 20.7% were risk-loving. This result is in line with previous studies showing that the majority of respondents were risk-averse [91], and only a small proportion of participants were risk-loving [92]. In addition, with regard to food security experiences, 29.2% of participants stated that they experienced a food shortage during the lockdown, and 60.7% of them experienced rising food prices. As for COVID-19 experiences, the results showed that 71.7% of respondents stated that they did not have symptoms, so did not opt for a test. Only 1.5% of respondents tested positive for the COVID-19 virus. 21.7% of consumers did not know due to no access to a test. 37.2% of respondents did not know anyone who had been diagnosed or died due to the COVID-19 virus.

Table 2. Results of the independent variables included in the logit model.

Variables	Percentage (%)	Scales
Knowledge		
Subjective knowledge level	77.26	1–100%
Objective knowledge level	67.44	1–100%
Risk preference		
Risk-loving	20.7	
Risk-neutral	13.6	
Risk-averse	65.7	
Experiences		
Food security experiences		
Experienced food shortages	Yes 29.2; No 70.8	
Experienced higher food prices	Yes 60.7; No 39.3	
Experienced neither	Yes 28.4; No 71.6	
COVID-19 experiences		
Q. Have you contracted the COVID-19 virus?		
Yes. I tested positive for the COVID-19 virus.	1.5	
No, I had the symptoms, but the test result was negative.	5.1	
No. I did not have the symptoms, so I did not opt for a test.	71.7	
I don't know. I had the symptoms but did not have access to tests.	21.7	
Q. Do you know someone who has been diagnosed or died due to the COVID-19 virus?		
Members of my family	Yes 19.0; No 81.0	
Friends	Yes 26.4; No 73.6	
Neighbors	Yes 14.3; No 85.7	
Friends of my friends	Yes 25.6; No 74.4	
Colleagues	Yes 6.6; No 93.4	
No, I don't know any person	Yes 37.2; No 62.8	
Variables	**Mean (SD)**	**Scales**
Concern level about COVID-19	4.77 (1.70)	7-point Likert scale
Food security risk perception		
The probability of food shortages in the next 6 months	2.34 (1.49)	7-point Likert scale
The probability of higher food prices in the next 6 months	5.01 (1.61)	7-point Likert scale
Risk perception of COVID-19		
The severity of one's health condition will be if they contract COVID-19	6.04 (2.40)	11-point Likert scale
The probability of contracting COVID-19	2.65 (0.95)	5-point Likert scale
Trust in information sources		
Government	2.52 (1.27)	5-point Likert scale
Social media	2.70 (1.09)	
Health professionals (e.g., doctor)	4.27 (0.82)	
Family, friends, and colleagues	2.91 (1.04)	
Scientists	4.13 (0.91)	
News (e.g., papers, TV, radio)	1.90 (0.94)	

SD: standard deviation.

Participants' concern level about COVID-19 was above average (4.77 > 3.5 points on a 7-point scale), which is consistent with research showing that levels of concern about COVID-19 are relatively high in Spain [93]. The probability of consumers perceiving food shortages in the next six months was below average (2.34 < 3.5 points on a 7-point scale). With respect to the probability of facing higher food prices in the next six months, this was perceived to be above average (5.01 > 3.5 points on a 7-point scale). The news reported that in Spain, fruit and vegetables have become between 25% and 30% more expensive due to the increase in transport costs during the COVID-19 pandemic [94], which has increased consumers' perceived food price (food security) risk. In addition, this was supported by the result of "experiences" in this research (as shown earlier), which showed that 60.7% of consumers experienced a higher food price during the lockdown, increasing their food security risk perceptions. Consumers' experiences of food insecurity will increase their risk perception because direct exposure to risk events usually enhances consumers' memories and imaginations of hazards [95].

As for the severity of the perceived risk, this was above average (6.04 > 5.5 points on an 11-point scale). Regarding the probability of contracting COVID-19 in the next six months, the results indicated that consumers assessed their risk of being infected as high (2.65 > 2.5 points on a 5-point Likert scale). In both cases, the scores were slightly above the average, indicating a slightly higher perceived risk regarding COVID-19. These outcomes converge with the findings that Spain was the second country with the highest risk perception of COVID-19 among ten countries across Europe, America, and Asia [93]. Consumers' trust level in information sources from the highest to lowest was health professionals, scientists, family (friends and colleagues), social media, government, and news. This is in line with a study which concluded that consumers stated information from experts or scientists was the most reliable [2].

4.2. *Results of Consumers' Food Preferences and Sustainable Behavior*

According to Table 3, the majority of the respondents stated that they did not increase total food consumption (63.8%) or purchase more food with sustainable attributes (55.2%) when compared to the situation before the lockdown. However, the majority of respondents (52.6%) stated that they increased food expenditure during the lockdown.

Table 3. Behavioral changes during the lockdown.

Category	Percentage		
	Total Food Consumption	**Expenditure**	**Sustainable Food**
Increase (Y = 1)	36.0%	52.6%	37.8%
Did not increase (Y = 0)	63.8%	47.4%	55.2%
Missing	0.2%	Null	7.0%

4.2.1. Changes in Total Food Consumption (C) during the Lockdown

As reported in Table 4, the percentage of the model's correct classification was 75.2%, and the Hosmer–Lemeshow's goodness of fit was equal to 0.353, leading us to accept the null hypothesis that there was no significant difference between the observed and model-predicted values [96]. The OR of gender was equal to 1.394, meaning that females were 1.394 times more likely to increase food consumption than males during the lockdown. One possible reason was that many food-away-from-home establishments were closed because of the shutdown restrictions during COVID-19 in Spain, such that an increasing number of working women had to cook at home, where they tended to consume more food. Another reason may be that women were more prone to depression, stress, and anxiety than men, resulting in more emotional eating [97].

Table 4. Logit model of total food consumption **(C)**.

Significant Variables	Reference Category	Beta (B)	p-Value	Exp (B)
Gender Female	Male	0.332	0.063	1.394
Age 40–59 years old More than 60 years old	18–39 years old	−0.622 −0.977	0.003 0.001	0.537 0.376
Monthly household income Income (before the lockdown) >3000 euros	<999 euros	1.086	0.021	2.963
Employment status ERTE (partial or total) Sick leave Unemployed Unable to work	Student	−1.061 −2.142 −1.020 −1.979	0.080 0.017 0.087 0.023	0.346 0.117 0.361 0.138
Place of residence Living in rural place	Urban	−0.437	0.077	0.646
Risk preference Risk-averse	Risk-loving	−0.365	0.085	0.694
Experiences Did not experience food shortages or price increase "I know a friend of my friends has been diagnosed or died due to COVID-19"	Experienced Do not know	−0.785 0.564	0.026 0.011	0.456 1.759
Shopping places Specialized food stores (before the lockdown) Farmer's market/open markets (before the lockdown)	Supermarkets	−0.750 −1.480	0.021 0.052	0.473 0.228
Trust in information sources Health professionals were perceived to be a little trustworthy	Not at all	−3.078	0.042	0.046
Food security risk perception A little unlikely to face food shortages in the next 6 months	Very unlikely	0.643	0.003	1.903
Risk perception of COVID-19 Somewhat serious if contracting in the next 6 months Very serious if contracting in the next 6 months	Not at all	1.595 1.596	0.003 0.012	4.930 4.934
Percentage of correct classification Hosmer–Lemeshow's goodness of fit			75.2% 0.353	

In addition, people aged 40–59 years and more than 60 years old were less likely to increase total food consumption than those aged 18–39 years when compared to the situation before the lockdown. This is in line with a study which showed that older people consumed less than younger people during the COVID-19 lockdown [98]. The results also demonstrated a positive and significant association between income and total food consumption. This indicated that households whose monthly income before the lockdown was more than 3000 euros were 2.963 times more likely to increase total food consumption than those less than 999 euros. Not surprisingly, more income in a household denoted a stronger purchasing power to provide food for their family members, such that they were more likely to increase total food consumption during the COVID-19 lockdown. People whose current employment status was ERTE (partial or total), on sick leave, unemployed, or unable to work were less likely to increase their food consumption during the lockdown. It was expected that these people's jobs were suspended or they were unable to work, such that their sources of income were cut off by COVID-19, and they were less likely to increase their consumption level. However, there was little change in income (no income) for students before and during the lockdown. Results also indicated that people who live in rural places were less likely to consume more food than those living in urban places. This may be related to several reasons. Firstly, population flow is more frequent in urban areas than that in rural places, resulting in a higher risk of contracting COVID-19 for consumers who live in urban areas. Consequently, people living in urban places may feel worried, anxious, or negative about themselves; thus, they tended to display emotional eating behavior to avoid these negative feelings by turning their attention to food during the lockdown [99]. Secondly, consumers living in urban areas usually have a higher income than those living in rural places; that is, they have a stronger purchasing power and consumption power.

As for consumers' stated risk preference, the results showed that risk-averse people were less likely to increase their total food consumption than risk-loving persons. A previous study indicated that risk-averse respondents may seek out more insurance after a disaster [100]; thus, risk-averse people may focus on health insurance or save money to make themselves feel more secure and use it when there is a health threat in the future. Respondents who did not experience food shortages or higher food prices or did not know someone who had been diagnosed or died due to COVID-19 were less likely to consume more food than those who experienced these situations. This could be explained by the fact that subjects who experienced food shortages or higher food prices or knew someone who had been diagnosed or died due to COVID-19 were more likely to be anxious [101]; thus, they were prone to emotional eating (over-eating). Regarding shopping places, people who went to specialized food stores and farmers' markets to purchase food before the lockdown were less likely to consume more food than those who went to supermarkets. This may be because specialized food stores and farmers' markets only sell food, while supermarkets have a wider variety of not only food products but also other necessities, such as toilet paper, shampoo, and pet supplies. Therefore, in order to reduce the number of visits to stores and reduce the risk of infection, consumers who used to buy food from specialized food stores and farmers' markets may have preferred to buy food from supermarkets during the lockdown, such that those who went to supermarkets consumed more food.

Results also showed that consumers were less likely to increase their food consumption when they perceived a higher trust level in health professionals (e.g., doctors) during the lockdown. Trust in reliable scientific information contributes to reducing unnecessary scares and inappropriate risk perceptions [55]. Hence, consumers who trust health professionals could reduce their risk perception and were less likely to panic buy and consume food. Regarding risk perception of COVID-19, this category demonstrated that consumers who perceived a higher risk of COVID-19 were more likely to increase their total food consumption than those who perceived a lower risk during the lockdown. This may be because if consumers thought the situation was serious, they were worried about themselves and tended to display emotional eating behavior. As for food security risk perception, this category revealed that consumers who perceived a higher risk of food shortages in the next six months were more likely to increase total food consumption than those perceiving the lowest food security risk. It was not surprising that people with a higher food security risk perception tended to stockpile food products to reduce the food security risk; thus, they turned to increase food consumption.

4.2.2. Changes in the Total Food Expenditure (E) during the Lockdown

In Table 5, the percentage of correct classification was 70.3%, and the value of Hosmer–Lemeshow's goodness of fit was 0.311, indicating that the model presented an acceptable goodness of fit. The results demonstrated that females were less likely to spend more on food than males during the lockdown. The data from the National Statistics Institute in Spain showed that the unemployment rates of females and males in the first quarter of 2020 in Spain were 16.24% and 12.79%, respectively. In the second quarter, they stood at 16.72% (females) and 14.13% (males) [102], indicating that females had a higher likelihood of being unemployed than males during the lockdown. Hence, females were more cautious about their income and less likely to increase food expenditure. Another potential reason was that females were the main meal preparers and "food gatekeepers" in the household [103]. As a result, they were more familiar with the characteristics (e.g., the price and the quality) of food products and always knew what food to buy, such that females were less likely to increase food expenditure. Conversely, males were not usual food buyers and not familiar with food products; therefore, males may have increased their expenditure on food.

Table 5. Logit model of food expenditure (E).

Significant Variables	Reference Category	Beta (B)	*p*-Value	Exp (B)
Gender				
Female	Male	−0.458	0.008	0.632
Age				
40–59 years old	18–39 years old	−0.572	0.006	0.564
More than 60 years old		−0.675	0.015	0.509
Employment status				
Sick leave	Student	−1.617	0.054	0.199
Unable to work		−1.485	0.060	0.226
Family structure				
There are children aged 7–12 years in the household	No	0.797	0.079	2.218
Experiences				
Experienced food shortages during the lockdown	Did not	0.524	0.017	1.688
Did not have symptoms, so did not test	Tested positive	−1.265	0.078	0.282
Did not know anyone who has been diagnosed or died due to COVID-19	Knew someone	−0.784	0.002	0.457
Shopping places				
Buy food on retailers' websites during the lockdown	Supermarkets	1.520	0.015	4.574
Mood				
Feel a little reassured	None of this feeling	0.794	0.004	2.213
Feel moderately reassured		0.582	0.044	1.789
Feel moderately angry		−0.859	0.017	0.424
Feel a great deal of angry		−0.722	0.095	0.486
Risk preference				
Risk-neutral	Risk-loving	−0.505	0.066	0.604
Risk-averse		−0.528	0.009	0.590
Trust in information sources				
Government information regarding COVID-19 was perceived to be a little trustworthy	Not trustworthy at all	−0.425	0.092	0.654
News information regarding COVID-19 was perceived to be very trustworthy		−1.021	0.030	0.360
Food security risk perception				
A little unlikely to face food shortages in the next 6 months	Very unlikely	0.543	0.036	1.722
Risk perception of COVID-19				
A little unlikely to contract COVID-19	Very unlikely	0.819	0.004	2.268
Financial risk perception				
Feel threatened moderately about financial situation	Not at all	−0.836	0.033	0.434
Feel threatened considerably about financial situation		−0.981	0.035	0.375
Feel threatened a great deal about financial situation		−1.502	0.009	0.223
Knowledge regarding COVID-19				
A higher level of objective knowledge		0.944	0.075	2.570
Percentage of correct classification			70.3%	
Hosmer–Lemeshow's goodness of fit			0.311	

People aged 40–59 years and more than 60 years old were less likely to increase expenditure than those aged 18–39 years when compared with the situation before the lockdown. The elderly were at a high risk of death due to COVID-19, which may have increased their worry and further affected their appetite [104]. Therefore, their cost was not likely to increase compared to younger people during the COVID-19 lockdown. Results also indicated that respondents whose employment status was sick leave and unable to work were less likely to spend more on food during the lockdown, which may be related to the interruption of their income. In addition, households with children aged 7–12 years were 2.218 times more likely to increase food expenditure than those without children. It was expected that primary schools were closed during the lockdown, such that children aged 7–12 years had to stay at home, resulting in more expenditure. Participants who experienced food shortages during the COVID-19 lockdown were 1.688 times more likely to increase their food expenditure than those who did not face food shortages. If consumers had experienced food shortages, they were likely to perceive that future food supplies may also be limited. Therefore, they spent more and stockpiled more food to reduce food security risks. In addition, consumers who tested positive or knew someone who had been diagnosed or died due to COVID-19 were more likely to increase food expenditure. This may be attributed to the fact that these people perceived a higher risk of contracting COVID-19. They therefore tended to buy more food per visit and reduce the number of shopping trips, thus reducing the risk of infection and consequently spending more on food. As for shopping places, consumers who bought food on retailers' websites during the lockdown were 4.574 times more likely to spend more on food than those who bought food in supermarkets. This is consistent with a study which found a significant increase in online shopping due to COVID-19 [7]. It was expected that consumers tended to shop online rather than in supermarkets to minimize store visits, aiming to reduce the risk of infection.

In addition, our results demonstrated that consumers with a positive mood (reassured) were more likely to increase food expenditure, while those with a negative mood (angry) were less likely. This outcome is supported by Mehrabian and Riccioni, who concluded that a positive mood was associated with high appetite levels [46]. Therefore, people with a positive mood during the lockdown tended to purchase more food and increase food expenditure, while a negative mood decreased consumers' appetite; thus, they were less likely to increase food expenditure. With regard to risk preference, the results implied that risk-neutral and risk-averse people were less likely to increase their food expenditure than those who were risk-loving during the lockdown. This may be related to risk-averse people's aversion to uncertainty, i.e., risk-averse consumers prefer certainty to uncertainty more than risk-loving ones. Due to the COVID-19 outbreak, they may tend to reduce food expenditure and save more money to prevent insufficient money when uncontrollable situations arise in the future. The findings also revealed that consumers were less likely to spend more on food when they perceived greater trust in government and news information regarding COVID-19 during the lockdown. This is supported by a study which demonstrated that higher trust in the national government had positive effects, such as reducing the likelihood of respondents' fears and worry of food shortages [105]. Consequently, these consumers perceived a lower food security risk and were less likely to stock up on food and increase food expenditure.

As for consumers' risk perceptions, the results indicated that the higher the COVID-19 risk and food security risk the consumers perceived, the more expenditure was seen. This is in line with a study which showed that consumers tend to purchase more stock goods when they perceive a higher risk, and this also indicates that a high risk perception during the COVID-19 pandemic will cause the intention to buy goods, leading to a higher probability of increasing food expenditure [106]. Another study also demonstrated that risk perception of the COVID-19 pandemic has positively affected consumers' behavior regarding the tendency to maintain food stocks [107]. The results also showed that consumers would not increase food expenditure when they perceived a higher financial risk, which highlighted previous research showing that risk perception negatively affected attitude and purchasing behavior [108]. This was expected, because when consumers feel threatened about their current financial situation, that is, perceiving a higher financial risk, they are more cautious about spending money. Additionally, consumers with a higher objective knowledge level regarding COVID-19 were found to have a higher likelihood of increasing food expenditure. It was expected that the more knowledge consumers had, the more severity about COVID-19 they perceived, such that they were more likely to increase expenditure to stock up on food.

4.2.3. Changes in Purchasing Food with Sustainable Attributes (S) during the Lockdown

As shown in Table 6, the fit was acceptable as indicated by Hosmer–Lemeshow's goodness of fit measures and the percentage of correct classification. The result showed that households with 5 members were 2.551 times more likely to purchase more food with sustainable attributes than those with 1 member when compared with the situation before the lockdown. This is supported by a study which indicated that consumers living in larger households were more likely to purchase organic food products [23].

Table 6. Logit model of purchasing food with sustainable attributes **(S)**.

Significant Variables	Reference Category	Beta (B)	*p*-Value	Exp (B)
Household size Households with 5 members	1 member	0.936	0.066	2.551
Risk preference Risk-averse	Risk-loving	−0.403	0.058	0.668
Shopping places Specialized food stores (before the lockdown)	Supermarkets	−0.710	0.028	0.492
Mood Feel considerably reassured Feel moderately angry	None of this feeling	0.773 −0.953	0.036 0.010	2.166 0.386
Trust in information sources Government information regarding COVID-19 was perceived to be very trustworthy	Not at all	0.481	0.095	1.618
Food security risk perception A little unlikely to face food shortages in the next 6 months A little likely to face food shortages in the next 6 months	Very unlikely	0.369 1.152	0.082 0.064	1.446 3.163
Risk perception of COVID-19 A little unlikely to contract COVID-19	Very unlikely	0.748	0.015	2.113
Financial risk perception Feel threatened moderately about financial situation Feel threatened a great deal about financial situation	Not at all	−0.675 −1.125	0.093 0.051	0.509 0.325
Percentage of correct classification Hosmer–Lemeshow's goodness of fit			73.0% 0.095	

In addition, risk-averse consumers were less likely to increase their purchases of food with sustainable attributes during the lockdown. This outcome converges with the finding that risk-averse respondents avoided buying more sustainable food during the lockdown in China [109]. It may relate to the uncertainty consumers feel when uncertain about food with sustainable attributes (e.g., whether organic certification can be trusted); they may therefore prefer the certainty of conventional products to the uncertainty that may come from sustainable ones [63]. The results also indicated that people who used to purchase food from specialized food stores (before the lockdown) were less likely to buy more food with sustainable attributes than those who usually went to supermarkets. Similar to the previous explanation, one possible reason was that specialized food stores only have food, while supermarkets have a more complete variety (e.g., food, alcohol, toilet paper, and pet supplies). As a consequence, consumers who used to purchase food from specialized food stores may be inclined to buy food (including food with sustainable attributes) and other necessities from the supermarkets during the lockdown to minimize trips to the store and reduce the risk of infection. Additionally, consumers with a positive mood (reassured) were more likely to purchase more food with sustainable attributes while those with a negative mood (angry) were less likely. One possible explanation was that positive emotions make consumers perceive sustainable food (e.g., organic food) as more attractive, and they are eager to purchase and consume healthy food [110].

According to the results, consumers with a higher trust level in government were more likely to increase their purchasing of food with sustainable attributes. This is supported by a study indicating that in public health emergencies, people who have high trust in government health agencies were more likely to follow health recommendations (including food choice recommendations) made by the government [90], and they regard sustainable food (e.g., organic food) as healthier food. Thus, they are more likely to purchase more food with sustainable attributes. The results also implied that consumers with higher risk perceptions of COVID-19 and food security were more likely to buy more food with sustainable attributes. Similarly, consumers in Spain perceived these products were healthier than conventional ones [111], which contributes to improving their immunity and reducing the risk of infection. The results also demonstrated that respondents who perceived a higher financial risk were less likely to purchase more food products with sustainable attributes when compared with the situation before the lockdown. Not surprisingly, food products with sustainable attributes were more expensive than conventional food [112]. Consumers tended to buy less sustainable food (expensive) when they perceived a higher financial risk, and they would spend money more carefully during the COVID-19 pandemic. The results of food security risk perception and financial risk perception are similar to the previous research conducted in China [109], but we did not find the effects of gender and age on the purchases of food with sustainable attributes in this research.

4.3. Overall Discussion

Overall, the majority of respondents stated that they did not increase food consumption (63.8%) or purchase more food with sustainable attributes (55.2%) during the lockdown. This is supported by a recent study which showed that 74% of respondents in Spain did not increase their food intake [113]. However, the majority of them (52.6%) stated that they increased food expenditure during the lockdown. This may be due to the fact that most Spanish participants reduced their food purchase frequency, which led to increased expenditure for each food purchase occasion [113].

Our results showed that females tended to consume more food but with less expenditure on food than males during the lockdown. Females were more likely to be depressed, stressed, and anxious, which can lead to emotional eating [97,114]. As previously explained, on the one hand, women were more likely to be unemployed during the COVID-19 pandemic compared to men, causing them to experience financial pressure [102]. On the other hand, women were the main food buyers and gatekeepers in the household [32,103]; therefore, they were more aware of food and did not engage in more spending. The literature review concluded that women were more likely to buy foods with sustainable attributes [31,32] because they were more health conscious and perceived sustainable foods as healthier [29]. However, this research did not identify a relationship between gender and purchase of sustainable food during the lockdown in Spain.

In this research, family structure only affected food expenditure, which is in line with an early study that showed households with children increased their food expenditure during the lockdown [39]. Nevertheless, we did not find that this factor influenced the other eating and purchasing behaviors defined in this study. From the literature review, household size positively influenced food expenditure [39], but our result indicated that household size was only a statistically significant factor affecting purchasing behavior towards sustainable attributes. Our results indicated that age was an indicator related to total food consumption and expenditure when compared to the situation before the lockdown, i.e., older people were less likely to increase food intake and expenditure. This can be explained by the fact that older people need fewer calories than younger people; thus, they buy less food and consume less [36]. This may also be related to the high risk of death in older adults due to COVID-19, which may increase their worry and fear, further affecting their appetite [104].

Mood was found to be associated with expenditure and purchasing food with sustainable attributes. This may be because, on the one hand, positive emotions make consumers perceive sustainable food (e.g., organic food) as more attractive, and they are more likely to be eager to purchase and consume healthy food [110]. On the other hand, positive emotion was correlated with high appetite levels [46], and it has been a neglected trigger for eating more food [47]. Hence, people who have positive emotions during the lockdown tended to buy more food and increase their food expenses, while negative emotions decreased consumers' appetite, and therefore, they were less likely to increase their food spending. However, we did not find a relationship between mood and total food consumption.

In addition, consumers' risk perceptions and trust in information sources were crucial factors in understanding consumers' food preferences and sustainable behavior during the lockdown. To be specific, consumers increased their food consumption, food expenditure, and purchased more sustainable food when they perceived a higher risk of COVID-19 and food security. However, consumers were less likely to increase expenditure and sustainable food when they perceived higher financial risks. This was in accordance with our expectation that, as explained earlier, food with sustainable attributes was more expensive than conventional food, and these people were more careful in spending their money [112]; thus, they were less likely to increase purchases of food with sustainable attributes and food expenditure. The results of food security risk perceptions and financial risk perceptions were comparable to previous studies conducted in China [109]. In addition, consumers' trust level in information from health professionals and scientists was higher than that from the government and news. Similar findings were found in a Chinese study, where health professionals were the most trusted source about COVID-19 [115]. This is also consistent with a previous study investigating perceived trust in general health information which showed that health professionals were identified as the most trusted sources [116]. This suggested that health professionals were the most trusted source of information, both for general health information and specific disease (e.g., COVID-19) information.

The results did not identify significant impacts of subjective knowledge, concerns, or stated health status on food preferences and sustainable behavior defined in this study. These results allow the government and stakeholders to deepen their understanding of consumers' preferences and sustainable behaviors during the lockdown in order to develop realistic policies and strategies.

5. Conclusions

This study explored trends in food preferences and sustainable purchasing behavior of Spanish consumers during the COVID-19 lockdown and the factors influencing them. Our empirical results gave some insights to the government, retailers, and other stakeholders that help them to adapt to the new COVID-19 situation.

5.1. Practical Implications

Firstly, based on the result of the increased expenditure on the retailers' websites, retailers should design a more visually attractive and convenient website, taking advantage of this opportunity to retain customers. Secondly, the Spanish government should make efforts to design more effective information to communicate with people and should enhance the quality and level of detail of the information that they share in such an emergency. This is because consumers reported low trust in government and news while reporting high trust in health professions and scientists, inspiring health professions and scientists to share more reliable and trustworthy information about COVID-19 and recommendations of food choices and consumption. Thirdly, households with children aged 7–12 years were more likely to increase food expenditure. As a result, retailers could carry out promotion activities (e.g., children's related food can be given as a gift if consumers spend a certain amount of money in the store), so as to attract families with children. Finally, consumers who live with large households and those who often go to the supermarket to buy food were more likely to purchase more food with sustainable attributes, reminding retailers to focus on these people by using this argument to first place and highlight sustainable items (e.g., organic items) in hotlines on the shelves.

5.2. Limitations and Future Research

Despite the contributions of this study, it has some limitations. Firstly, the data are based on stated rather than revealed behavior. Self-report items may be a limitation with respect to data quality, e.g., social desirability bias and lack of memory, inspiring future research with a focus on consumers' revealed behavior. Secondly, this research explored consumers' behavior before and during the lockdown but did not measure changes after the lockdown. Therefore, further research could explore whether this change in consumption and purchasing behavior is long-term in this global crisis, and can also explore other consumption and purchasing behaviors. Finally, the online survey excluded those who were unfamiliar with and did not have access to the internet. Nevertheless, because of its low cost and rapidity, it still gave valid data for this study, representing the population of Spain. Future studies could try to conduct face-to-face surveys after this outbreak to obtain a more comprehensive sample.

Author Contributions: Conceptualization, S.L., Z.K., D.R. and J.M.G.; methodology, S.L., Z.K. and D.R.; data curation, S.L., Z.K. and D.R.; formal analysis, S.L.; software, S.L.; visualization, S.L.; writing—original draft preparation, S.L.; validation, Z.K.; writing—review & editing, Z.K. and D.R.; supervision, Z.K.; project administration, Z.K. and D.R.; questionnaire design, S.L., Z.K. and D.R.; investigation, S.L., Z.K., D.R. and J.M.G.; funding acquisition, J.M.G. All authors have read and agreed to the published version of the manuscript.

Funding: This research received no external funding.

Institutional Review Board Statement: The study was conducted according to the guidelines of the Declaration of Helsinki, and approved by the Ethics Committee of the Centre for Agro-food Economy and Development (CREDA) (protocol code 2020-1 and date of approval 6 May 2020).

Informed Consent Statement: Informed consent was obtained from all subjects involved in the study.

Data Availability Statement: The data presented in this study are available on request from the corresponding author.

Acknowledgments: The authors are thankful to the anonymous reviewers and the editor for their valuable comments, which helped to considerably improve the manuscript. S.L. thanks the China Scholarship Council (CSC) for funding her Ph.D. scholarship.

Conflicts of Interest: The authors declare no conflict of interest.

References

1. Chakraborty, I.; Maity, P. COVID-19 outbreak: Migration, effects on society, global environment and prevention. *Sci. Total Environ.* **2020**, *728*, 138882. [CrossRef] [PubMed]
2. Laguna, L.; Fiszman, S.; Puerta, P.; Chaya, C.; Tárrega, A. The impact of COVID-19 lockdown on food priorities. Results from a preliminary study using social media and an online survey with Spanish consumers. *Food Qual. Prefer.* **2020**, *86*, 104028. [CrossRef] [PubMed]
3. Coibion, O.; Gorodnichenko, Y.; Weber, M. *The Cost of the Covid-19 Crisis: Lockdowns, Macroeconomic Expectations, and Consumer Spending*; National Bureau of Economic Research: Cambridge, MA, USA, 2020.
4. Wanberg, C.R. The Individual Experience of Unemployment. *Annu. Rev. Psychol.* **2012**, *63*, 369–396. [CrossRef] [PubMed]
5. D'Alessandro, D.; Gola, M.; Appolloni, L.; Dettori, M.; Fara, G.M.; Rebecchi, A.; Settimo, G.; Capolongo, S. COVID-19 and Living space challenge. Well-being and Public Health recommendations for a healthy, safe, and sustainable housing. *Acta Biomed. Atenei Parm.* **2020**, *91*, 61–75.

6. Scarmozzino, F.; Visioli, F. Covid-19 and the Subsequent Lockdown Modified Dietary Habits of Almost Half the Population in an Italian Sample. *Foods* **2020**, *9*, 675. [CrossRef]
7. Nuijten, S. Pandemic Impacts on Consumer food Behaviour. Available online: https://www.eitfood.eu/media/news-pdf/COVID-19_Study_-_European_Food_Behaviours_-_Report.pdf (accessed on 6 April 2021).
8. Głąbska, D.; Skolmowska, D.; Guzek, D. Population-Based Study of the Changes in the Food Choice Determinants of Secondary School Students: Polish Adolescents' COVID-19 Experience (PLACE-19) Study. *Nutrients* **2020**, *12*, 2640. [CrossRef]
9. Scacchi, A.; Catozzi, D.; Boietti, E.; Bert, F.; Siliquini, R. COVID-19 Lockdown and Self-Perceived Changes of Food Choice, Waste, Impulse Buying and Their Determinants in Italy: QuarantEat, a Cross-Sectional Study. *Foods* **2021**, *10*, 306. [CrossRef]
10. Rodríguez-Pérez, C.; Molina-Montes, E.; Verardo, V.; Artacho, R.; García-Villanova, B.; Guerra-Hernández, E.J.; Ruíz-López, M.D. Changes in Dietary Behaviours during the COVID-19 Outbreak Confinement in the Spanish COVIDiet Study. *Nutrients* **2020**, *12*, 1730. [CrossRef]
11. Ben-Hassen, T.; El-Bilali, H.; Allahyari, M.S. Impact of COVID-19 on Food Behavior and Consumption in Qatar. *Sustainability* **2020**, *12*, 6973. [CrossRef]
12. Yuen, K.F.; Wang, X.; Ma, F.; Li, K.X. The Psychological Causes of Panic Buying Following a Health Crisis. *Int. J. Environ. Res. Public Health* **2020**, *17*, 3513. [CrossRef]
13. Lins, S.; Aquino, S. Development and initial psychometric properties of a panic buying scale during COVID-19 pandemic. *Heliyon* **2020**, *6*, e04746. [CrossRef] [PubMed]
14. MAPA Spanish Ministry of Agriculture, Fisheries and Food (MAPA). Press Releases, March/April, Madrid, Spain. 2020. Available online: https://www.mapa.gob.es/es/prensa/ultimas-noticias (accessed on 6 April 2021).
15. Cranfield, J.A.L. Framing consumer food demand responses in a viral pandemic. *Can. J. Agric. Econ. Can. d'Agroecon.* **2020**, *68*, 151–156. [CrossRef]
16. Chauhan, V.; Shah, H. An Empirical Analysis into Sentiments, Media Consumption Habits, and Consumer Behaviour during the Coronavirus (COVID-19) Outbreak. *Purakala UGC Care J.* **2020**, *31*, 353–378. [CrossRef]
17. Cha, W.M.; Chintagunta, P.K.; Dhar, S.K. Food Purchases during the Great Recession. *Kilts Cent. Mark. Chic. Booth Nielsen Dataset Pap. Ser.* **2015**, *1*, 8. [CrossRef]
18. Baker, S.R.; Farrokhnia, R.A.; Meyer, S.; Pagel, M.; Yannelis, C. How Does Household Spending Respond to an Epidemic? Consumption during the 2020 COVID-19 Pandemic. *Rev. Asset Pricing Stud.* **2020**, *10*, 834–862. [CrossRef]
19. World Economic Forum. These Charts Show How COVID-19 Has Changed Consumer Spending Around the World. Available online: https://www.weforum.org/agenda/2020/05/coronavirus-covid19-consumers-shopping-goods-economics-industry/ (accessed on 6 April 2021).
20. Martin-Neuninger, R.; Ruby, M. What Does Food Retail Research Tell Us About the Implications of Coronavirus (COVID-19) for Grocery Purchasing Habits? *Front. Psychol.* **2020**, *11*, 1–4. [CrossRef] [PubMed]
21. Di Renzo, L.; Gualtieri, P.; Pivari, F.; Soldati, L.; Attinà, A.; Cinelli, G.; Leggeri, C.; Caparello, G.; Barrea, L.; Scerbo, F.; et al. Eating habits and lifestyle changes during COVID-19 lockdown: An Italian survey. *J. Transl. Med.* **2020**, *18*, 229. [CrossRef]
22. Mankoff, J.; Matthews, D.; Fussell, S.R.; Johnson, M. Leveraging Social Networks To Motivate Individuals to Reduce their Ecological Footprints. In Proceedings of the 2007 40th Annual Hawaii International Conference on System Sciences (HICSS'07), Big Island, HI, USA, 3–6 January 2007; p. 87.
23. Tsakiridou, E.; Mattas, K.; Tzimitra-Kalogianni, I. The Influence of Consumer Characteristics and Attitudes on the Demand for Organic Olive Oil. *J. Int. Food Agribus. Mark.* **2006**, *18*, 23–31. [CrossRef]
24. Rahmani, D.; Kallas, Z.; Pappa, M.; Gil, J.M. Are Consumers' Egg Preferences Influenced by Animal-Welfare Conditions and Environmental Impacts? *Sustainability* **2019**, *11*, 6218. [CrossRef]
25. Escobar, C.; Kallas, Z.; Gil, J. Consumers' wine preferences in a changing scenario. *Br. Food J.* **2018**, *120*, 18–32. [CrossRef]
26. Li, S.; Kallas, Z. Meta-analysis of consumers' willingness to pay for sustainable food products. *Appetite* **2021**, *163*, 105239. [CrossRef]
27. Romeo-Arroyo, E.; Mora, M.; Vázquez-Araújo, L. Consumer behavior in confinement times: Food choice and cooking attitudes in Spain. *Int. J. Gastron. Food Sci.* **2020**, *21*, 100226. [CrossRef]
28. Chen, P.-J.; Antonelli, M. Conceptual Models of Food Choice: Influential Factors Related to Foods, Individual Differences, and Society. *Foods* **2020**, *9*, 1898. [CrossRef] [PubMed]
29. Lea, E.; Worsley, T. Australians' organic food beliefs, demographics and values. *Br. Food J.* **2005**, *107*, 855–869. [CrossRef]
30. Kriwy, P.; Mecking, R.-A. Health and environmental consciousness, costs of behaviour and the purchase of organic food. *Int. J. Consum. Stud.* **2011**, *36*, 30–37. [CrossRef]
31. Memery, J.; Angell, R.J.; Megicks, P.; Lindgreen, A. Unpicking motives to purchase locally-produced food: Analysis of direct and moderation effects. *Eur. J. Mark.* **2015**, *49*, 1207–1233. [CrossRef]
32. Onyango, B.M.; Hallman, W.K.; Bellows, A.C. Purchasing organic food in US food systems: A study of attitudes and practice. *Br. Food J.* **2007**, *109*, 399–411. [CrossRef]
33. Aertsens, J.; Mondelaers, K.; Verbeke, W.; Buysse, J.; Van Huylenbroeck, G. The influence of subjective and objective knowledge on attitude, motivations and consumption of organic food. *Br. Food J.* **2011**, *113*, 1353–1378. [CrossRef]
34. Gracia Royo, A.; De-Magistris, T. Organic food product purchase behaviour: A pilot study for urban consumers in the South of Italy. *Span. J. Agric. Res.* **2007**, *5*, 439–451. [CrossRef]

35. Zepeda, L.; Li, J. Characteristics of Organic Food Shoppers. *J. Agric. Appl. Econ.* **2007**, *39*, 17–28. [CrossRef]
36. Garcia, T.; Grande, I. Determinants of food expenditure patterns among older consumers. The Spanish case. *Appetite* **2010**, *54*, 62–70. [CrossRef]
37. Banovic, M.; Barreira, M.M.; Fontes, M.A. The Portuguese Household Food Expenditure: 1990, 1995 and 2000. *New Medit* **2006**, *5*, 25.
38. Davies, A.; Titterington, A.J.; Cochrane, C. Who buys organic food? A profile of the purchasers of organic food in Northern Ireland. *Br. Food J.* **1995**, *97*, 17–23. [CrossRef]
39. Babalola, D.A.; Isitor, S.U. Analysis of the determinants of food expenditure patterns among urban households in Nigeria: Evidence from Lagos State. *IOSR J. Agric. Vet. Sci.* **2014**, *7*, 71–75.
40. Hsu, S.-Y.; Chang, C.-C.; Lin, T.T. An analysis of purchase intentions toward organic food on health consciousness and food safety with/under structural equation modeling. *Br. Food J.* **2016**, *118*, 200–216. [CrossRef]
41. Januszewska, R.; Pieniak, Z.; Verbeke, W. Food choice questionnaire revisited in four countries. Does it still measure the same? *Appetite* **2011**, *57*, 94–98. [CrossRef]
42. Canetti, L.; Bachar, E.; Berry, E.M. Food and emotion. *Behav. Process.* **2002**, *60*, 157–164. [CrossRef]
43. Cardi, V.; Leppanen, J.; Treasure, J. The effects of negative and positive mood induction on eating behaviour: A meta-analysis of laboratory studies in the healthy population and eating and weight disorders. *Neurosci. Biobehav. Rev.* **2015**, *57*, 299–309. [CrossRef]
44. Lyman, B. The Nutritional Values and Food Group Characteristics of Foods Preferred during Various Emotions. *J. Psychol.* **1982**, *112*, 121–127. [CrossRef]
45. Mennella, J.A.; Pepino, M.Y.; Lehmann-Castor, S.M.; Yourshaw, L.M. Sweet preferences and analgesia during childhood: Effects of family history of alcoholism and depression. *Addiction* **2010**, *105*, 666–675. [CrossRef]
46. Mehrabian, A.; Riccioni, M. Measures of Eating-Related Characteristics for the General Population: Relationships with Temperament. *J. Pers. Assess.* **1986**, *50*, 610–629. [CrossRef]
47. Evers, C.; Adriaanse, M.; De Ridder, D.T.; Huberts, J.C.D.W. Good mood food. Positive emotion as a neglected trigger for food intake. *Appetite* **2013**, *68*, 1–7. [CrossRef]
48. Wansink, B. Environmental Factors that Increase the Food Intake and Consumption Volume of Unknowing Consumers. *Annu. Rev. Nutr.* **2004**, *24*, 455–479. [CrossRef]
49. Boggiano, M.M.; Wenger, L.E.; Mrug, S.; Burgess, E.E.; Morgan, P.R. The Kids-Palatable Eating Motives Scale: Relation to BMI and binge eating traits. *Eat. Behav.* **2015**, *17*, 69–73. [CrossRef]
50. Garbarino, E.; Johnson, M.S. The different roles of satisfaction, trust, and commitment in customer relationships. *J. Mark.* **1999**, *63*, 70–87. [CrossRef]
51. Pornpitakpan, C. The Persuasiveness of Source Credibility: A Critical Review of Five Decades' Evidence. *J. Appl. Soc. Psychol.* **2004**, *34*, 243–281. [CrossRef]
52. Hoque, M.Z.; Alam, M. What determines the purchase intention of liquid milk during a food security crisis? The role of perceived trust, knowledge, and risk. *Sustainability* **2018**, *10*, 3722. [CrossRef]
53. Thomas, M.S.; Feng, Y. Consumer risk perception and trusted sources of food safety information during the COVID-19 pandemic. *Food Control* **2021**, *130*, 108279. [CrossRef]
54. Lobb, A.; Mazzocchi, M.; Traill, W. Modelling risk perception and trust in food safety information within the theory of planned behaviour. *Food Qual. Prefer.* **2007**, *18*, 384–395. [CrossRef]
55. Faour-Klingbeil, D.; Osaili, T.M.; Al-Nabulsi, A.A.; Jemni, M.; Todd, E.C. The public perception of food and non-food related risks of infection and trust in the risk communication during COVID-19 crisis: A study on selected countries from the Arab region. *Food Control* **2021**, *121*, 107617. [CrossRef]
56. Arafat, S.Y.; Kar, S.K.; Marthoenis, M.; Sharma, P.; Apu, E.H.; Kabir, R. Psychological underpinning of panic buying during pandemic (COVID-19). *Psychiatry Res.* **2020**, *289*, 113061. [CrossRef]
57. Islam, T.; Pitafi, A.H.; Aryaa, V.; Wang, Y.; Akhtar, N.; Mubarik, S.; Xiaobei, L. Panic buying in the COVID-19 pandemic: A multi-country examination. *J. Retail. Consum. Serv.* **2020**, *59*, 102357. [CrossRef]
58. Rohrmann, B. Risk perception, risk attitude, risk communication, risk management: A conceptual appraisal. In Proceedings of the International Emergency Management Society Annual Conference, Prague, Czech Republic, 17–19 June 2008.
59. Taylor-Gooby, P. Varieties of risk. *Health Risk Soc.* **2002**, *4*, 109–111. [CrossRef]
60. Ko, H.; Jung, J.; Kim, J.; Shim, S.W. Cross-Cultural Differences in Perceived Risk of Online Shopping. *J. Interact. Advert.* **2004**, *4*, 20–29. [CrossRef]
61. Hallowell, M. Safety risk perception in construction companies in the Pacific Northwest of the USA. *Constr. Manag. Econ.* **2010**, *28*, 403–413. [CrossRef]
62. Pieniak, Z.; Verbeke, W.; Scholderer, J.; Brunsø, K.; Olsen, S.O. Impact of consumers' health beliefs, health involvement and risk perception on fish consumption: A study in five European countries. *Br. Food J.* **2008**, *110*, 898–915. [CrossRef]
63. Smith, S.; Paladino, A. Eating clean and green? Investigating consumer motivations towards the purchase of organic food. *Australas. Mark. J.* **2010**, *18*, 93–104. [CrossRef]
64. Lusk, J.L.; Coble, K. Risk Perceptions, Risk Preference, and Acceptance of Risky Food. *Am. J. Agric. Econ.* **2005**, *87*, 393–405. [CrossRef]

65. Brucks, M. The Effects of Product Class Knowledge on Information Search Behavior. *J. Consum. Res.* **1985**, *12*, 1–16. [CrossRef]
66. Dodd, T.H.; Laverie, D.A.; Wilcox, J.F.; Duhan, D.F. Differential Effects of Experience, Subjective Knowledge, and Objective Knowledge on Sources of Information used in Consumer Wine Purchasing. *J. Hosp. Tour. Res.* **2005**, *29*, 3–19. [CrossRef]
67. Ali, A.; Sohaib, M.; Iqbal, S.; Hayat, K.; Khan, A.U.; Rasool, M.F. Evaluation of COVID-19 Disease Awareness and Its Relation to Mental Health, Dietary Habits, and Physical Activity: A Cross-Sectional Study from Pakistan. *Am. J. Trop. Med. Hyg.* **2021**, *104*, 1687–1693. [CrossRef]
68. Clements, J.M. Knowledge and Behaviors toward COVID-19 among US Residents during the Early Days of the Pandemic: Cross-Sectional Online Questionnaire. *JMIR Public Health Surveill.* **2020**, *6*, e19161. [CrossRef]
69. Musche, V.; Bäuerle, A.; Steinbach, J.; Schweda, A.; Hetkamp, M.; Weismüller, B.; Kohler, H.; Beckmann, M.; Herrmann, K.; Tewes, M.; et al. COVID-19-Related Fear and Health-Related Safety Behavior in Oncological Patients. *Front. Psychol.* **2020**, *11*, 1984. [CrossRef]
70. Latip, M.S.A.; Newaz, F.T.; Ramasamy, R.; Tumin, S.A.; Noh, I. How do food safety knowledge and trust affect individual's green considerations during the covid-19 pandemic in malaysia? *Malays. J. Consum. Fam. Econ.* **2020**, *24*, 261–285.
71. Zhu, H.; Deng, F. How to Influence Rural Tourism Intention by Risk Knowledge during COVID-19 Containment in China: Mediating Role of Risk Perception and Attitude. *Int. J. Environ. Res. Public Health* **2020**, *17*, 3514. [CrossRef]
72. Torres, M.A.O.; Kallas, Z.; Herrera, S.I.O. Analysis of Farmers' Stated Risk Using Lotteries and Their Perceptions of Climate Change in the Northwest of Mexico. *Agronomy* **2018**, *9*, 4. [CrossRef]
73. Harrison, G.W.; Lau, M.I.; Rutström, E.E. Estimating Risk Attitudes in Denmark: A Field Experiment. *Scand. J. Econ.* **2007**, *109*, 341–368. [CrossRef]
74. Bults, M.; Beaujean, D.J.M.A.; de Zwart, O.; Kok, G.; van Empelen, P.; van Steenbergen, J.E.; Richardus, J.H.; Voeten, H.A.C.M. Perceived risk, anxiety, and behavioural responses of the general public during the early phase of the Influenza A (H1N1) pandemic in the Netherlands: Results of three consecutive online surveys. *BMC Public Health* **2011**, *11*, 2. [CrossRef]
75. Rosi, A.; van Vugt, F.T.; Lecce, S.; Ceccato, I.; Vallarino, M.; Rapisarda, F.; Vecchi, T.; Cavallini, E. Risk Perception in a Real-World Situation (COVID-19): How It Changes From 18 to 87 Years Old. *Front. Psychol.* **2021**, *12*, 646558. [CrossRef] [PubMed]
76. Brug, J.; Aro, A.R.; Oenema, A.; De Zwart, O.; Richardus, J.H.; Bishop, G.D. SARS Risk Perception, Knowledge, Precautions, and Information Sources, The Netherlands. *Emerg. Infect. Dis.* **2004**, *10*, 1486–1489. [CrossRef] [PubMed]
77. Xu, M.L.; Leung, S.O. Effects of varying numbers of Likert scale points on factor structure of the Rosenberg Self-Esteem Scale. *Asian J. Soc. Psychol.* **2018**, *21*, 119–128. [CrossRef]
78. Gross, R.; Schoeneberger, H.; Pfeifer, H.; Preuss, H.-J. The four dimensions of food and nutrition security: Definitions and concepts. *SCN News* **2000**, *20*, 20–25.
79. Gardner, M.P.; Wansink, B.; Kim, J.; Park, S.-B. Better moods for better eating? How mood influences food choice. *J. Consum. Psychol.* **2014**, *24*, 320–335. [CrossRef]
80. Dillard, J.P.; Peck, E. Affect and persuasion: Emotional responses to public service announcements. *Communic. Res.* **2000**, *27*, 461–495. [CrossRef]
81. An, L.C.; Wallner, L.; Kirch, M.A.; Grainger, R.; Attai, D. Online Social Engagement by Cancer Patients: A Clinic-Based Patient Survey. *JMIR Cancer* **2016**, *2*, e10. [CrossRef] [PubMed]
82. Power, M.; Doherty, B.; Pybus, K.; Pickett, K. How COVID-19 has exposed inequalities in the UK food system: The case of UK food and poverty. *Emerald Open Res.* **2020**, *2*, 11. [CrossRef]
83. Boylan, J.; Seli, P.; Scholer, A.A.; Danckert, J. Boredom in the COVID-19 pandemic: Trait boredom proneness, the desire to act, and rule-breaking. *Pers. Individ. Differ.* **2021**, *171*, 110387. [CrossRef]
84. Mostafa, M.M. Antecedents of Egyptian consumers' green purchase intentions: A hierarchical multivariate regression model. *J. Int. Consum. Mark.* **2006**, *19*, 97–126. [CrossRef]
85. Aday, S.; Aday, M.S. Impact of COVID-19 on the food supply chain. *Food Qual. Saf.* **2020**, *4*, 167–180. [CrossRef]
86. Jung, H.; Park, M.; Hong, K.; Hyun, E. The Impact of an Epidemic Outbreak on Consumer Expenditures: An Empirical Assessment for MERS Korea. *Sustainability* **2016**, *8*, 454. [CrossRef]
87. Adelaar, T.; Chang, S.; Lancendorfer, K.M.; Lee, B.; Morimoto, M. Effects of Media Formats on Emotions and Impulse Buying Intent. *J. Inf. Technol.* **2003**, *18*, 247–266. [CrossRef]
88. Thai, C.L.; Gaysynsky, A.; Falisi, A.; Chou, W.-Y.S.; Blake, K.; Hesse, B.W. Trust in Health Information Sources and Channels, Then and Now: Evidence from the Health Information National Trends Survey (2005–2013). *eHealth Curr. Evidence Promises Perils Futur. Dir.* **2018**, *15*, 43–67.
89. Camporesi, S.; Vaccarella, M.; Davis, M. Investigating Public trust in Expert Knowledge: Narrative, Ethics, and Engagement. *J. Bioethical Inq.* **2017**, *14*, 23–30. [CrossRef] [PubMed]
90. Freimuth, V.S.; Musa, D.; Hilyard, K.; Quinn, S.C.; Kim, K. Trust During the Early Stages of the 2009 H1N1 Pandemic. *J. Health Commun.* **2014**, *19*, 321–339. [CrossRef]
91. Weber, E.U.; Milliman, R.A. Perceived Risk Attitudes: Relating Risk Perception to Risky Choice. *Manag. Sci.* **1997**, *43*, 123–144. [CrossRef]
92. Harrison, G.W.; Lau, M.; Rutström, E.; Sullivan, M.B. Eliciting Risk and Time Preferences Using Field Experiments: Some Methodological Issues. *Res. Exp. Econ.* **2005**, 125–218. [CrossRef]

93. Dryhurst, S.; Schneider, C.R.; Kerr, J.; Freeman, A.L.J.; Recchia, G.; van der Bles, A.M.; Spiegelhalter, D.; van der Linden, S. Risk perceptions of COVID-19 around the world. *J. Risk Res.* **2020**, *23*, 994–1006. [CrossRef]
94. Informacion Producir Frutas y Verduras Cuesta Ahora un 25% Más Que Antes de la Crisis del Covid-19—Información. Available online: https://www.informacion.es/economia/2020/04/18/producir-frutas-verduras-cuesta-ahora-4692057.html (accessed on 22 February 2021).
95. Kasperson, R.E.; Renn, O.; Slovic, P.; Brown, H.S.; Emel, J.; Goble, R.; Kasperson, J.X.; Ratick, S. The Social Amplification of Risk: A Conceptual Framework. *Risk Anal.* **1988**, *8*, 177–187. [CrossRef]
96. Maharjan, K.L.; Joshi, N.P. Determinants of household food security in Nepal: A binary logistic regression analysis. *J. Mt. Sci.* **2011**, *8*, 403–413. [CrossRef]
97. Thompson, S.H. Gender and Racial Diferences in Emotional Eating, Food Addiction Symptoms, and Body Weight Satisfaction among ndergraduates. *J. Diabetes Obes.* **2015**, *2*, 93–98. [CrossRef]
98. Emerson, K.G. Coping with being cooped up: Social distancing during COVID-19 among 60+ in the United States. *Rev. Panam. Salud Pública* **2020**, *44*, e81. [CrossRef] [PubMed]
99. Nguyen-Rodriguez, S.T.; Unger, J.; Spruijt-Metz, D. Psychological Determinants of Emotional Eating in Adolescence. *Eat. Disord.* **2009**, *17*, 211–224. [CrossRef] [PubMed]
100. Cameron, L.; Shah, M. Risk-Taking Behavior in the Wake of Natural Disasters. *J. Hum. Resour.* **2015**, *50*, 484–515. [CrossRef]
101. Liu, M.; Zhang, H.; Huang, H. Media exposure to COVID-19 information, risk perception, social and geographical proximity, and self-rated anxiety in China. *BMC Public Health* **2020**, *20*, 1649. [CrossRef] [PubMed]
102. Spanish Statistical Office Unemployment Rates by CCAA. Available online: https://www.ine.es/infografias/tasasepa/desktop/tasas.html?t=0&lang=en (accessed on 22 February 2021).
103. Solgaard, H.S.; Yang, Y. Consumers' perception of farmed fish and willingness to pay for fish welfare. *Br. Food J.* **2011**, *113*, 997–1010. [CrossRef]
104. Vives, J.M.M.; Jornet-Gibert, M.; Cámara-Cabrera, J.; Esteban, P.L.; Brunet, L.; Delgado-Flores, L.; Camacho-Carrasco, P.; Torner, P.; Marcano-Fernández, F. Spanish HIP-COVID Investigation Group Mortality Rates of Patients with Proximal Femoral Fracture in a Worldwide Pandemic. *J. Bone Jt. Surg. Am. Vol.* **2020**, *102*, e69. [CrossRef]
105. Jeżewska-Zychowicz, M.; Plichta, M.; Królak, M. Consumers' Fears Regarding Food Availability and Purchasing Behaviors during the COVID-19 Pandemic: The Importance of Trust and Perceived Stress. *Nutrients* **2020**, *12*, 2852. [CrossRef]
106. Long, N.N.; Khoi, B.H. An Empirical Study about the Intention to Hoard Food during COVID-19 Pandemic. *Eurasia J. Math. Sci. Technol. Educ.* **2020**, *16*, em1857. [CrossRef]
107. Borsellino, V.; Kaliji, S.; Schimmenti, E. COVID-19 Drives Consumer Behaviour and Agro-Food Markets towards Healthier and More Sustainable Patterns. *Sustainability* **2020**, *12*, 8366. [CrossRef]
108. Tham, K.W.; Dastane, O.; Johari, Z.; Ismail, N.B. Perceived Risk Factors Affecting Consumers' Online Shopping Behaviour. *J. Asian Financ. Econ. Bus.* **2019**, *6*, 246–260. [CrossRef]
109. Li, S.; Kallas, Z.; Rahmani, D. Did the COVID-19 lockdown affect consumers' sustainable behaviour in food purchasing and consumption in China? *Food Control* **2021**. [CrossRef]
110. Köster, E.P.; Mojet, J. From mood to food and from food to mood: A psychological perspective on the measurement of food-related emotions in consumer research. *Food Res. Int.* **2015**, *76*, 180–191. [CrossRef]
111. Gil Roig, J.M.; Gracia Royoz, A.; Sánchez García, M. Market segmentation and willingness to pay for organic products in Spain. *Int. Food Agribus. Manag. Rev.* **2000**, *3*, 207–226.
112. Bhaskaran, S.; Polonsky, M.; Cary, J.; Fernandez, S. Environmentally sustainable food production and marketing. *Br. Food J.* **2006**, *108*, 677–690. [CrossRef]
113. Vidal-Mones, B.; Barco, H.; Diaz-Ruiz, R.; Fernandez-Zamudio, M.-A. Citizens' Food Habit Behavior and Food Waste Consequences during the First COVID-19 Lockdown in Spain. *Sustainability* **2021**, *13*, 3381. [CrossRef]
114. Pak, H.; Süsen, Y.; Nazlıgül, M.D.; Griffiths, M. The Mediating Effects of Fear of COVID-19 and depression on the association between intolerance of uncertainty and emotional eating during the COVID-19 Pandemic in Turkey. *Int. J. Ment. Health Addict.* **2021**, *19*, 1–15.
115. Lu, L.; Liu, J.; Yuan, Y.C.; Burns, K.S.; Lu, E.; Li, D. Source Trust and COVID-19 Information Sharing: The Mediating Roles of Emotions and Beliefs About Sharing. *Health Educ. Behav.* **2021**, *48*, 132–139. [CrossRef] [PubMed]
116. Dutta-Bergman, M. Trusted Online Sources of Health Information: Differences in Demographics, Health Beliefs, and Health-Information Orientation. *J. Med. Internet Res.* **2003**, *5*, e21. [CrossRef]

foods

MDPI

Article

Changes in Eating Behaviour during SARS-CoV-2 Pandemic among the Inhabitants of Five European Countries

Paulina Górska [1,*], Ilona Górna [1], Izabela Miechowicz [2] and Juliusz Przysławski [1]

1 Department of Bromatology, Poznań University of Medical Sciences, 60-354 Poznan, Poland; igorna@ump.edu.pl (I.G.); jotespe@ump.edu.pl (J.P.)
2 Department of Computer Science and Statistics, Poznań University of Medical Sciences, 60-806 Poznan, Poland; iza@ump.edu.pl
* Correspondence: paulina.gorska@student.ump.edu.pl

Abstract: Psychological factors and restrictions imposed due to the pandemic may influence eating behaviours and physical activity. With the above thesis in mind, questionnaire-based surveys were conducted amongst residents of five European countries: Poland, Italy, Spain, Portugal and Great Britain (England and Scotland). A specially devised, structured questionnaire was used to conduct anonymous internet surveys between 28 April and 16 July 2020. It contained questions pertaining to sociodemographic data, eating behaviours, the impact of the pandemic on the diet and physical activity. The questionnaire was made available to internet users in Poland, Italy, Spain, Great Britain (England and Scotland), and Portugal. The questionnaire was translated by native speakers into five languages: Polish, English, Spanish, Italian and Portuguese. Survey results were then analysed using StatSoft's Statistica v. 13 software and Cytel's StatXact v. 9.0.0. Age was the parameter that impacted changing eating behaviours to the largest extent during the pandemic. It was also found that during the pandemic, regular consumption of meals was most dependent on various factors. The negative impact of the pandemic within this scope was most profound amongst women, city residents regardless of gender and people over 35 years of age. A change in the frequency of consumption of selected product groups during the pandemic was also observed. Reduced consumption of meat and fish was identified. Especially among people under 35 living in Portugal, almost half—45.5% ($p = 0.0210$) declared lower consumption of meat, and more than half—54.5% ($p = 0.011$) reported lower consumption of fish. An analysis of the obtained results also showed an increase in the consumption of products with lower nutritional values, particularly amongst people under 35 years of age and also amongst residents of Great Britain (regardless of age). Moreover, the results showed that the pandemic may have had an impact on the weight reduction diet. A negative impact was declared by 16.5% of people, compared to 9.7% who said that the pandemic facilitated the use of the weight reduction diet ($p = 0.006$). The results of our survey also showed a decrease in the level of physical activity among people over 35 living in Poland (69.6%, $p = 0.0497$) and people living in Portuguese cities (72.73%, $p = 0.0245$). Our survey results showed that the impact of the pandemic on eating behaviours was particularly profound when it came to meal consumption regularity. Changes to the consumption of products with lower nutritional values, which may decrease immunity, have also been found during the pandemic. Our results showed that the problem associated with consuming products with lower nutritional values was particularly evident amongst people under 35. Considering the global character of SARS-CoV-2 transmission, further research is necessary to determine its impact on the diet, nutritional status and physical activity.

Keywords: eating behaviour; consumer preference; COVID-19

Citation: Górska, P.; Górna, I.; Miechowicz, I.; Przysławski, J. Changes in Eating Behaviour during SARS-CoV-2 Pandemic among the Inhabitants of Five European Countries. *Foods* **2021**, *10*, 1624. https://doi.org/10.3390/foods10071624

Academic Editors: Derek V. Byrne and Christopher John Smith

Received: 5 May 2021
Accepted: 11 July 2021
Published: 13 July 2021

1. Introduction

Towards the end of 2019, several pneumonia cases of unknown aetiology [1,2] were reported in Wuhan, the capital of the Chinese Hubei Province. The virus causes the onset of a disease, which the World Health Organisation named "coronavirus 2019" (COVID-19) [3]. Approximately two months after the CDC identified the new disease, the World Health Organisation characterised the spread of the disease as a pandemic [4].

The scale with which the virus is able to spread is exceptional. This has forced the governments of many countries to freeze many sectors of the economy, social and cultural life, which is why it is so important to find an answer to how such an unusual situation affects our eating behaviours and physical activity required to stay healthy. Moreover, because we cannot forecast when the pandemic will end, this might profoundly impact public health.

Social isolation is one of the most effective means used to control the pandemic [5]. However, despite its benefits in curbing the spread of the virus, it also has negative consequences. Data acquired during the SARS-CoV epidemic have shown that quarantine is associated with a higher risk of mental disorders, exhaustion, anger, sleep problems and the occurrence of depression and post-traumatic stress symptoms [6–8]. Psychological factors may influence eating behaviours. In a survey conducted during the SARS-CoV-2 pandemic, 42.7% of the respondents indicated heightened anxiety as the leading cause for changing eating behaviours [9]. Stress may contribute to increased consumption of products with lower nutritional values, amongst other behaviours [10]. Restrictions imposed due to the pandemic by governments in their respective countries may also influence eating behaviours by altering the individual economic situations [9].

Surveys carried out amongst Italian residents indicated that during the pandemic, approximately 21.2% of the respondents consumed more fresh fruit and vegetables and less alcohol (36.8% of the respondents). 19.5% of the respondents reported an increase in body weight. Increased consumption was particularly evident for products such as ice creams and desserts (42.5%), as well as savoury snacks (23.5%) [9]. Other surveys found that the frequency of snacking between meals and the number of main meals also increased [11]. Reduced levels of physical activity were also noted [12].

The aim of the study was to assess changes in eating behaviour and physical activity during a pandemic in a group of people living in various European countries. Apart from identifying differences between the various countries, we wanted to check whether other factors such as place of residence (city or village), gender and age are significant. Compared to other studies that have looked at changes in eating behaviour during a pandemic, the aim of the study was also to see how the pandemic affects the weight reduction diet. Moreover, the design of the questionnaire allowed for a subjective assessment of changes in eating behaviour before and during the pandemic, not only for the assessment of eating behaviour during it.

2. Materials and Methods

2.1. Study Design and Participants

A specially devised, structured and anonymous internet questionnaire was used as the research tool. From the research point of view, the respondents who answered the questionnaire did not have to be nationals of the country in question but had to reside there during the pandemic. Only people older than 18 took part in the survey. Participants aged <18 years old were excluded. Age was verified using a question in the questionnaire. Data were collected via an online survey running over three months (April–July 2020). Respondents were recruited on social networks through several groups on Facebook (for example, on groups that bring together students and people living in various places in Poland, Italy, Portugal, Great Britain—Scotland and England, Spain) and through posts on LinkedIn, Instagram and Twitter. Native speakers translated the questionnaire into five languages: Polish, English, Spanish, Italian and Portuguese. The internet survey was conducted in agreement with the national and international regulations and the Declaration of Helsinki. All participants were informed about the objectives and requirements of the study. They filled out the questionnaire by connecting to Google Forms, and then the questionnaires were downloaded as Microsoft Excel sheets with a guarantee of anonymity.

The questionnaire did not include questions about personal data, and all questionnaires were downloaded as a Microsoft Excel sheet. The Poznan University of Medical Sciences Bioethical Committee declared that no ethical clearance was needed.

2.2. Measures

The questionnaire used for the survey contained questions pertaining to sociodemographic data (gender, age, place of residence), eating behaviours, the impact of the pandemic on the diet and physical activity. Respondents were asked to make a subjective assessment of changes by comparing the period before the pandemic to the period during the pandemic.

2.2.1. Meal Consumption Regularity, Eating Behaviours, Frequency of Food Consumption

The questionnaire included questions on how regularly meals were consumed and on the frequency of consumption of selected product groups: milk and dairy products, cereal products, eggs, meat and meat-based products, fish, fats (vegetable oils; butter; margarine; cream; other animal fats, e.g., lard; mayonnaise and dressings e.g., salad dressings), vegetables, fruit and pulses. Respondents were also asked whether the pandemic impacted consumption levels of products with lower nutritional values: sweets, savoury snacks, fizzy soft drinks, energy drinks, fast foods and instant meals. The questionnaire also included a question on changes to the frequency of alcohol consumption. The questionnaire was developed based on the nutritional recommendations published by the WHO and the CDC during the pandemic [13,14].

2.2.2. Impact of the Pandemic on the Weight Reduction Diet

One of the research objectives was to find an answer to whether the pandemic had affected the weight reduction diet. Respondents were asked to make a subjective comparison of the ease of adherence to dietary rules before and during the pandemic. In a closed question, respondents could indicate a number of factors that made an impact. It was assumed that a positive impact might be the result of having more time to prepare correctly balanced meals and/or to find out more about healthy nutrition, improved conditions for eating meals at regular times, and/or no temptation and pressure within the scope of consuming low nutritional values meals during family/social meetings. On the other hand, a negative impact may be the result of deteriorated conditions for eating meals at regular times, less time to prepare well-balanced meals, increased desire to eat products with lower nutritional values (e.g., sweets, savoury snacks) and/or more difficult access to medical facilities and dieticians.

2.2.3. Physical Activity

Individuals taking part in the survey were asked whether their physical activity level during the pandemic decreased, increased or had not changed.

2.3. Statistical Analyses

Survey results were then analysed using StatSoft's Statistica v. 13 software and Cytel's StatXact v. 9.0.0. The Shapiro–Wilk test was used to test the distribution of variables for normality. To compare two groups of normally distributed variables with the same variance, the Student's t-test was used for unrelated individuals and the Mann–Whitney test was used where the variables were not normally distributed. To compare a larger number of groups (between countries), as the normality criteria were not satisfied, the Kruskal–Wallis test followed by Dunn's Multiple Comparison Test were applied. To test for dependencies among categorical variables, the Chi-squared test for independence, Fisher's exact test or the Fisher–Freeman–Halton test were used. The statistical power of the study was between 0.66 and 0.98. Therefore, an $\alpha = 0.05$ significance level was adopted. Results were considered to be statistically significant for $p < 0.05$.

3. Results

3.1. Demographic Characteristics

A total of 279 people took part in the survey. Participant characteristics can be seen in Table 1.

Table 1. Participant characteristics.

Variables	*n* (%)
N_{total}	279 (100.00)
Country	
Poland	100 (36.00)
Great Britain (England and Scotland)	56 (20.00)
Spain	58 (21.00)
Italy	31 (11.00)
Portugal	34 (12.00)
Age	
<35 years	160 (57.35)
>35 years	119 (42.65)
Gender	
female	226 (81.00)
male	53 (19.00)

3.2. Meal Consumption Regularity

It was observed that there was a statistically significant dependence (p = 0.0045) between meal consumption regularity during the pandemic and the country of residence (Figure 1).

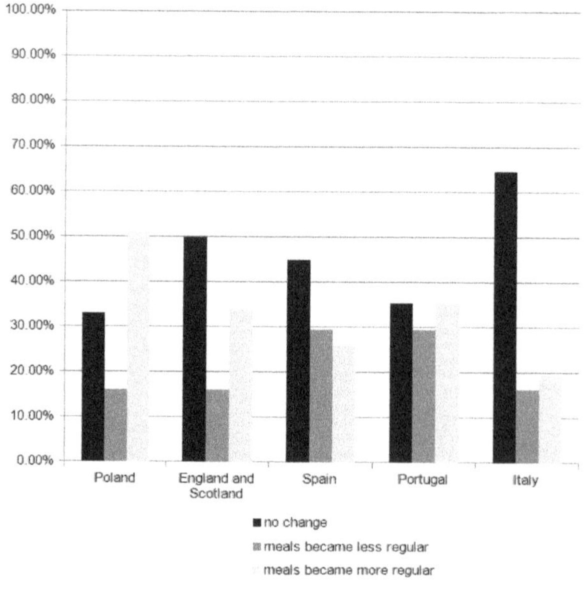

Figure 1. Meal consumption regularity during the pandemic per country.

Within the group comprising people residing in Poland, it was observed that the pandemic contributed to improved meal consumption regularity for more than half (56.1%) of the women (p = 0.0262). Concerning the place of residence, a statistically significant dependence (p < 0.001) was found for people residing in Portugal, where residents of cities were more likely to declare that the pandemic contributed to decreased meal consumption regularity. On the other hand, more than half of those living in the country (66.7%) stated that their meals became more regular. The respondents' age was also a significant factor within the scope of meal consumption regularity. That relationship is shown in Table 2.

Table 2. Meal consumption regularity during the pandemic acc. to age.

	Poland (n = 100)	England and Scotland (n = 56)	Spain (n = 58)	Portugal (n = 34)	Italy (n = 31)
	Percentage of Surveyed (%)				
No change					
<35 years	22.08 *	53.85	47.22	18.18	60.00
>35 years	69.57 *	46.67	40.91	43.48	66.67
Meals became less regular					
<35 years	16.88 *	19.23	25.00	45.45	20.00
>35 years	13.04 *	13.33	36.36	21.74	14.29
Meals became more regular					
<35 years	61.04 *	26.92	27.78	36.36	20.00
>35 years	17.39 *	40.00	22.73	34.78	19.05

* Significant differences, p-values < 0.05.

3.3. Eating Behaviours

Great Britain (England and Scotland) residents were most likely to declare increased consumption of products with lower nutritional values. As compared to respondents from other countries, they were more likely to state that during the pandemic, they ate more savoury snacks (25.9%), fast food and instant meals (12.3%) and consumed more energy drinks (7.4%). Furthermore, 25.9% of respondents from Great Britain indicated increased alcohol consumption during the pandemic. Increased consumption of sweets was most frequently indicated by residents of Portugal (41.2%). In comparison, residents of Spain showed increased consumption of fizzy soft drinks (13.8%) and fried dishes (13.8%). At the same time, residents of Poland (28%) were most likely to declare that they paid more attention to the principles of healthy eating. Assuming that the pandemic could have not affected the respondents' eating behaviours, the questionnaire did include such an answer. Residents of Portugal were most likely to select that answer (38.2%). These results were statistically significant ($p < 0.05$).

A statistically significant dependence ($p < 0.05$) between gender, the place of residence (village or city) and eating behaviours during the pandemic was not found in any of the countries in question. However, such dependence was found for age in a group of people living in Poland. Participants under 35 years of age more often declared that during the pandemic, they increased their consumption of savoury snacks (24.68%). In the group of people above 35 years of age, 4.35% reported increased consumption of savoury snacks ($p = 0.0375$).

3.4. Frequency of Food Consumption

Respondents were asked how the frequency of consumption of given product groups changed during the pandemic. It was observed that among the respondents from all the countries, more people declared an increase in the consumption of cereal products (with the highest percentage recorded in the case of inhabitants of Portugal—32.25%, $p = 0.4784$). In Portugal, people up to 35 years of age were more likely to declare a higher consumption of cereal products than people over 35 ($p = 0.0339$). No statistically significant dependence between the frequency of consumption and gender, age or place of residence was found ($p > 0.05$).

In the case of milk and dairy products, the inhabitants of Poland (28.00%) and Portugal (23.53%) more often declared increased consumption of this type of product. In contrast, in Spain, a more significant percentage of respondents indicated that their consumption decreased (25.86%, $p = 0.08959$). Further analysis showed increased milk and dairy product consumption among male respondents from Great Britain—England, and Scotland ($p = 0.0045$). Age was another factor for which a statistically significant relationship was identified ($p = 0.0269$)—23.1% of people up to 35 years of age indicated higher consumption of milk and dairy products while 6.7% of respondents above 35 years of age indicated that answer. The situation was similar in Spain ($p = 0.0161$) and Italy ($p = 0.0370$), where people up to 35 years of age declared higher consumption of this product group more often.

It was observed that among the respondents from all the countries, more people declared an increase in the consumption of eggs (with the highest percentage recorded in the case of inhabitants of Portugal—32.29%, $p = 0.0756$). No statistically significant dependence between the frequency of consumption and gender, age or place of residence was found ($p > 0.05$).

Frequency of consumption of meat and meat products was also analysed. It was observed that among the respondents from all the countries, more people declared a decrease in the consumption of meat and meat products (with the highest percentage recorded in the case of inhabitants of Spain—31.03%, $p = 0.4837$). Moreover, it was shown that people up to 35 years of age who live in Portugal were more likely to declare decreased meat and meat product consumption during the pandemic—45.5% of respondents in this age group declared decreased consumption ($p = 0.0210$).

In the case of fish, respondents from all countries more often declared a reduction in their consumption (with the highest percentage in Italy—29.03%, $p = 0.5217$). Further analysis of the results showed a statistically significant relationship between the age of those living in Portugal and the frequency of fish consumption during the pandemic ($p = 0.0011$). More than half (54.5%) of people up to 35 years of age declared that their consumption of those products decreased.

The frequency of consumption of fats was also analysed. The increase in their consumption during the pandemic was most often declared by the inhabitants of Italy (19.35%), while an overall decrease was reported by the participants from Spain (27.59%, $p = 0.2209$). In Portugal ($p = 0.0011$) and Italy ($p = 0.0105$), the frequency was found to be age-dependent. In both countries, people up to 35 years old increased their consumption during the pandemic.

It was observed that among the respondents from all the countries, more people declared an increase in the consumption of vegetables (with the highest percentage recorded in the case of inhabitants of Spain—43.10%, $p = 0.1189$) and fruits (with the highest percentage recorded in the case of inhabitants of Spain—39.66%, $p = 0.2244$). For vegetables and fruits, significant differences were observed between age and frequency of consumption of vegetables ($p = 0.0368$) and fruit ($p = 0.0012$) during the pandemic. This consumption increased in the over 35 age group for both vegetables and fruit. Whereas in Portugal, respondents in the under 35 age group were more likely to declare increased consumption of fruit ($p = 0.0492$).

Pulses constitute a significant group of foodstuffs. The increase in their consumption during the pandemic was most often declared by the inhabitants of Spain (25.86%), while a decrease was reported by the partipants from Great Britain (21.43%, $p = 0.1682$). It was shown that in the group of respondents from Poland, 22.5% of women increased their consumption of pulses during the pandemic. Correspondingly, 11.1% of men declared higher consumption ($p = 0.0178$).

3.5. Impact of the Pandemic on the Weight Reduction Diet

It was observed that the pandemic might have an impact on the weight reduction diet. A negative impact was declared by 16.5% of people, compared to 9.7% who said that the pandemic facilitated the use of the weight reduction diet ($p = 0.006$). Individuals who declared a negative impact of the pandemic were most likely to indicate increased craving for unhealthy products (69.6%) as the leading cause. Out of the factors that had the most significant positive impact on their diets during the pandemic, respondents were most likely to indicate having more time to prepare healthy meals (74.1%) as the leading factor. The impact of the pandemic on the diets of residents of various countries is shown in Table 3.

Table 3. Impact of the pandemic on the weight reduction diet of residents of various countries.

	Poland	England and Scotland	Spain	Portugal	Italy	*p*
	Percentage of Surveyed (%)					
Positive Impact	8.00	3.57	20.69	2.94	12.90	
Negative Impact	9.00	17.86	13.79	32.35	25.81	0.0064
No impact	18.00	25.00	17.24	11.76	25.81	
I'm not on the weight reduction diet	65.00	53.57	48.28	52.95	35.48	

Further analysis showed that for respondents from Portugal, there was a dependence between the place of residence and the impact of the pandemic on diet ($p = 0.0055$). Half of those living in cities declared it has a negative impact. Furthermore, respondents were most likely to indicate increased craving for unhealthy products as the leading cause for changes in diet ($p = 0.0356$). Furthermore, the obtained results made it possible to conclude that the impact of the pandemic on the weight reduction diet may also be age-dependent. This is shown in Table 4.

Table 4. Impact of the pandemic on the weight reduction diet acc. to age.

	Poland	England and Scotland	Portugal	Spain	Italy
	Percentage of Surveyed (%)				
		<35 years			
I'm not on the weight reduction diet	66.23	73.08	45.45	55.56	60.00
Positive Impact	9.09	3.85	9.09	16.67	30.00
Negative impact	9.09	11.54	36.36	8.33	10.00
No impact	15.58	11.54	9.09	19.44	0.00
		>35 years			
I'm not on the weight reduction diet	60.87	36.67	56.52	36.36	23.81
Positive Impact	4.35	3.33	0.00	27.27	4.76
Negative Impact	8.07	23.33	30.43	22.73	33.33
No impact	26.09	36.67	13.04	13.64	38.10
p-value	0.6629	0.0301	0.6192	0.2654	0.0068

3.6. Physical Activity

It was observed that age could be a significant factor influencing the level of physical activity. In the group of respondents living in Poland, 69.6% of people over 35 stated that their level of physical activity had decreased ($p = 0.0497$). For those living in Portugal, the place of residence was significant. People living in cities were more likely to report reduced levels of physical activity ($p = 0.0245$).

4. Discussion

4.1. Principal Findings

4.1.1. Country Differences

The results of the research conducted showed that problems with adherence to the principles of a healthy diet may differ from the country. In the case of respondents from Spain and Portugal, problems with the regularity of eating meals were more frequent than in other respondents. In addition, the problem was the increase in the consumption of products such as sweets (in the case of Portuguese residents) and sweet sodas and fried foods (in the group of people living in Spain). However, the greatest problem with the consumption of low-nutrient products was observed among people from Great Britain. Compared to respondents from other countries, they more often declared an increase in consumption of savoury snacks, fast food, instant meals, energy drinks and alcohol. On the other hand, only 16.07% of respondents in Great Britain indicated that the regularity of eating meals had decreased, and 50.00% said it was unchanged. Different results were obtained among respondents from Poland, who stated that they adhered to the principles of a healthy diet more often, and the regularity of their meals had improved. The results obtained in the group of people from Poland showed that the pandemic can positively and negatively impact eating behaviour. A positive effect was also observed concerning the consumption of vegetables, fruits and legumes among the inhabitants of all countries. Similarly to the research by Rodríguez-Pérez et al. (2020), Renzo et al. (2020) and Ruiz-Roso et al. (2020), we also noted an increased consumption of vegetables and fruit and pulses (especially in Spain) [15–17]. This is in line with World Health Organisation (WHO) guidelines [13]. These are products with high nutritional values, characterised by a high content of nutrients that boost immunity. For fruit and vegetables, the advantage is that they can be consumed quickly. Pulses require longer preparation times, but the advantage is that they can be stored for a longer period [12]. Concerning meat and fish, we observed lower consumption (among the respondents from all the countries), as did Renzo et al. (2020) [15]. This may be due to the short shelf life and more time required to prepare them for meat and fish.

More than half of the respondents (53.8%) reported decreased physical activity during the pandemic. The inhabitants of Portugal most often declared a decrease in their levels of physical activity. Among the respondents from Portugal, negative changes were also observed in the meal consumption regularity and the frequency of consumption of sweets. This can be an important signal for health promoters in the country, as lower physical activity levels combined with poor eating habits can lead to weight gain and obesity.

Such differences show that the negative impacts of a pandemic may be different around the world. The problems with adherence to the principles of a healthy diet may differ depending on the country. One of the reasons may be that the number of cases and deaths from COVID-19 have varied from country to country, which could affect the level of stress and anxiety among residents. Moreover, when comparing changes in eating habits between individual countries, it should be considered that they may also be caused by factors not related to the pandemic. Other influences that may also be of importance here include cultural and religious factors and culinary traditions.

4.1.2. Age Differences

Pursuant to the research, age was found to be the factor that impacted changing eating behaviours to the largest extent. For younger people (<35 years), the pandemic has had a positive impact on the weight reduction diet and regularity of meals and has increased the frequency of consumption of cereal products, eggs, vegetables, fruit and fats. At the same time, it contributed to increased consumption of products with reduced nutritional value and reduced consumption of fish and meat.

Renzo et al. (2020) found that quarantine contributed to increasing appetites amongst young people. This might explain why meals were consumed more often [15]. At the same time, an increased appetite can be a reason for snacking and eating unhealthy snacks. Our research results confirmed this, as they demonstrated that people under 35 consumed more savoury snacks, fast food and instant meals. A study by Ammar et al. (2020) delivered similar results. Therein, subjects over 18 years of age were more likely to be willing to eat unhealthy snacks during a pandemic and eat at night [12]. Ruiz-Roso et al. (2020) obtained different results within this scope. They found lower consumption of fast food products during quarantine amongst Spanish teenagers. This could be due to the fact that they were still under the care of older individuals [16].

When comparing two age groups: <35 years and >35 years, it should be taken into account that apart from emotional factors (related to, for example, increased levels of anxiety and stress) or those resulting from the introduced restrictions, the influence on changing eating habits could also have included a change of life situation during the pandemic. The group of people <35 years could include students who returned to their family home due to the introduction of online learning. This could impact the regularity of meals and changes in the frequency of consumption of selected groups of products due to the family home diet. On the other hand, among respondents >35 years, there could be people whose level of anxiety and stress influencing eating habits was increased due to the deterioration of the financial situation. Moreover, the eating habits in this age group may have been influenced by the provision of 24 h care for children due to the closure of kindergartens and schools. Therefore, more research is needed to help explain the causes of changes in eating habits in different age groups.

4.1.3. Place of Residence (Village or City)

The study results showed that there was a dependence between the place of residence, eating behaviour and physical activity. This tendency was observed in Portugal, where compared to people living in the countryside, respondents living in cities experienced more significant difficulties in adhering to the recommendations of the weight reduction diet and regular eating. Moreover, people living in cities declared a reduction in physical activity.

One of the reasons for this may be that in places with a more significant number of inhabitants, the stress related to the risk of infection and the feeling of the restrictions introduced may be more important.

4.2. Study Strengths and Weaknesses

The timing of the research problem is one of the strengths of this study. As the SARS-CoV-2 pandemic began at the beginning of 2020, few studies on its impact on eating behaviours and physical activity have been published thus far. Another strength was that the research group, consisting of people from different countries, was formed when the most stringent restrictions were being introduced. Knowing the differences that exist between countries provides an opportunity for international exchange of experiences.

Another strength was using an online questionnaire supported by an eating behaviours assessment questionnaire in different language versions, making it possible to form the study group from different countries quickly. This is particularly important during the pandemic, as the situation is changing rapidly. Furthermore, such a format reduces the risk of infection resulting from direct contact.

Another strength of the research was the question on the impact of the pandemic on the weight reduction diet that was implemented before the pandemic. Continuing a particular diet is essential for maintaining proper body weight.

When analysing the weaknesses of the study, the fact that the online questionnaire has certain limitations should be taken into account. These include, among other things, difficulties with validating it and verifying the entered data and no opportunity to consult possible doubts with the investigator.

Lack of inclusion and exclusion criteria was also a weakness of the research, which could affect the results. Some people may have had to deal with eating disorders, alcohol addiction or a job that did not force them to change their lifestyle during the pandemic. The number of respondents is also a weakness—quite few and varied from country to country.

5. Conclusions

The survey results showed differences between countries in terms of changes in eating behaviour and levels of physical activity during the pandemic. However, it should be considered that they may have resulted from cultural, regional or culinary differences specific to a given place. Therefore, when planning further research aimed at a more detailed analysis, it should be considered. Perhaps a good solution would be to add questions to the questionnaire that consider these factors by people who know the specifics of a given place.

Despite some differences between countries, the study results showed that a worrying trend in all countries is reducing the consumption of omega-3-rich fish (even in countries like Spain and Portugal). One of the reasons for this may be their short use-by date. This is a signal for both health promotion professionals and food producers. Perhaps a way to increase consumption of these products during a pandemic would be to promote ways of preparing fish and prepared products based on them that allow them to be stored longer.

The results of our research showed that special attention should also be paid to the under 35 age group. Considering that these are people who easily use digital technologies, it is worth considering their use in promoting healthy eating and regular physical activity. For this purpose, information and communication technologies (ICT) can be used increasingly during a pandemic. The use of ICT makes it possible to analyse data (including health status, eating behaviour, physical activity) and then implement a personalised and multi-dimensional intervention. ICT can be used in various ways to promote a healthy lifestyle. For young people who often use mobile applications, it is worth considering the preparation of an application that will allow them to monitor diet and physical activity. By entering data on the meals consumed and physical activity, the user will receive personalised feedback prepared in cooperation with dietitians or personal trainers. These could include clues about where the nutritional error is being made and how to fix it. The application may also have recipes for healthy meals and suggestions for exciting forms of physical activity at home. For added value, it may be possible to calculate the nutritional value of the meals consumed. It is also worth considering introducing gamification, which is more and more often used in mobile applications. The user who performs specific tasks (e.g., performs the right amount of exercise on a given day, eats a certain amount of healthy products) receives points similar to in the game, which encourages compliance with the principles of a healthy lifestyle. The use of ICT to promote healthy eating and physical activity during a pandemic should not be limited to young people. ICT can also be used in the elderly, but some age-related restrictions should be taken into account (e.g., by introducing personalised feedback in the form of a voice) [18].

We have not seen a virus spreading on such a large scale for many years. That is why it is important to determine where the problem may lie with observing healthy eating principles and regular physical activity. Through its impact on the functioning of the immune system, amongst others, a healthy lifestyle may be an essential element in the prevention of COVID-19.

Author Contributions: P.G., I.G. and J.P. designed the study. P.G. and I.G. conducted the data collection. I.M. conducted statistical analysis. P.G., I.G. and J.P. contributed in the interpretation of the data. P.G. wrote the first draft of the manuscript. P.G., I.G. and J.P. wrote, reviewed and edited. All authors contributed to the critical revision final manuscript. All authors have read and agreed to the published version of the manuscript.

Funding: This research received no external funding.

Institutional Review Board Statement: Not applicable.

Informed Consent Statement: Not applicable.

Data Availability Statement: Data can be accessible upon request to corresponding author.

Conflicts of Interest: The authors declare no conflict of interest.

References

1. Huang, C.; Wang, Y.; Li, X.; Ren, L.; Zhao, J.; Hu, Y.; Zhang, L.; Fan, G.; Xu, J.; Gu, X.; et al. Clinical features of patients infected with 2019 novel coronavirus in Wuhan, China. *Lancet* **2020**, *395*, 497–506. [CrossRef]
2. Wang, D.; Hu, B.; Hu, C.; Zhu, F.; Liu, X.; Zhang, J.; Wang, B.; Xiang, H.; Cheng, Z.; Xiong, Y.; et al. Clinical characteristics of 138 hospitalised patients with 2019 novel coronavirus-infected pneumonia in Wuhan, China. *JAMA* **2020**, *11*, 1061–1069. [CrossRef] [PubMed]
3. Notice of the National Health Commission of the People's Republic of China on Revising the English Name of Novel Coronavirus Pneumonia. 2020. Available online: http://www.nhc.gov.cn/yzygj/s7653p/202002/33393aa53d984ccdb1053a52b6bef810.shtml (accessed on 29 February 2020).
4. Zarocostas, J. What next for the coronavirus response? *Lancet* **2020**, *395*, 401. [CrossRef]
5. Manuell, M.E.; Cukor, J. Mother Nature versus human nature: Public compliance with evacuation and quarantine. *Disasters* **2011**, *35*, 417–442. [CrossRef] [PubMed]
6. Trabelsi, K.; Ammar, A.; Masmoudi, L.; Boukhris, O.; Chtourou, H.; Bouaziz, B.; Brach, M.; Bentlage, E.; How, D.; Ahmed, M.; et al. Globally altered sleep patterns and physical activity levels by confinement in 5056 individuals: ECLB COVID-19 international online survey. *Biol. Sport* **2020**, 495–506. [CrossRef]
7. Bai, Y.; Lin, C.C.; Lin, C.Y.; Chen, J.Y.; Chue, C.M.; Chou, P. Survey of stress reactions among health care workers involved with the SARS outbreak. *Psychiatr. Serv.* **2004**, *55*, 1055–1057. [CrossRef] [PubMed]
8. Richter, I.; Gabe-Thomas, E.; Maharja, C.; Nguyen, T.H.; Van Nguyen, Q.; Praptiwi, R.; Pahl, S. Virtual Capacity Building for International Research Collaborations in Times of COVID-19 and #Flygskam. *Front. Commun.* **2020**, *5*, 146.
9. Scarmozzino, F.; Visioli, F. Covid-19 and the Subsequent Lockdown Modified Dietary Habits of Almost Half the Population in an Italian Sample. *Foods* **2020**, *5*, 675. [CrossRef] [PubMed]
10. Groesz, L.M.; McCoy, S.; Carl, J.; Saslow, L.; Stewart, J.; Adler, N.; Laraia, B.; Epel, E. What is eating you? Stress and the drive to eat. *Appetite* **2012**, *2*, 717–721. [CrossRef] [PubMed]
11. Ammar, A.; Brach, M.; Trabelsi, K.; Chtourou, H.; Boukhris, O.; Masmoudi, L.; Bouaziz, B.; Bentlage, E.; How, D.; Ahmed, M.; et al. Effects of COVID-19 Home Confinement on Eating Behaviour and Physical Activity: Results of the ECLB-COVID19 International Online Survey. *Nutrients* **2020**, *6*, 1583. [CrossRef] [PubMed]
12. Ammar, A.; Trabelsi, K.; Brach, M.; Chtourou, H.; Boukhris, O.; Masmoudi, L.; Bouaziz, B.; Bentlage, E.; How, D.; Ahmed, M.; et al. Effects of home confinement on mental health and lifestyle behaviours during the COVID-19 outbreak: Insight from the ECLB-COVID19 multicenter study. *Biol. Sport* **2021**, *1*, 9–21. [CrossRef] [PubMed]
13. World Health Organization. Food and Nutrition Tips during Self-Quarantine. Available online: https://www.euro.who.int/en/health-topics/health-emergencies/coronavirus-covid-19/publications-and-technical-guidance/food-and-nutrition-tips-during-self-quarantine (accessed on 24 May 2020).
14. Centers for Disease Control and Prevention. Food and Coronavirus Disease 2019 (COVID-19). Available online: https://www.cdc.gov/coronavirus/2019-ncov/daily-life-coping/food-and-COVID-19.html (accessed on 25 May 2020).
15. Di Renzo, L.; Gualtieri, P.; Pivari, F.; Soldati, L.; Attinà, A.; Cinelli, G.; Leggeri, C.; Caparello, G.; Barrea, L.; Scerbo, F.; et al. Eating habits and lifestyle changes during COVID-19 lockdown: An Italian survey. *J. Transl. Med.* **2020**, *18*, 229. [CrossRef] [PubMed]
16. Ruiz-Roso, M.B.; de Carvalho Padilha, P.; Mantilla-Escalante, D.C.; Ulloa, N.; Brun, P.; Acevedo-Correa, D.; Arantes Ferreira Peres, W.; Martorell, M.; Aires, M.T.; de Oliveira Cardoso, L.; et al. Covid-19 Confinement and Changes of Adolescent's Dietary Trends in Italy, Spain, Chile, Colombia and Brazil. *Nutrients* **2020**, *12*, 1807. [CrossRef] [PubMed]
17. Rodríguez-Pérez, C.; Molina-Montes, E.; Verardo, V.; Artacho, R.; García-Villanova, B.; Guetrra-Hernández, E.J.; Ruíz-López, M.D. Changes in Dietary Behaviours during the COVID-19 Outbreak Confinement in the Spanish COVIDiet Study. *Nutrients* **2020**, *12*, 1730. [CrossRef] [PubMed]
18. Ammar, A.; Bouaziz, B.; Trabelsi, K.; Glenn, J.; Zmijewski, P.; Müller, P.; Chtourou, H.; Jmaiel, M.; Chamari, K.; Driss, T.; et al. Applying digital technology to promote active and healthy confinement lifestyle during pandemics in the elderly. *Biol. Sport* **2021**, *38*, 391–396. [CrossRef]

foods

MDPI

Article

Explaining Chinese Consumers' Green Food Purchase Intentions during the COVID-19 Pandemic: An Extended Theory of Planned Behaviour

Xin Qi * and Angelika Ploeger

Specialized Partnerships in Sustainable Food Systems and Food Sovereignty, University of Kassel, 37213 Kassel, Germany; a.ploeger@uni-kassel.de
* Correspondence: x.qi@uni-kassel.de; Tel.: +49-561-804-1717; Fax: +49-561-804-1713

Abstract: The outbreak of the COVID-19 pandemic has strongly influenced consumers' habits and behaviours, creating a more sustainable and healthier era of consumption. Hence, there is a potential for further expanding the green food sector in China. The theory of planned behaviour (TPB) is one widely used framework to explain consumers' food choices. Considering consumers' internal norms, their perceptions of green food attributes, and the shifting consumer behaviour, our study has extended the TPB framework (E-TPB) by adding constructs of moral attitude, health consciousness, and the impact of COVID-19 (IOC). The results of structural equation modelling among 360 functional samples revealed that the E-TPB model has a superior explanatory and predictive power, compared with the original TPB model regarding Chinese consumers' green food buying intentions in the current and post-pandemic periods. The path analysis demonstrated that attitude, perceived behavioural control, moral attitude, health consciousness, and IOC have significant positive effects on green food purchase intentions. However, the association between subjective norm and purchase intention varies within the TPB and E-TPB models, which showed a non-significant impact in E-TPB. These findings can generate more suitable managerial implications to promote green food consumption in China during the current and post-pandemic periods.

Keywords: green food; purchase intention; TPB; E-TPB; COVID-19; Chinese consumer

Citation: Qi, X.; Ploeger, A. Explaining Chinese Consumers' Green Food Purchase Intentions during the COVID-19 Pandemic: An Extended Theory of Planned Behaviour. *Foods* **2021**, *10*, 1200. https://doi.org/10.3390/foods10061200

Academic Editor: Derek V. Byrne

Received: 3 May 2021
Accepted: 25 May 2021
Published: 26 May 2021

1. Introduction

In March 2020, the World Health Organization (WHO) announced the outbreak of the highly transmittable Coronavirus (COVID-19) as a pandemic [1], which is considered to be the third pandemic in the 21st century [2]. After one year's development, there have been more than 138 million confirmed cases of COVID-19 globally and more than 751 million vaccine doses administered in April 2021 [1]. Although the situation is expected to improve in the next several years, the COVID-19 pandemic undoubtedly has widespread effects on society and consumers, hinting towards the dynamic changes in the market [3,4]. Some studies have found that there has been increased consumption of unhealthy food during the initial 'lockdown' period [5,6]. However, many studies have recently reported that consumers are increasingly concerned about the health and safety aspects of their food consumption and want to protect and strengthen their immune systems through their food diets [7–10]. In a 2021 survey from Accenture, an investigation involving more than 3000 consumers in 15 countries has shown that the pandemic is likely to create a more sustainable and healthier consumption era over the following 10 years [11]. This development can permanently alter consumer behaviours and cause lasting structural changes to products and industries [11]. Therefore, enterprises and marketers in the organic and green food sectors need to accelerate their business objectives and capabilities to match the shifting consumption patterns for their products and services during the current pandemic and post-pandemic periods.

Beginning in the 1990s, green food has been one of the most successful eco-labelling innovations in the Chinese food production industry [12]. Green food primarily refers to a full range of safe and premium edible agricultural products and related processed products, and are required to be grown in an ecologically sound environment, produced based on green food production standards, adopt the wholesome quality control, and granted the right to have a 'green food' certification [13]. There are two different standards for green food: Grade A and Grade AA. Grade A allows food producers to use limited chemical pesticides, chemical fertilisers, and other chemical inputs. Grade AA has stricter standards that exclude the guidelines mentioned above and are equivalent to Chinese organic food production standards [13]. After 30 years of development, Chinese green food production and consumption have experienced rapid growth and continued to expand in scale. According to official statistics in 2020, there have been 19,321 green food enterprises, 43,729 green-food-labelled products, and USD 66.52 billion of domestic sales, with a growing average of 9.2% annually [14]. Meanwhile, many Chinese families pay increasing attention to their health, quality lifestyles, environmental protection, and food security, and prefer to consume safe and green food products [15,16]. Despite undergoing exponential development and resulting in a booming trend, green food sales account for an extremely low percentage of the total food sales in China (i.e., less than 1%). Thus, there is an excellent prospect for the further expansion of green food consumption in China.

Since the Chinese green food industry faces an upward demand for sustainable consumption and growth potential, stakeholders in the green food sector need to understand the effective mechanisms of consumers' green food purchase intentional behaviours during the current and post-pandemic periods in more detail. The theory of planned behaviour (TPB), proposed by Ajzen [17], is one of the most widely used theories predicting consumers' purchase intentions of environmentally friendly food products [18,19]. According to Ajzen [17], the TPB is open for modification by incorporating new constructs or altering the path. Moreover, some nonnegligible limitations of applying TPB exist, and therefore, various refinements and extensions of TPB theory have been suggested to improve its effectiveness and applicability [20,21]. It is necessary to use multidisciplinary approaches for better understanding consumers' food preferences and acceptances in different contexts and eating scenarios [22]. Consequently, certain important factors related to consumer behaviours during a pandemic should be investigated and validated to establish an updated green food purchase intention model.

Researchers and theorists have recurrently criticised the TPB for its insufficient consideration of other human behavioural constructs such as moral and ethical concerns [20]. Moral attitude is considered another salient behavioural factor in purchasing environmentally friendly food products since these behaviours are commonly perceived as pro-environmental actions [23]. Moreover, recent studies that investigated the influence of the COVID-19 pandemic on food consumption show that many consumers have been increasingly concerned about the health aspects [7–9]. Health concerns are considered as one of the significant drivers promoting consumers' attitudes and intentions towards purchasing environmentally friendly food products [24]. Lastly, the emergence of the pandemic has strongly affected the global food systems, such as a collapse in a growing demand for global agri-food products, the severe disruptions of domestic and international food supply chains, the shortage of labour for food production enterprises, and the shifting consumption pattern [8,25,26]. Currently, no studies have established a green food purchase intention model with an integration of COVID-19 pandemic influences and incorporating salient factors among Chinese consumers. Therefore, it is necessary to adjust former models by integrating important and new factors that account for green food consumption and COVID-19 to understand consumers' green food consumption better during a pandemic.

Hence, this paper aims to explore an appropriate model to explain and predict Chinese consumers' green food purchase intentions during the current and post-pandemic periods. Based on the TPB, we have proposed an extended theory of planned behaviour (E-TPB) model and applied a structural equation modelling (SEM) approach to conduct model comparisons and examine the performance of each construct. Therefore, the outcome of this research can contribute to offering significant practical implications for researchers and marketers in the Chinese green food industry. The present study can generate new insight into the effects of the COVID-19 pandemic crisis on green food consumption. Moreover, marketers can use our findings to develop innovative marketing strategies to promote green food consumption in China further.

The remainder of this paper is organised as follows: Section 2 outlines the conceptual model developed for the present study and includes hypotheses to be examined; Section 3 explains the research methodology, which includes data collection, measurement, and data analysis; Section 4 displays descriptive statistics and SEM; Section 5 discusses of results and implications; Section 6 provides conclusions and includes research limitations and suggestions for future work.

2. Theoretical Framework and Development of Hypotheses

2.1. Theoretical Framework

The present study has adopted TPB and proposed an E-TPB model by adding three constructs (i.e., moral norm, health consciousness, and the impact of COVID-19), to explain and predict Chinese consumers' green food purchase intentions during the current and post-pandemic periods. The conceptual framework is represented in Figure 1.

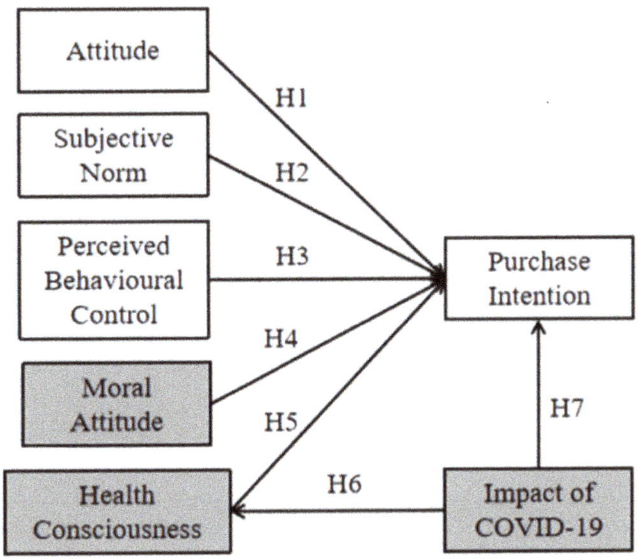

Figure 1. Research model: the white blocks are variables in the standard TPB model; the grey blocks and the white blocks are variables in the E-TPB model; H1, Hypothesis 1; H2, Hypothesis 2; H3, Hypothesis 3; H4, Hypothesis 4; H5, Hypothesis 5; H6, Hypothesis 6; H7, Hypothesis 7.

2.2. Development of Hypotheses

2.2.1. TPB

According to the TPB, there are three factors that collectively lead to the formation of an individual's intentional behaviour: attitude, subjective norm, and perceived behavioural control [17].

- Attitude

Attitude assesses the extent to which people favourably or unfavourably evaluate the subject in question [17]. A subject can be a product, a person, or any other physical or intangible stimulus. There are substantial empirical studies on attitudes affecting consumers' food choices towards environmentally friendly food products. Previous research emphasises a strong positive correlation between attitude and purchase intention towards organic food [27,28], green food [19], and sustainably sourced food [29]. Therefore, the present study introduces Hypothesis 1 (H1):

Hypothesis 1 (H1). *Chinese consumers' attitudes towards green food products significantly influence their green food purchase intentions.*

- Subjective Norm

Subjective norm relates to perceived social influences or stresses to engage or disengage in a given behaviour [17]. Subjective norm also reveals the individuals' beliefs about how their reference groups would view them if they perform a certain behaviour [28]. According to Scalco et al. [30], the most important social influences related to consumers' environmentally friendly food purchases are from their families, friends, colleagues, and other reference groups. Previous research indicated that subjective norm is a positive driver of consumers' behavioural intentions to indulge in their food choices [31,32]. However, other scholars [19,33] argue the efficacy of subjective norm in explaining consumer food choices and agree that more examinations into the role of the subjective norm are needed. Based on the aforementioned discussion, Hypothesis 2 (H2) is proposed as the following:

Hypothesis 2 (H2). *Subjective norm has a significant impact on Chinese consumers' green food purchase intentions.*

- Perceived Behavioural Control (PBC)

PBC refers to an individual's ability to control their behaviour independently [17]. Previous studies highlight PBC as a salient factor of intention during sustainable food consumption [34,35]. Studies from Yadav and Pathak [27] and Carfora et al. [36] have confirmed that the most significant impact of PBC on consumers' buying intentions towards organic food products is mainly due to unavailability issues. Since some attributes of green and organic food products are similar, findings from consumer studies about organic food can serve as a reference for green food studies. Hence, Hypothesis 3 (H3) is presented as follows:

Hypothesis 3 (H3). *Perceived behavioural control is significantly related to Chinese consumers' intentions to buy green food.*

2.2.2. Incorporating Additional Constructs in the TPB

- Moral Attitude

Moral attitude refers to a person's self-evaluation resulting from their expected compliance with their moral principles [37]. Moral attitude is considered to be a significant driver that impacts organic and green food consumption since consumers realise that their sustainable purchases can affect other people's well-being. Thus, they perceive a sense of responsibility for their purchases and look for opportunities to fulfil them [29,38]. Studies from Dowd and Burke [29] and Yazdanpanah and Forouzani [39] have inserted moral norms into the TPB model. Both studies reported that when adding moral norms into the original model, there are significant increases in the fidelity and explanatory capability of the model. Hence, Hypothesis 4 (H4) is proposed based on the above discussion:

Hypothesis 4 (H4). *The moral attitude among Chinese consumers positively influences consumers' intentions to purchase green food.*

- Health Consciousness

Health consciousness is defined as 'the degree to which health concerns are integrated into a person's daily activities' [40], which reveals a person's willingness to conduct health behaviours [41]. According to Paul and Rana [42], people who are more concerned about their health have frequent positive attitudes towards buying organic products since they are commonly considered as a healthier choice, compared to conventionally grown food varieties [24,43]. Many studies have found health consciousness as a significant motivator for consumer decisions towards environmentally friendly food products [44–47] and a crucial factor that strongly influences consumers' willingness to pay for premium products [48]. Thus, Hypothesis 4 (H4) is proposed as follows:

Hypothesis 5 (H5). *Health consciousness among Chinese consumers positively influences consumers' intentions towards buying green food.*

2.2.3. Incorporating the Impact of COVID-19 (IOC) into the TPB Framework

The ongoing coronavirus pandemic crisis has caused a severe global health emergency and, consequently, has led to a shift in food systems as well as the way people purchase and consume their food [49]. The study from Meixner and Katt [3] has assessed the IOC on consumers' perceptions about food safety issues. Their findings suggest that food safety concerns are becoming increasingly important due to COVID-19. Moreover, the latest studies [8,50] reported that the COVID-19 pandemic could lead people's behaviours and lifestyles towards a sustainable and healthier direction. People tend to consume more environmentally friendly food products due to an increase in their food safety concerns [8,50]. With its specific attributes, green food meets the current demand. Therefore, we have added the IOC construct into the standard TPB model, and we aim to investigate its influence on consumers' green food purchase intentions and health consciousness. Accordingly, Hypothesis 6 (H6) and Hypothesis 7 (H7) are proposed:

Hypothesis 6 (H6). *The impact of COVID-19 is significantly related to Chinese consumers' health consciousness.*

Hypothesis 7 (H7). *The impact of COVID-19 is significantly related to Chinese consumers' green food purchase intentions.*

3. Methodology

3.1. Data Collection

As part of the study, an online survey was applied to collect data and analyse the developed research framework. The geographic location used within the research was the Chinese mainland, a country containing the world's largest food consumer group. Data were collected using a questionnaire survey platform (i.e., www.wenjuan.com (accessed on 24 May 2021)). After a brief pilot study involving 15 consumers, the initial questionnaire was adjusted and refined to improve comprehension and readability. The online questionnaire was distributed via WeChat, i.e., the most widely used mobile messaging application among Chinese people. Participants could answer the questionnaire by accessing the WeChat app with their smartphones. In addition, a certain number of 'red packets', which is an electronic monetary function in WeChat, were enabled to attract more consumers. The target group of this survey included consumers over the age of 20 due to the age category accounting for the majority of Chinese green consumers [51]. Therefore, the participant's age was asked at the beginning of the online survey and was used for filtering. The survey would only continue if the age requirement was met. The survey was available to WeChat active users from 12 to 19 April 2021. A total of 398 questionnaires were returned and 38 of the 398 respondents were excluded due to the straight-line answer pattern and failure to complete the survey questions. A valid sample of 360 respondents (i.e., response rate = 90.4%) was used as a research dataset. According to Kline [52], there should be at least 10 cases per measurement as an acceptable sample size for conducting an experimental investigation. This study contains 19 measured items with a required minimum of 190 responses. Thereby, a total of 360 valid questionnaires was considered a valid sampling and sufficient for further data analysis. Table 1 provides an overview of the demographics of the samples.

Table 1. Demographics of samples (n = 360).

Demographics Variables		Frequency	Percent (%)
Gender	Male	166	46.1
	Female	194	53.9
Age	20–30	124	34.4
	31–40	87	24.2
	41–50	83	23.0
	51–60	42	11.7
	>60	24	6.7
Marital Status	Married with a child or children	149	41.4
	Married	80	22.2
	Single	115	32.0
	Other	16	4.4
Education	Junior school or below	57	15.9
	High school or technical secondary school	124	34.4
	University or above	179	49.7
Monthly Income (RMB)	<4500	98	27.2
	4500–9000	174	48.3
	>9000	88	24.5

3.2. Measurements

The scales used in this study were adopted from previous studies and modified to employ valid measurement instruments. This study considered the particularity of the Chinese language and culture. The contents were screened by two academic experts and back-translated by two local language experts to ensure accuracy and data integrity. Besides demographic questions, the other responses on the statements were obtained on a seven-point Likert scale (i.e., 1 for strongly disagree and 7 for strongly agree). The measurement items and their sources of adoption are shown in Table 2.

3.3. Data Method

SPSS Statistics version 24 and AMOS version 24 were applied to analyse the data and test the hypothetical associations between the constructs in the research model. Firstly, data were examined by descriptive statistics (i.e., means and standard deviations) to analyse the characteristics of participants and visualise the responses received. Secondly, confirmatory factor analysis (CFA) was conducted to measure the validity and reliability of measurement items within the proposed models. Next, we applied SEM to evaluate the model fit and hypothesis testing between TPB and E-TPB models. Finally, the indicators and hypothesis testing results of the two frameworks were compared and summarised based on data analysis results.

Table 2. Measurement of constructs.

Constructs	Items	Measurement Items	Adopted From
Purchase Intention (PI)	PI1 PI2	I prefer to choose green food products if they are available for purchase. In the near future, I will try to buy green food.	Yazdanpanah and Forouzani [39]
Attitude (AT)	AT1 AT2 AT3	I think purchasing green food is a good concept. I believe buying green food is pleasant. I believe buying green food is of importance.	Wang et al. [53]
Subjective Norm (SN)	SN1 SN2	Most people I value believe I should purchase green food. Most people I value will purchase green food rather than non-green food.	Han et al. [54]
Perceived Behavioural Control (PBC)	PBC1 PBC2 PBC3	If I want to, I can easily buy green food. I have all resources for buying green food. Buying green food is entirely up to me.	Han et al. [54]
Moral Attitude (MA)	MA1 MA2 MA3	If I purchase green food rather than non-green food, it feels like a personal contribution to something better. If I purchase green food rather than non-green food, it feels like I'm doing the morally right thing. If I purchase green food rather than non-green food, I feel like I'm being a better person.	Arvola et al. [55]
Health Consciousness (HC)	HC1 HC2 HC3	I chose food carefully to ensure good health. I consider myself a health-conscious consumer. I often think about health-related issues.	Yadav and Pathak [27]
Impact of COVID-19 (IOC)	IOC1 IOC2 IOC3	I perceive the COVID-19 pandemic has influenced me personally. I perceive the COVID-19 pandemic will shift my consumption pattern. I perceive the COVID-19 pandemic will change society.	Meixner and Katt [3]

4. Results

4.1. Descriptive Statistics

In general, participants in this investigation expressed positive purchase intentions for green food (Mean = 5.645; SD = 0.872). For other determinants, participant responses were higher for PBC (Mean = 6.061; SD = 0.887), moral attitude (Mean = 5.909; SD = 0.827), health consciousness (Mean = 5.777; SD = 0.773), subjective norm (Mean = 5.750; SD = 0.910), and IOC (Mean = 5.381; SD = 1.007). Surprisingly, consumers' answers for the attitude construct (Mean = 4.678; SD = 1.047) were neutral. Figure 2 presents the responses from all participants in this survey. Although some curves displayed slight fluctuations, most curves in general showed a homogeneous response pattern for all constructs.

4.2. Measurement Model

Table 3 presents the results from the reliability and validity analysis of each measurement. All of the Cronbach's α values were higher than the threshold of 0.7 [56], indicating that the questionnaire's data have adequate reliability. In regard to convergent validity, all variables presented high composite reliability (CR), with scores ranging from 0.795 to 0.904, and were above the recommended standard of 0.6 [57]. In addition, the values of factors loading for all variables ranged from 0.719 to 0.944 and exceeded the acceptable value of 0.6 [58]. Additionally, the values of AVE (0.563 to 0.758) were above the acceptable limit of 0.5 [56]. Therefore, the convergent validity of the measurements was satisfied. Regarding the discriminant validity, the value of the square root of AVE was estimated for each variable and compared with its correlation value. As shown in Table 4, most construct pairs satisfied this requirement, except in one case (i.e., moral attitude and health consciousness). Therefore, the chi-square ($\Delta\chi^2$) difference test was performed regarding this lone problematic case. The chi-square difference test exhibited highly significant differences

($\Delta\chi^2 = 182.139$, $p < 0.001$). Therefore, discriminant validity was confirmed, which indicates that all variables used in the study were distinctively different.

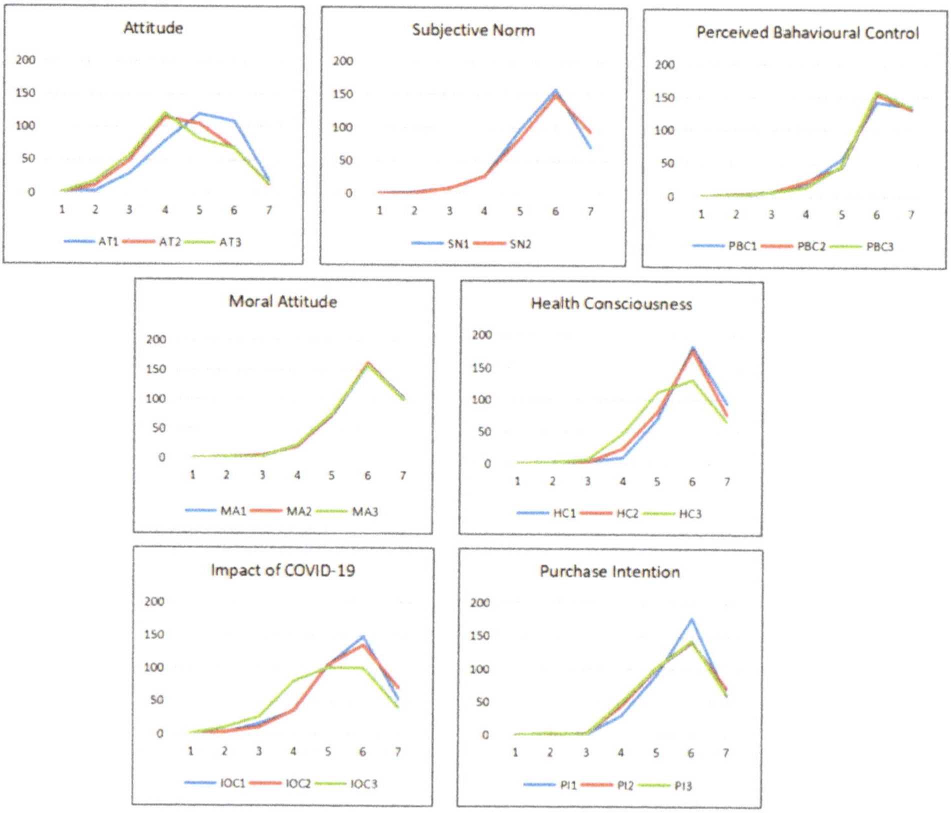

Figure 2. The overview of responses ($n = 360$): X-axis, seven-point scale (1 = strongly disagree, 2 = disagree, 3 = somewhat disagree, 4 = neither agree nor disagree, 5 = somewhat agree, 6 = agree, 7 = strongly agree); Y-axis, number of responses; AT, attitude; SN, subjective norm; PBC, perceived behavioural control; MA, moral attitude; HC, health consciousness; IOC, impact of COVID-19; PI, purchase intention.

4.3. Structural Model

The goodness-of-fit indices of the structural model are presented in Table 5. Regarding the original TPB framework, the structural TPB model demonstrated a good fit to the sample data, with $\chi^2/df = 2.533$, GFI = 0.956, TLI = 0.970, IFI = 0.980, CFI = 0.980, and RMSEA = 0.065. As for the proposed extended framework (i.e., E-TPB model), its goodness-of-fit indices ($\chi^2/df = 2.870$; GFI = 0.893; TLI = 0.938; IFI = 0.951; CFI = 0.950; RMSEA = 0.068) also showed satisfactory fit indices. Although the value of GFI (0.893) was slightly smaller than the suggested level ($\geq$0.9), the structural model of E-TPB still can be accepted due to the good performance of other indices. Finally, the E-TPB model was compared with the standard TPB model. Our findings show that the E-TPB model has a better explanatory power ($R^2 = 0.68$), in comparison to the original TPB ($R^2 = 0.49$), for measuring Chinese consumers' green food intentional purchases during the COVID-19 pandemic period. Notably, the extended model can explain 68% of the total variance in this study.

Table 3. Reliability and validity analysis.

Constructs	Factor Loadings		CR		SMC		AVE		Cronbach's α	√AVE
	TPB	E-TPB	TPB	E-TPB	TPB	E-TPB	TPB	E-TPB		
PI			0.904	0.901			0.758	0.752	0.902	0.867
PI1	0.840	0.838			0.706	0.703				
PI2	0.870	0.866			0.757	0.751				
PI3	0.901	0.897			0.813	0.805				
AT			0.893	0.894			0.738	0.740	0.888	0.859
AT1	0.733	0.743			0.538	0.553				
AT2	0.944	0.942			0.891	0.888				
AT3	0.887	0.883			0.786	0.780				
PBC			0.900	0.900			0.750	0.750	0.899	0.866
PBC1	0.864	0.864			0.747	0.746				
PBC2	0.886	0.887			0.785	0.787				
PBC3	0.847	0.846			0.717	0.716				
SN			0.830	0.834			0.710	0.716	0.830	0.843
SN1	0.826	0.789			0.682	0.623				
SN2	0.859	0.900			0.739	0.810				
MA				0.841				0.640	0.849	0.800
MA1		0.845				0.714				
MA2		0.830				0.688				
MA3		0.719				0.516				
HC				0.795				0.563	0.792	0.750
HC1		0.778				0.605				
HC2		0.737				0.544				
HC3		0.736				0.541				
IOC				0.891				0.731	0.887	0.855
IOC1		0.905				0.818				
IOC2		0.844				0.712				
IOC3		0.814				0.662				

Note: PI, purchase intention; AT, attitude; PBC, perceived behavioural control; SN, subjective norm; MA, moral attitude; HC, health consciousness; IOC, impact of COVID-19; CR, composite reliability; SMC, squared multiple correlation; AVE, average variance extracted; √AVE, square root of average variance extracted.

Table 4. Correlation matrix for discriminant validity.

	SN	IOC	HC	MA	PBC	AT	PI
SN	0.843						
IOC	0.458	0.855					
HC	0.599	0.569	***0.750***				
MA	0.755	0.529	***0.771***	0.800			
PBC	0.670	0.432	0.561	0.630	0.866		
AT	0.478	0.527	0.421	0.440	0.376	0.859	
PI	0.550	0.693	0.686	0.700	0.550	0.594	0.867

Note: The diagonal elements represent the square root of AVE; off-diagonal elements show the correlations between constructs; values in italics boldface indicate that values for the shared variance are larger than the square root of AVE values; SN, subjective norm; IOC, impact of COVID-19; HC, health consciousness; MA, moral attitude; PBC, perceived behavioural control; AT, attitude; PI, purchase intention.

4.4. Hypotheses Testing

The path analysis results of the TPB and E-TPB models are presented in Table 6, including standardised parameter estimates, *t*-values, significance levels, and the results of each hypothesis. For the original variables of the TPB framework, the constructs of attitude ($\beta = 0.395$, $t = 7.373$, $p < 0.001$; $\beta = 0.237$, $t = 4.806$, $p < 0.001$, respectively) and PBC ($\beta = 0.284$, $t = 4.153$, $p < 0.001$; $\beta = 0.122$, $t = 2.111$, $p < 0.05$, respectively) had a significant impact on Chinese consumers' green food purchase intentions in both TPB and E-TPB models. Therefore, hypotheses H1 and H3 are justified. However, the path analysis results of subjective norm ($\beta = 0.188$, $t = 2.498$, $p < 0.05$) were significant in the TPB model, but the results of subjective norm in the E-TPB model ($\beta = -0.119$, $t = -1.478$, $p > 0.05$) were nonsignificant and negatively related to consumers' buying intentions of green food products. Thus, H2 is partially supported. Regarding the extended constructs in the E-TPB model, all the structural model results presented in Table 6 show the significant relationships between the three additional variables (i.e., moral norm, health consciousness, and IOC) and purchase intention, albeit with different levels of significance. Moral attitude ($\beta = 0.318$, $t = 3.352$, $p < 0.001$) and health consciousness ($\beta = 0.154$, $t = 2.023$, $p < 0.05$) affected consumers' purchase intentions towards green food. Hence, H4 and H5 are supported. Regarding the role of IOC, there were significantly strong and positive associations between IOC and health consciousness ($\beta = 0.600$, $t = 9.579$, $p < 0.001$), as well as purchase intention ($\beta = 0.315$, $t = 4.950$, $p < 0.001$). Accordingly, H6 and H7 are supported.

Table 5. Goodness-of-fit indices and explanatory power of two models.

Models	χ^2/df	GFI	TLI	IFI	CFI	RMSEA	R^2
Thresholds	>1 and <5 *	≥0.9 *	≥0.9 *	≥0.9 *	≥0.9 *	≤0.08 *	
TPB	2.533	0.956	0.970	0.980	0.980	0.065	0.49
E-TPB	2.870	0.893	0.938	0.951	0.950	0.068	0.68

Note: * Source from Bagozzi and Yi [57]; GFI, goodness-of-fit index; NFI, normative fit index; TLI, Tucker–Lewis index; CFI, comparative fit index; IFI, incremental fit index; RMSEA, root mean square error approximation.

Table 6. Hypotheses test results.

Hypothesised Path	Standardised Estimate		*t*-Value		Result
	TPB	E-TPB	TPB	E-TPB	
H1: AT → PI	0.395	0.237	7.373 ***	4.806 ***	Support
H2: SN → PI	0.188	−0.119	2.498 *	−1.478	Partly support
H3: PBC → PI	0.284	0.122	4.153 ***	2.111 *	Support
H4: MA → PI		0.318		3.352 ***	Support
H5: HC → PI		0.154		2.023 *	Support
H6: IOC → HC		0.600		9.579 ***	Support
H7: IOC → PI		0.315		4.950 ***	Support

Note: *** $p < 0.001$; ** $p < 0.01$; * $p < 0.05$; AT, attitude; PI, purchase intention; SN, subjective norm; PBC, perceived behavioural control; MA, moral attitude; HC, health consciousness; IOC, impact of COVID-19.

5. Discussion

The present study explores Chinese consumers' green food purchase intentions by developing and applying an extended model adapted to COVID-19 pandemic influences. A new E-TPB model was proposed by extending the original TPB model, incorporating three salient variables (i.e., moral norm, health consciousness, and IOC) into the framework. The results of our empirical investigation revealed better applicability in the E-TPB model than the TPB model and identified several key factors relating to Chinese consumers' green food purchase intentions.

Regarding the original TPB constructs' impact on green food intentional purchases, attitude and PBC were found to have a significant positive effect on Chinese consumers' green food intentional behaviour. Consumers' attitudes play a significant role in driving consumers' intentions towards purchasing green food. When consumers have a positive attitude towards green food products, their intentions to buy green food increase. This result is consistent with former studies [19,27–29] involving environmentally friendly food products. Thus, green food enterprises need to make an effort to increase consumers' positive perceptions and attitudes of green food products, such as highlighting the benefits of branded food and conducting promotional campaigns to increase consumers' beliefs and knowledge. PBC is also an important aspect that is directly associated with consumers' green food purchase intentions, which conforms with conclusions of previous studies [19,34,59]. Specifically, the present study applies a quantitative approach to validate the results of recent qualitative research [7] that the PBC is the influential antecedent of an intentional purchase of green food products during the COVID-19 pandemic. Hence, marketers in the Chinese green food sector should increase varieties and expand the supply channels of green food products. Online shopping is a trend for current food consumption in China, especially during the ongoing COVID-19 pandemic. Marketers in the green food industry can cooperate with a takeout platform (e.g., Eleme app) and retail podium (e.g., Freshhema app) to increase availability and convenience. Interestingly, the correlation of the association between consumers' subjective norms and their green food buying intentions differed within the TPB and E-TPB models, which was significant in TPB but was negative and not significant in the E-TPB. A possible explanation of this incongruence is due to the unstable, poor predictive power of subjective norms, and variation in different contexts [60,61], especially regarding organic and green food purchases [32,39]. Qi and Ploeger [19] have substituted the subjective norm into the factors of face consciousness and group conformity when investigating Chinese consumers' behaviour. The results from their study [19] showed that the replacement greatly improved the predictive power of explaining consumers' intentional behaviour of green food products in the Chinese context.

In the extended model, the analysed results supported findings from studies involving the purchase of environmentally friendly food products [27,39] and demonstrated that consumers' moral attitude towards green food is a significant positive driver of intentional purchases. In particular, moral attitude showed a significant effect since it resulted in a larger contribution to the explanatory power of the proposed E-TPB framework. Our findings suggest that more Chinese consumers feel it is a moral norm to buy green food products as their purchase intentions increase. Therefore, marketers can highlight concepts related to moral imperatives in their marketing strategies to influence consumers to gain positive feelings in purchasing green food. As expected, health consciousness emerges as one significant driver of green food purchase intention as well, which correlates with previous findings that the consumers' health concern is one of the primary determinants influencing their environmentally friendly food behavioural intentions [62–64]. Thus, marketers in the Chinese green food industry should disseminate its health-related benefits and make it a primary objective while communicating with consumers. In regard to the IOC, our results indicated that there is a significant impact on consumers' health consciousness and their purchase intentions during the pandemic. Our findings show that the pandemic has shifted an individual's consumption pattern and structure, which is consistent with recent studies [4,49,50]. The pandemic has greatly increased an individual's safety and health concerns, and people are increasingly focused on health benefits, which also supports findings from Meixner and Katt [3]. Thus, facing the rise in willingness and existing challenges, companies in the green food industry should quickly adjust their production, inspection, supply, and marketing strategies to better respond to the pandemic. For example, companies can provide information about virus and safety inspections with their product packages, increase online sales channels, prevent the upswing in prices, and strengthen promotional activities, especially in highlighting the benefits of green food products.

Finally, in terms of comparing the overall performance between the standard and extended frameworks, our results have validated the effectiveness of TPB and demonstrated the superior performance of E-TPB in regard to explaining and predicting Chinese consumers' green food purchase intentions. Notably, the explanatory power difference between the original TPB model (R^2 = 49%) and the E-TPB model (R^2 = 68%) in predicting consumers' intentions to buy green food products was higher than 19%. Hence, the E-TPB model is more appropriate for explaining and predicting Chinese consumers' green food purchase intentions in the current and post-pandemic periods.

6. Conclusions

The present study has revealed that after the outbreak of COVID-19, the E-TPB model has exhibited better explanatory power in predicting Chinese consumers' purchase intentions towards green food products when compared with the original TPB model. The findings from our investigation have reinforced existing evidence that factors including attitude, PBC, moral attitudes, health consciousness, and IOC have played significant roles in the intentional processes of buying green food during a pandemic crisis. In addition, our work is among the first attempt to explore the impacts of the COVID-19 pandemic on consumers' green food purchase behaviour by distributing an online survey. Additionally, the above findings have potentially mapped a pathway to expand the green food market in China further. Our findings explored a newly developed model to gain a better understanding of how different and new factors affect consumers' behavioural intentions towards green food purchases during a pandemic. In addition, certain limitations should be noted for further research. Firstly, our study has investigated an intentional stage, not the actual purchase behaviour of green food. Since the correlation between behavioural intentions and actual or observed behaviours is not always perfect [65], a further study can extend our framework to a final purchase behaviour phase to substantiate current research findings. Secondly, we used an online survey platform to collect data. This approach may result in sample bias because consumers without internet access were not included in our samples. The education levels among our respondents could have been over-represented in our investigation. Thereby, our findings could not be considered representative of the whole population. Accordingly, a future investigation should enlarge the sample size and investigate more diverse populations from different backgrounds. Thirdly, recent studies [66–68] have reported that there is an increasing need from consumers to receive food information services. Therefore, further research should consider the important role of food information and incorporate it within our framework and then apply the framework to green food research. Finally, since the pandemic is constantly changing and there exist large uncertainties, further research can update and modify our proposed model according to shifting consumer behaviours and consumption patterns.

Author Contributions: Conceptualisation, X.Q. and A.P.; methodology, X.Q. and A.P.; formal analysis, X.Q.; investigation, X.Q.; resources, X.Q. and A.P.; writing—original draft preparation, X.Q.; writing—review and editing, A.P.; visualisation, X.Q. and A.P.; supervision, A.P.; validation, X.Q. and A.P. All authors have read and agreed to the published version of the manuscript.

Funding: This research received no external funding. The costs of the open access publication were covered by open access publications funds from the University of Kassel, which is financed by the German Research Foundation (DFG) and the Library of the University of Kassel.

Data Availability Statement: The datasets generated and/or analysed during the current study are not publicly available due data are not public but are available from the corresponding author on reasonable request.

Acknowledgments: The authors are thankful to the anonymous reviewers and the editor for their valuable comments. The authors are thankful to Ryan M. Kelly (https://orcid.org/0000-0003-2322-2848) of Rykell Scientific Editorial, located in Los Angeles, CA, USA, for technical editing and proofreading this manuscript.

Conflicts of Interest: The authors declare no conflict of interest.

References

1. World Health Organization. WHO Coronavirus (COVID-19) Dashboard. Available online: https://covid19.who.int/ (accessed on 22 April 2021).
2. Perlman, S. Another decade, another coronavirus. *N. Engl. J. Med.* **2020**, *382*, 760–762. [CrossRef]
3. Meixner, O.; Katt, F. Assessing the Impact of COVID-19 on Consumer Food Safety Perceptions—A Choice-Based Willingness to Pay Study. *Sustainability* **2020**, *12*, 7270. [CrossRef]
4. Mehta, S.; Saxena, T.; Purohit, N. The new consumer behaviour paradigm amid covid-19: Permanent or transient? *J. Health Manag.* **2020**, *22*, 291–301. [CrossRef]
5. Khan, M.; Menon, P.; Govender, R.; Samra, A.; Nauman, J.; Ostlundh, L.; Al Kaabi, J. Systematic review of the effects of pandemic confinements on body weight and their determinants. *Br. J. Nutr.* **2021**. [CrossRef] [PubMed]
6. Zhang, X.; Oluyomi, A.; Woodard, L.; Raza, S.A.; Adel Fahmideh, M.; El-Mubasher, O.; Byun, J.; Han, Y.; Amos, C.I.; Badr, H. Individual-Level Determinants of Lifestyle Behavioral Changes during COVID-19 Lockdown in the United States: Results of an Online Survey. *Int. J. Environ. Res. Public Health* **2021**, *18*, 4364. [CrossRef] [PubMed]
7. Qi, X.; Yu, H.; Ploeger, A. Exploring Influential Factors Including COVID-19 on Green Food Purchase Intentions and the Intention–Behaviour Gap: A Qualitative Study among Consumers in a Chinese Context. *Int. J. Environ. Res. Public Health* **2020**, *17*, 7106. [CrossRef]
8. Xie, X.; Huang, L.; Li, J.J.; Zhu, H. Generational Differences in Perceptions of Food Health/Risk and Attitudes toward Organic Food and Game Meat: The Case of the COVID-19 Crisis in China. *Int. J. Environ. Res. Public Health* **2020**, *17*, 3148. [CrossRef]

9. Scacchi, A.; Catozzi, D.; Boietti, E.; Bert, F.; Siliquini, R. COVID-19 Lockdown and Self-Perceived Changes of Food Choice, Waste, Impulse Buying and Their Determinants in Italy: QuarantEat, a Cross-Sectional Study. *Foods* **2021**, *10*, 306. [CrossRef]
10. Rodríguez-Pérez, C.; Molina-Montes, E.; Verardo, V.; Artacho, R.; García-Villanova, B.; Guerra-Hernández, E.J.; Ruíz-López, M.D. Changes in Dietary Behaviours during the COVID-19 Outbreak Confifinement in the Spanish COVIDiet Study. *Nutrients* **2020**, *12*, 1730. [CrossRef] [PubMed]
11. Accenture. COVID-19 Increasing Consumers' Focus on "Ethical Consumption". Available online: https://newsroom.accenture.com/news/covid-19-increasing-consumers-focus-on-ethical-consumption-accenture-survey-finds.htm (accessed on 22 April 2021).
12. Paull, J. Green Food in China. *Elem. J. Bio Dyn. Tasma.* **2008**, *91*, 48–53.
13. Green Food Development Center. Available online: http://www.greenfood.agri.cn/ztzl/zspj/lsspzs/jczs/ (accessed on 26 May 2021).
14. Green Food Development Center. Available online: http://www.greenfood.agri.cn/dtyw/gzywx/202104/t20210425_7655830.htm (accessed on 26 May 2021).
15. Ayyub, S.; Wang, X.; Asif, M.; Ayyub, R.M. Antecedents of trust in organic foods: The mediating role of food related personality traits. *Sustainability* **2018**, *10*, 3597. [CrossRef]
16. Yu, X.; Gao, Z.; Zeng, Y. Willingness to pay for the "Green Food" in China. *Food Policy* **2014**, *45*, 80–87. [CrossRef]
17. Ajzen, I. The theory of planned behavior. *Organ. Behav. Hum. Decis. Process.* **1991**, *50*, 179–211. [CrossRef]
18. Wang, Y.F.; Wang, C.J. Do psychological factors affect green food and beverage behaviour? *An application of the theory of planned behaviour. Br. Food J.* **2016**, *118*, 2171–2199.
19. Qi, X.; Ploeger, A. Explaining consumers' intentions towards purchasing green food in Qingdao, China: The amendment and extension of the theory of planned behavior. *Appetite* **2019**, *133*, 414–422. [CrossRef] [PubMed]
20. Conner, M.; Armitage, C.J. Extending the theory of planned behavior: A review and avenues for further research. *J. Appl. Sol. Psychol.* **1998**, *28*, 1429–1464. [CrossRef]
21. Nejad, L.M.; Wertheim, E.H.; Greenwood, K.M. Predicting dieting behavior by using, modifying, and extending the theory of planned behavior. *J. Appl. Sol. Psychol.* **2004**, *34*, 2099–2132. [CrossRef]
22. Byrne, D.V. Current Trends in Multidisciplinary Approaches to Understanding Consumer Preference and Acceptance of Food Products. *Foods* **2020**, *9*, 1380. [CrossRef] [PubMed]
23. Manstead, A.S.R. The role of moral norm in the attitude-behavior relationship. In *Attitudes, Behavior and Social Context: The Role of Norms and Group Membership*; Terry, D.J., Hogg, M.A., Eds.; Lawrence Erlbaum Associates, Inc.: Mahwah, NJ, USA, 2000; pp. 11–30.
24. Lin, L.; Zhou, D.; Ma, C. Green food industry in China: Development, problems and policies. *Renew. Agric. Food Syst.* **2010**, *25*, 69–80. [CrossRef]
25. Cranfield, J.A.L. Framing consumer food demand responses in a viral pandemic. *Can. J. Agric. Econ. Can. D'agroecon.* **2020**, *68*, 151–156. [CrossRef]
26. Carducci, B.; Keats, E.C.; Ruel, M.; Haddad, L.; Bhutta, Z.A. Food systems, diets and nutrition in the wake of covid-19. *Nat. Food* **2021**, *2*, 68–70. [CrossRef]
27. Yadav, R.; Pathak, G.S. Intention to purchase organic food among young consumers: Evidences from a developing nation. *Appetite* **2016**, *96*, 122–128. [CrossRef] [PubMed]
28. Al-Swidi, A.; Huque, S.M.R.; Hafeez, M.H.; Shariff, M.N.M. The role of subjective norms in theory of planned behavior in the context of organic food consumption. *Br. Food J.* **2014**, *116*, 1561–1580. [CrossRef]
29. Dowd, K.; Burke, K.J. The influence of ethical values and food choice motivations on intentions to purchase sustainably sourced foods. *Appetite* **2013**, *69*, 137–144. [CrossRef] [PubMed]
30. Scalco, A.; Noventa, S.; Sartori, R.; Ceschi, A. Predicting organic food consumption: A meta-analytic structural equation model based on the theory of planned behavior. *Appetite* **2017**, *112*, 235–248. [CrossRef]
31. Shen, Y.-C.; Chen, H.-S. Exploring Consumers' Purchase Intention of an Innovation of the Agri-Food Industry: A Case of Artificial Meat. *Foods* **2020**, *9*, 745. [CrossRef] [PubMed]
32. Sultan, P.; Tarafder, T.; Pearson, D.; Henryks, J. Intention-behaviour gap and perceived behavioural control-behaviour gap in theory of planned behaviour: Moderating roles of communication, satisfaction and trust in organic food consumption. *Food Qual. Prefer.* **2020**, *81*, 103838. [CrossRef]
33. Stranieri, S.; Ricci, E.C.; Banterle, A. Convenience food with environmentally sustainable attributes: A consumer perspective. *Appetite* **2017**, *116*, 11–20. [CrossRef]
34. Qi, X.; Ploeger, A. An integrated framework to explain consumers' purchase intentions toward green food in the Chinese context. *Food Qual. Prefer.* **2021**, *92*, 104229. [CrossRef]
35. Zhou, Y.; Thogersen, J.; Ruan, Y.; Huang, G. The moderating role of human values in planned behavior: The case of Chinese consumers' intention to buy organic food. *J. Consum. Mark.* **2013**, *30*, 335–344. [CrossRef]
36. Carfora, V.; Cavallo, C.; Caso, D.; Del Giudice, T.; De Devitiis, B.; Viscecchia, R.; Nardone, G.; Cicia, G. Explaining consumer purchase behavior for organic milk: Including trust and green self-identity within the theory of planned behavior. *Food Qual. Prefer.* **2019**, *76*, 1–9. [CrossRef]
37. Schwartz, S.H. Normative influences on altruism. *Adv. Exp. Soc. Psychol.* **1977**, *10*, 222–280.

38. Olsen, N.V.; Sijtsema, S.J.; Hall, G. Predicting consumers' intention to consume ready-to-eat meals. The role of moral attitude. *Appetite* **2010**, *55*, 534–539. [CrossRef] [PubMed]
39. Yazdanpanah, M.; Forouzani, M. Application of the theory of planned behaviour to predict Iranian students' intention to purchase organic food. *J. Clean. Prod.* **2015**, *107*, 342–352. [CrossRef]
40. Jayanti, R.K.; Burns, A.C. The antecedents of preventive health care behavior: An empirical study. *J. Acad. Mark. Sci.* **1998**, *26*, 6–15. [CrossRef]
41. Becker, M.H.; Maiman, L.A.; Kirscht, J.P.; Haefner, D.P.; Drachman, R.H. The health belief model and prediction of dietary compliance: A field experiment. *J. Health Soc. Behav.* **1977**, *18*, 348–366. [CrossRef]
42. Paul, J.; Rana, J. Consumer behavior and purchase intention for organic food. *J. Consum. Mark.* **2012**, *29*, 412–422. [CrossRef]
43. Lea, E.; Worsley, T. Australians' organic food beliefs, demographics and values. *Br. Food J.* **2005**, *107*, 855–869. [CrossRef]
44. Katt, F.; Meixner, O. Is It All about the Price? An Analysis of the Purchase Intention for Organic Food in a Discount Setting by Means of Structural Equation Modeling. *Foods* **2020**, *9*, 458. [CrossRef]
45. Ismael, D.; Ploeger, A. The Potential Influence of Organic Food Consumption and Intention-Behavior Gap on Consumers' Subjective Wellbeing. *Foods* **2020**, *9*, 650. [CrossRef]
46. Soroka, A.; Wojciechowska-Solis, J. Consumer Motivation to Buy Organic Food Depends on Lifestyle. *Foods* **2019**, *8*, 581. [CrossRef] [PubMed]
47. Rizzo, G.; Borrello, M.; Dara Guccione, G.; Schifani, G.; Cembalo, L. Organic Food Consumption: The Relevance of the Health Attribute. *Sustainability* **2020**, *12*, 595. [CrossRef]
48. Kriwy, P.; Mecking, R.-A. Health and environmental consciousness, costs of behaviour and the purchase of organic food. *Int. J. Consum. Stud.* **2012**, *36*, 30–37. [CrossRef]
49. Alaimo, L.S.; Fiore, M.; Galati, A. How the Covid-19 Pandemic Is Changing Online Food Shopping Human Behaviour in Italy. *Sustainability* **2020**, *12*, 9594. [CrossRef]
50. Ben Hassen, T.; El Bilali, H.; Allahyari, M.S. Impact of COVID-19 on Food Behavior and Consumption in Qatar. *Sustainability* **2020**, *12*, 6973. [CrossRef]
51. AliResearch; Alibaba Foundation. Green Consumer Reports of China. Available online: http://i.aliresearch.com/fle/20160803/20160803103534.pdf (accessed on 22 April 2021).
52. Kline, R.B. *Principals and Practice of Structural Equation Modeling*, 3rd ed.; The Guilford Press: New York, NY, USA, 2011.
53. Wang, Y.; Wiegerinck, V.; Krikke, H.; Zhang, H. Understanding the purchase intention towards remanufactured product in closed-loop supply chains: An empirical study in China. *Int. J. Phys. Distr. Log.* **2013**, *43*, 866–888. [CrossRef]
54. Han, H.; Hsu, L.T.J.; Sheu, C. Application of the theory of planned behavior to green hotel choice: Testing the effect of environmental friendly activities. *Tour. Manage.* **2010**, *31*, 325–334. [CrossRef]
55. Arvola, A.; Vassallo, M.; Dean, M.; Lampila, P.; Saba, A.; Lahteenmaki, L.; Shepherd, R. Predicting intentions to purchase organic food: The role of affective and moral attitudes in the theory of planned behavior. *Appetite* **2008**, *50*, 443–454. [CrossRef] [PubMed]
56. Hair, J.F.; Black, W.C.; Babin, B.J.; Anderson, R.E. *Multivariate Data Analysis*; Pearson: Harlow, UK, 2014.
57. Bagozzi, R.P.; Yi, Y. On the evaluation of structural equation models. *J. Acad. Mark. Sci.* **1988**, *16*, 74–94. [CrossRef]
58. Chin, W.W.; Gopal, A.; Salisbury, W.D. Advancing the theory of adaptive structuration: The development of a scale to measure faithfulness of appropriation. *Inform. Syst. Res.* **1997**, *8*, 342–367. [CrossRef]
59. Li, S.; Jaharuddin, N. Identifying the key purchase factors for organic food among Chinese consumers. *Front. Bus. Res. China* **2020**, *14*, 25. [CrossRef]
60. Armitage, C.J.; Conner, M. Efficacy of the theory of planned behaviour: A meta-analytic review. *Br. J. Soc. Psychol.* **2001**, *40*, 471–499. [CrossRef] [PubMed]
61. Saba, A.; Di Natale, R. A study on the mediating role of intention in the impact of habit and attitude on meat consumption. *Food Qual. Prefer.* **1998**, *10*, 69–77. [CrossRef]
62. Hoek, A.C.; Pearson, D.; James, S.W.; Lawrence, M.A.; Friel, S. Shrinking the food-print: A qualitative study into consumer perceptions, experiences and attitudes towards healthy and environmentally friendly food behaviours. *Appetite* **2017**, *108*, 117–131. [CrossRef]
63. Kareklas, I.; Carlson, J.; Muehling, D.D. 'I eat organic for my benefit and yours': Egoistic and altruistic considerations for purchasing organic food and their implications for advertising strategists. *J. Advert.* **2014**, *43*, 18–32. [CrossRef]
64. Verain, M.C.D.; Sijtsema, S.J.; Dagevos, H.; Antonides, G. Attribute Segmentation and Communication Effects on Healthy and Sustainable Consumer Diet Intentions. *Sustainability* **2017**, *9*, 743. [CrossRef]
65. Belk, R.W. Issues in the intention-behavior discrepancy. *Res. Consum. Behav.* **1985**, *1*, 1–34.
66. Felicetti, A.M.; Volpentesta, A.P.; Ammirato, S. Analyzing app-based food information services: The case of Olive Oil sector. *VINE J. Inf. Knowl. Manag. Syst.* **2019**, *50*, 427–453. [CrossRef]
67. Volpentesta, A.P.; Felicetti, A.M. Research investigation on food information user's behaviour. In *Collaborative Networks of Cognitive Systems*; Camarinha-Matos, L.M., Afsarmanesh, H., Rezgui, Y., Eds.; Springer: Berlin/Heidelberg, Germany, 2018; pp. 190–202.
68. Wu, X.; Xiong, J.; Yan, J.; Wang, Y. Perceived quality of traceability information and its effect on purchase intention towards organic food. *J. Market. Manag.* **2021**. [CrossRef]

foods

MDPI

Article

COVID-19′s First Wave: Examination of Impact on Food Purchasing Behaviour in the Eurozone

Belén Gutiérrez-Villar [1], Rosa Melero-Bolaños [1,*] and Mariano Carbonero-Ruz [2]

[1] Departamento de Gestión Empresarial, Universidad Loyola Andalucía, Escritor Castilla Aguayo 4, 14014 Córdoba, Spain; belengut@uloyola.es

[2] Departamento de Métodos Cuantitativos, Universidad Loyola Andalucía, Escritor Castilla Aguayo 4, 14014 Córdoba, Spain; mcarbonero@uloyola.es

* Correspondence: rmelero@uloyola.es

Abstract: COVID-19 has had a negative impact on the living conditions of people in all countries worldwide. With a devastating economic crisis where many families are finding it difficult to pay bills and make ends meet, increases in prices of food basket staples can be very worrying. This study examines the relationship between the incidence of the pandemic during the first wave in 16 Eurozone countries with the variation experienced in food prices. We analysed the harmonised index of consumer food prices (included in HICP) and the classification of the degree of pandemic impact by country, the latter established with the index of deaths provided by the Johns Hopkins Center. The procedure used compared actual food prices during the first wave (March to June 2020) with those foreseeable in the absence of the pandemic. Time series analysis was used, dividing the research period into two phases. In both phases, the Holt–Winters model was applied for estimation and subsequent prediction. After a contrast using Kendall's tau correlation index, it was concluded that in the countries with the highest death rates during the first wave, there was a higher increase in food prices than in the least affected countries of the Eurozone.

Keywords: food prices; COVID-19; Eurozone; Holt–Winters model

Citation: Gutiérrez-Villar, B.; Melero-Bolaños, R.; Carbonero-Ruz, M. COVID-19′s First Wave: Examination of Impact on Food Purchasing Behaviour in the Eurozone. *Foods* **2021**, *10*, 1179. https://doi.org/10.3390/foods10061179

Academic Editor: Derek V. Byrne

Received: 4 May 2021
Accepted: 21 May 2021
Published: 24 May 2021

1. Introduction

The coronavirus pandemic (COVID-19) is one of the most significant challenges humanity has faced in recent times. Although the disease has affected the entire planet, it has not had the same incidence everywhere and at the same time since it was classified as a pandemic by the World Health Organization (WHO) on 11 March 2020 [1].

Since that fateful moment, humanity has already experienced two peaks in the number of deaths and infections. The first, known as the first wave of the pandemic, lasted from its beginning in March until June 2020; the second wave, whose effects began to be felt at the end of the summer, lasted until the end of 2020, although there have been spikes in cases in January 2021 that warn of the possibility of the beginning of a third wave [2,3].

Today, the coronavirus has affected almost every country on the planet, but its impact during the first wave of the disease was uneven. Among the main factors that, combined, could explain the differences in mortality figures caused by the virus are the speed and restrictive measures imposed by governments, along with factors such as demographics, culture, and the country's environment [4].

In addition, the suspension of activities and restrictions on movement that many governments have adopted to curb the pandemic have meant a sudden and abrupt change in the lives of citizens and businesses, changes in the structure of demand, and a drastic contraction of the global economy.

According to World Bank forecasts [5], the economic effects of the containment measures undertaken by governments are particularly profound in the countries most affected by the pandemic, leading to significant reductions in employment and incomes. The ILO [6]

estimates global income losses over the first three quarters of 2020 (without considering the implementation of income-sustaining measures) at 10.7 per cent compared to the same period in 2019. This reduction in income will not only be projected onto extreme poverty (less than USD 1.90 per day), which is estimated to increase by 20% in 2020 [7], but will also result in new groups facing this situation [8].

From the point of view of the structure of demand, measures relating to the closure of non-essential activities and home confinements have led to a change in the structure of food demand. Thus, closures or capacity constraints in restaurants and catering services in many countries have led to a decrease in the demand for food consumed outside the home [9], while hoarding of products due to panic over possible stock-outs at the beginning of the pandemic, coupled with increased purchases due to the need to eat at home, may have had the opposite effect, i.e., an increase in demand, in this case from households [10,11].

In terms of food availability, governments in most developed countries have placed a high priority on ensuring that food can reach consumers. Overall, it can be argued that, except for occasional moments at the beginning of the COVID-19 pandemic, global staple food supply chains appear to have held up reasonably well, with few instances of substantial supply disruptions, even in countries with strict social distancing requirements [7].

In this dystopian environment, with worsening living conditions for many individuals who have seen their freedoms curtailed across the board, with many jobs lost, and a clear decline in demand for restaurants, bars, and hotels, but with a guaranteed supply of food, some reports raise alarm bells about the rising price of food paid by citizens.

An empirical examination of FAO data by Mogues [12] shows a clear increase in prices paid by the final consumer. In the span of approximately three months between 14 February and 18 May 2020, global average prices for a variety of food commodities increased by between 2% and 9%, with half of the prices tracked increasing by 7% or more. Underlying these averages, however, are large variations between countries.

A macro-survey conducted by the consulting firm IPSOS [13] from 22 May to 5 June among citizens in 26 countries to find out to what extent they perceived a rise in the cost of living in the wake of the first wave of the pandemic points in the same direction. Among the main findings of the survey is that, globally, 60% of respondents believe that the price of food, goods, and services has risen sharply since the arrival of the coronavirus. In particular, the prices of food, groceries, and household products have risen the most for 63% of all respondents. The study also shows differences between countries, with 77% of Belgians and 59% of Germans, to cite the two extremes among European countries.

At the European level, Akter [14] examined the impact of the level of "stay-at-home" restrictions (high or low) and food prices in 31 European countries from January to May 2020. Their findings reveal that the severity of stay-at-home restrictions increased overall food prices by 1% in March 2020 compared to January and February 2020. They continued to increase in countries with high restrictions in April and stabilised in May.

In this environment, it should be a higher priority than ever to study food systems to make them more resilient [15]. Among the many effects that the pandemic shock has caused and the multiple impacts it may have had on food systems, this research focuses only on analysing how food prices have responded during the first wave of COVID-19 in the most important countries. The study focuses on the countries of the Eurozone and the results can be considered by governments to, in other similar situations, take measures aimed at mitigating the short-term inflationary effect that new waves of COVID-19 or other pandemic diseases can cause.

The objective of this research is to study whether there has been a relationship between the price level of food products and the impact of the coronavirus, measured by the death rate, in the countries of the Eurozone during the 1st wave of COVID-19. In the following paragraphs, we review some empirical evidence on the relationship between the incidence of COVID-19 and the evolution of food prices in general and in Europe in particular, during the first wave of COVID-19, which will support the research question on which this paper is built. In subsequent sections, the selected data and the methodology applied are presented and justified, followed by an analysis and a discussion and conclusions.

2. Background and Research Hypothesis

2.1. Mortality Differences between Countries

On the human level, the lives taken by the pandemic in the first wave rose, according to the data of confirmed deaths as of 30 June 2020, to a total of 511,700 people, of which approximately 23% occurred in Europe [16]. As is well known, the death toll is not evenly distributed among all countries. Table 1 shows the data for the European countries with the largest number of confirmed deaths in the 1st wave of the virus. Europe comprises 50 countries. Among the 10 European countries most affected by the pandemic, eight are part of the Eurozone.

Table 1. Ranking of the 10 European countries with the most cumulative confirmed COVID-19 deaths, 30 June 2020 [16].

Country	Cumulative Confirmed COVID-19 Deaths
United Kingdom	43,880
Italy	34,773
France	29,845
Spain	28,360
Belgium	9747
Netherlands	6132
Sweden	5515
Ireland	1736
Portugal	1576
Luxembourg	110

These figures on confirmed deaths are contested and it is very likely that the total number of COVID-19 deaths is higher than confirmed deaths due to differences in the types and number of tests performed and counted by countries, problems in the attribution of cause of death, and delays in reporting [17].

Thus, recent studies on mortality experienced between March and June 2020 compared to the same period in previous years show a clear increase in the number of deaths, suggesting that mortality attributable to COVID-19 during the first wave was higher than officially recognised. Among the EU member states for which data are available, the largest number of additional deaths in 2020 during weeks 10–26 compared to the four-year average from 2016 to 2019 was recorded in Spain (48,000) followed closely by Italy (46,000) and France (30,000) [18].

2.2. Evolution of Food Prices

Mead et al. [19], in one of the first studies carried out on the impact of the COVID-19 pandemic on food price indices, studied the volatility in import, export, production, and consumption prices in the months following the onset of the COVID-19 pandemic in the United States, concluding that the shift from institutional and restaurant food consumption to home food consumption created short-term shocks and disturbances in the economy, especially for perishable foods. These shocks affected the economy and affected the prices paid by consumers at the supermarket.

Ebrahimy et al. [20] studied inflation in 107 countries around the world. The only category that showed a noticeable increase in prices at the beginning of the pandemic was food: the price of meat, dairy products, and canned/frozen fruits and vegetables skyrocketed from the beginning of the confinement. The authors studied inflation behaviour for three categories of countries: advanced economies, emerging markets, and low-income countries. In advanced economies, inflation between March and July 2020 was negative in 54.3% of the countries analysed and in 45.9% of emerging market countries and 26.3%

of low-income countries. They also related inflation to mitigating policies, finding a low correlation between how strict the mitigating measures were and how high inflation was since March. This seems to be true for both advanced economies and emerging markets. The picture does not change much when looking at individual CPI components rather than the overall index.

If one confines the objective data to the European level and looks at the harmonised index of consumer prices (HICP) published by Eurostat [21], food prices were abnormally high, although a stable trend in headline inflation was observed in the period under study (March–June 2020), both for the EU as a whole and for the Eurozone countries (see Figure 1).

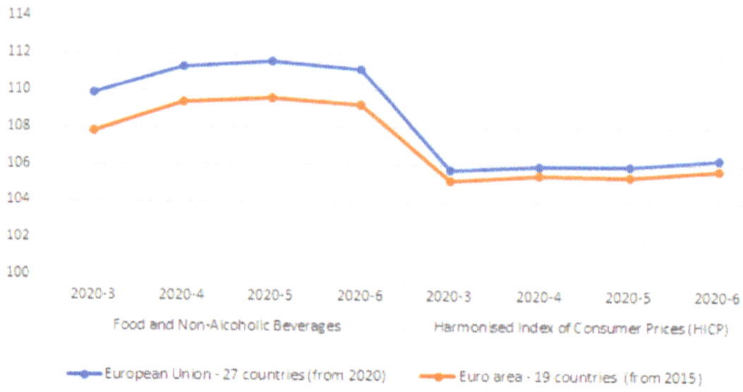

Figure 1. HICP versus food in EU (first wave COVID-19). Elaboration with Eurostat data [21].

Thus, food price increases in April were strong and reflected an unprecedented combination of supply and demand influences. However, the food HICP in May and June showed signs of normalisation as blockades eased and activity resumed [22].

Similar trends are pointed to by Akter's [14] study which, after examining the impact of the level of (high or low) "stay-at-home" restrictions and food prices in 31 European countries from January to May 2020, shows that the severity of stay-at-home restrictions increased overall food prices by 1% in March 2020 compared to January and February 2020. They continued to increase in countries with high restrictions in April and stabilised in May.

2.3. Research Hypothesis

This paper aims to delve deeper into the impact that the first wave of COVID-19 has had on food prices. As has been explained in the previous sections, the rise in prices is a fact confirmed by economic indicators and the opinions of the public. On the other hand, the figures also show differences between countries in terms of the incidence of the disease during the first wave.

To analyse the impact of the pandemic, we have focused our attention on the Eurozone. There are several reasons that led us to this choice: the high incidence rate of the disease and the priority consideration of health within the scope of action of the European Union, through the implementation of the European Health Union [23], which in practice is leading to a certain homogeneity in the adoption of measures against the disease.

On the other hand, the possibility of examining countries that, in addition to sharing currency, also share economic policy, seems a fundamental aspect for a study that analyses prices, since this common currency avoids effects and conversions related to economic decisions and exchange rates. For this reason, the analyses are confined to countries in the Eurozone.

Therefore, the objective of this research is to study whether there has been a relationship between the price level of food products and the impact of the coronavirus, measured by the death rate, in the countries of the Eurozone during the 1st wave of COVID-19.

Based on this objective, a single analytical hypothesis of difference between groups, stated as alternative hypothesis (H1), is formulated:

Hypothesis 1 (H1). *In the countries with the highest rate of COVID-19 deaths during the 1st wave, there has been a higher increase in food prices than that experienced in the least affected countries of the Eurozone.*

3. Materials and Methods

3.1. Confirmed Death Rates in Eurozone Countries

To measure the impact of the disease, the data of officially confirmed deaths in the countries of the Eurozone were selected, in proportion to the size of the country, to better adjust the incidence when making comparisons between countries. For example, if 1000 people died in Malta, with a population of approximately 514,654, that would have a much greater impact than the same number dying in the United States, with a population of 331 million. Therefore, the magnitude used in this study was the confirmed death rate per 100,000 inhabitants. Regarding the size limit, we have chosen to exclude extremely small countries (Malta, Cyprus, and Luxembourg) that are smaller than most of the capitals of the rest of the Eurozone countries.

Therefore, the 19 countries in the Eurozone are reduced to 16, which are those with more than one million inhabitants, with unequal effects of the disease (see Table 2).

Table 2. COVID-19 deaths per 100,000 inhabitants, 30 June 2020, in 16 countries of the Eurozone [1].

Country	Deaths per 100,000 Inhabitants	Country	Deaths per 100,000 Inhabitants
Belgium	84.10	Austria	7.82
Spain	60.64	Finland	5.93
Italy	57.50	Slovenia	5.34
France	45.72	Estonia	5.20
Netherlands	35.78	Lithuania	2.86
Ireland	35.15	Greece	1.84
Portugal	15.45	Latvia	1.59
Germany	10.73	Slovakia	0.51

[1] Table made with data from Johns Hopkins University [16].

We set 30 June 2020 as the date for the collection of data on the incidence of the disease, by which time most countries had already clearly begun to see a drop in the number of new cases or deaths and had begun to relax the restrictive measures adopted to curb the first wave of the pandemic.

3.2. The Food Price Index in the Eurozone

Based on previous studies, it can be stated that in EU member states, there are large differences in food prices between member states [20]. For example, in 2019, the price of an equivalent basket of food and non-alcoholic beverages was almost twice as high in the most expensive member state as in the cheapest one [21].

For this paper, the historical series of data for the Food and Non-Alcoholic Beverages category was compiled, taking the harmonised price indices for each of the 16 Eurozone countries with more than one million inhabitants, in the period from January 2013 to June 2020.

When the object of study was limited to European countries with more than one million inhabitants that have the euro as their currency, and time series were constructed by country (see Figure 2), differences between countries were again observed, year by year, and a clear upward trend in food prices was observed in most of the countries studied from 2018 onwards.

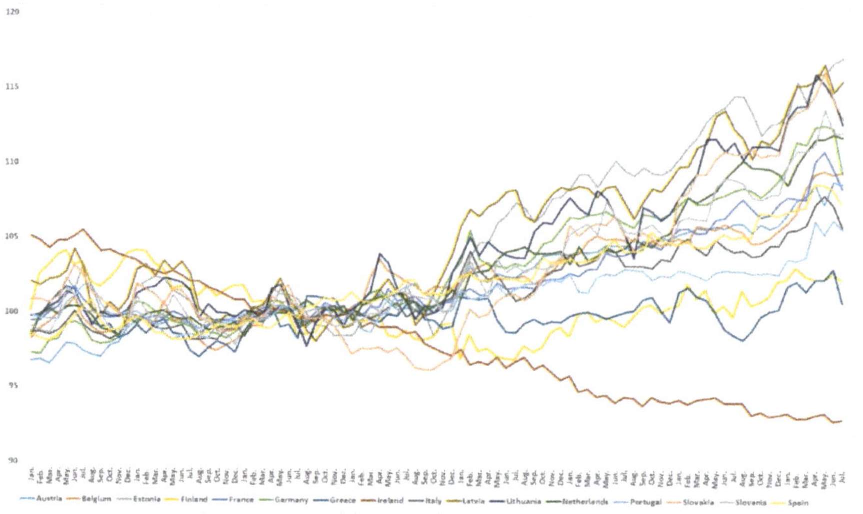

Figure 2. Monthly evolution of harmonised index of consumer prices food and non-alcoholic beverages prices (Eurozone) from January 2013 to July 2020 [21].

3.3. Data Analysis

The procedure used in this research was based on comparing actual food prices during the first wave (March to June 2020) with those that would have been expected in the absence of the pandemic. Time series analysis was used, dividing the research period into two phases. In both phases, the Holt–Winter model was applied for estimation and subsequent prediction.

The first phase, called the "learning phase", consisted of analysing the evolution of the historical series of food category prices in the countries studied, with the intention of predicting their subsequent behaviour. To investigate the price forecasts in each of the countries, the monthly data of the evolution of the HICP historical series have been used, taking January 2013 as the starting month and February 2020 as the final month of learning, the month before the declaration of the pandemic.

This represents a relatively long period of time. Then, if a high level of fit is obtained between the observed and estimated values during this phase, the model predictions for a short number of month (four in this case) should be considered adequate.

To estimate the price series, we have used the Holt–Winters (HW) seasonal algorithm, which is commonly used to capture the effect of seasonality in time series data [24]. This is a triple exponential smoothing procedure, which is especially interesting when, as is our case, we are dealing with short-term forecasting.

The HW algorithm is an iterative method that can be applied to both additive and multiplicative models, with or without stationarity. Among the possible specifications we have chosen, after a descriptive analysis of the price series (p_t), the additive formulation with a linear trend and annual seasonality is:

$$p_t = a_t + b_t\, t + s_t + u_t$$

The parameters on which the model depends $(a_t,\ b_t,\ s_t)$ were estimated recursively as a function of three quantities, $0 \leq \alpha, \beta, \gamma \leq 1$, according to the following equations:

$$a_t = \alpha(p_t - s_{t-T}) + (1-\alpha)(a_{t-1} + b_{t-1})$$

$$b_t = \beta(a_t - a_{t-1}) + (1-\beta)b_{t-1}$$

$$s_t = \gamma(p_t - a_t) + (1-\gamma)s_{t-T}$$

where T is the order of seasonality, the number of time periods until the pattern repeats.

If the series showed a high level of fit, the second phase of the study, called the "prediction phase", compared the situation of the countries according to the incidence of the epidemic in each of them. If the proposed hypothesis is true, in the countries most affected by the disease, the difference between actual prices during pandemic months (March to June 2020) and those estimated from the model built during the learning phase should be greater.

To obtain the predictions, the equations provided by the Holt–Winters method have been applied during these four months:

$$\widehat{p_{f+t}} = a_f + b_f t + s_{f+t-12}, t = 1, \ldots, 4$$

where the subscript f refers to the situation in February 2020.

Using these predictions, an index of the discrepancy between actual and estimated prices during the period of the first wave was calculated for each country, obtained as the average difference between the four values of each of the two series. A positive value of this measure indicates that prices rose more than they should have in the absence of the pandemic. Conversely, negative values of the index would point to an exceptional decline.

Finally, for testing the hypothesis, the Kendall's tau index was used to measure the association between the two variables: the incidence of the disease, measured by the death rate, and the price discrepancy index.

Kendall's tau is a measure of association between two variables. It is like the Pearson's correlation coefficient but used to compare the ranks (not the values) in a list. The Kendall's tau coefficient is defined as:

$$\tau = \frac{2(C-D)}{n(n-1)}$$

where C and D are the numbers of coincidences and discrepancies in the rank and n is the number of elements in the list.

4. Results and Discussion

4.1. Learning Phase

To study the past evolution of prices, there are two tasks to be carried out in this first learning phase: the estimation of the series and the measurement of the goodness of fit.

4.1.1. Estimation of the Equation

For the implementation of the HW method, we initialised the parameters a, b, and c estimating the series using a regression model. T = 12 has been taken by assuming that seasonality is annual, as usual with monthly data.

The optimal values of α, β, and γ have been calculated in such a way as to minimise the root mean square error (RMSE) of the learning model predictions. Tables 3 and 4 show the results for the 16 countries analysed:

Table 3. Optimal values of α, β, and γ [1].

Country	α	β	γ	Country	α	β	γ
Austria	1.000	0.000	0.000	Italy	0.891	0.000	1.000
Belgium	0.873	0.000	0.377	Latvia	0.890	0.042	0.649
Estonia	0.842	0.052	0.552	Lithuania	0.986	0.028	1.000
Finland	0.531	0.049	0.006	Netherlands	0.891	0.039	1.000
France	0.781	0.051	1.000	Portugal	0.984	0.000	1.000
Germany	1.000	0.012	0.975	Slovakia	0.952	0.073	1.000
Greece	1.000	0.000	0.000	Slovenia	0.747	0.035	0.801
Ireland	0.802	0.009	0.581	Spain	1.000	0.000	0.998

[1] Analysis performed with STATA.

Table 4. RMSE of the learning model predictions [1].

Country	RMSE	Country	RMSE
Austria	0.339	Italy	0.447
Belgium	0.448	Latvia	0.765
Estonia	0.739	Lithuania	0.713
Finland	0.806	Netherlands	0.420
France	0.395	Portugal	0.446
Germany	0.590	Slovakia	0.688
Greece	0.692	Slovenia	0.609
Ireland	0.282	Spain	0.445

[1] Analysis performed with STATA.

The maximum RMSE, obtained in Finland, is 0.806. As the values of all the series are around 100 (the base of the index was set at 100 points for the 2015 average), we can conclude that the RMSE is low in all cases, being a good indication of the achievement of a high degree of accuracy for the model predictions.

Although RMSE is one of the measures usually used as a fit indicator, we consider it convenient to complement it for two reasons: first, the fact that its minimisation has been the guide for estimating the model parameters will tend to reduce its value. Secondly, although, as we have pointed out, its values are very low with respect to the level of the series, there is no standardised reference for its valuation and, therefore, its value is not statistically verifiable.

4.1.2. Goodness of Fit

To measure the quality of the fit, and therefore the reliability of the predictions made with the estimated models, the coefficients of determination between each pair of real and estimated series during the training period (January 2013 to February 2020) have been calculated and statistically contrasted. The results are shown in Table 5:

Table 5. Coefficients of determination between each pair of real and estimated series during the training period [1].

Country	R^2	*p*-Value	Country	R^2	*p*-Value
Austria	0.986	0.0000	Italy	0.961	0.0000
Belgium	0.970	0.0000	Latvia	0.972	0.0000
Estonia	0.980	0.0000	Lithuania	0.970	0.0000
Finland	0.824	0.0000	Netherlands	0.985	0.0000
France	0.977	0.0000	Portugal	0.909	0.0000
Germany	0.975	0.0000	Slovakia	0.976	0.0000
Greece	0.693	0.0000	Slovenia	0.968	0.0000
Ireland	0.995	0.0000	Spain	0.970	0.0000

[1] Analysis performed with STATA.

With the exceptions of Greece and Finland, the coefficient of determination is in all cases above 0.9, indicating a very high level of agreement between the actual and estimated values. Even for these two countries, the value obtained is acceptable. As would be expected under these conditions, all the coefficients obtained are statistically significant, so we can conclude that the values estimated with the models obtained are, in all the countries examined, a good approximation of the true value of the harmonised price index for food.

4.2. Prediction Phase

The purpose of this phase is to evaluate the divergence between the prices that occurred between the first wave of COVID-19 (March to June 2020) and the forecasts obtained with the models validated in the training phase during the same period.

Figures 3 and 4 show the evolution of the two-price series (real and estimated) in the five most affected countries and the five least affected by the pandemic:

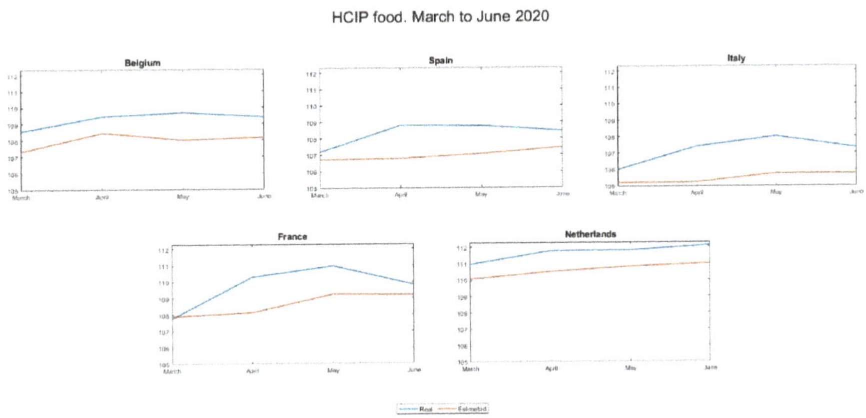

Figure 3. Evolution of the real and estimated price series in the five countries most affected by the pandemic.

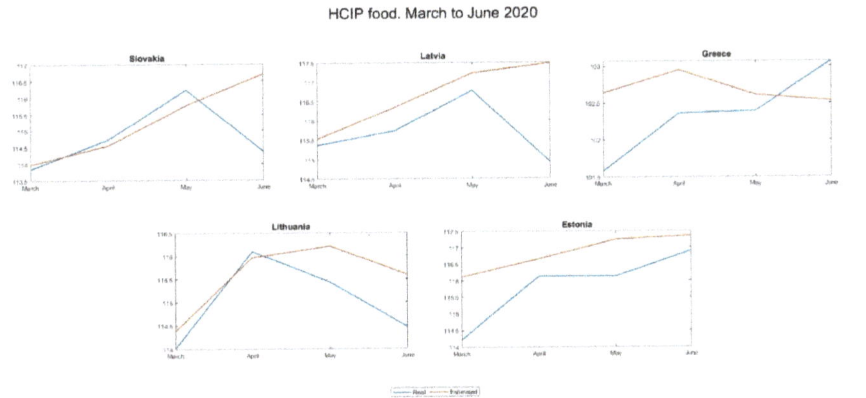

Figure 4. Evolution of the real and estimated price series in the five countries least affected by the pandemic.

The differences are manifest and seem to support the hypothesis put forward: in the countries with the highest incidence, prices grew more than expected, while in those less affected this was not the case.

To verify that this impression is correct, we calculated and show in Table 6 the index of discrepancy between actual and estimated prices, as defined in Section 3.3.

Table 6. Indicator of the discrepancy between actual and estimated prices during the first wave period [1].

Country	Index	Country	Index
Austria	0.339	Italy	0.707
Belgium	1.289	Latvia	−0.949
Estonia	−1.001	Lithuania	−0.534
Finland	−0.388	Netherlands	1.068
France	1.097	Portugal	1.580
Germany	0.245	Slovakia	−0.462
Greece	−0.331	Slovenia	0.583
Ireland	0.164	Spain	1.300

[1] Analysis performed with STATA.

This indicator makes it possible to establish a ranking among the countries investigated, headed by Italy, the country with the greatest divergence during the period, and Estonia, which, at the opposite pole, is the country where prices have grown the least with respect to expectations, and even below them.

This ranking, combined with the rate of deaths by COVID-19 as of 30 June 2020, was used to test the validity of the hypothesis of the work. If true, there should be a relationship between the position that countries occupy in both lists. To test whether this is true, we calculated Kendall's tau correlation index between both rankings, $\tau = 0.617$, and its statistical significance ($p = 0.001$) leads us to the conclusion that both rankings are positively related, in view of the sign of τ, and, therefore, there was a direct relationship, as our hypothesis states, between the incidence of COVID-19 and the rise in food prices during the first wave of the pandemic.

5. Conclusions

This research focused on analysing how food prices have responded during the first wave of COVID-19 in the most important countries of the Eurozone. Analysing the relationship between the price level of food products and the impact of the coronavirus, measured by the death rate, in the countries of the Eurozone during the 1st wave of COVID-19, this paper has shown that, during the first wave of the epidemic, there was a greater increase in food prices in the countries most affected by the disease than in those less affected.

Although there are many factors that could be related to this inflationary effect, the choice of countries from the same well-established economic environment (EU-16) and of a certain size (more than one million inhabitants) reduces the list, increasing the probability that the COVID-19 effect is significantly responsible for inflation.

Regarding the methodology, the combined use of a modelling and predictive tool (HW model) and an inferential tool (Kendall's tau) allows the construction of a hypothetical scenario (estimated prices in the case of no COVID-19) and its statistical comparison with the incidence of the pandemic in each country.

Governments should take these results into account and, in the face of other similar situations, take measures aimed at mitigating the short-term inflationary effect that new waves of COVID-19 or other pandemic diseases may cause. In this sense, it would be advisable for governments to consider measures that strengthen productive capacity on the one hand and help and mitigate the effects suffered by consumers on the other. To date, policymakers have provided massive fiscal support to protect firms, households, and vulnerable populations [1].

On the demand side, social measures have been introduced in the EU to help the most vulnerable consumers, which vary in form and amount depending on the country, but can be summarised as follows: unemployment payments, guaranteed minimum revenue, family support, housing subsidies, and emergency aid [25].

Without making an exhaustive analysis of the measures needed to curb food price increases caused by a pandemic, it seems advisable from the supply side to ensure that they do not occur. Thus, policymakers must act more quickly to weather non-economic disruptions in the supply of goods and services, such as the disruption of international supply chains or reduced labor efficiency due to COVID-19 mobility restriction measures. Other types of fiscal measures, such as the reduction in VAT rates implemented in Germany during the second half of 2020, do not seem to have had the expected effects on consumption [26].

In the case of support to producers, it is necessary to investigate the outcome of the recent package of exceptional measures, including private storage aid of the same product, temporary authorisation to self-organise market measures by operators, and flexibility in the implementation of market support programmes- to further support the agricultural and food sectors most affected by the coronavirus crisis approved by the EU in May 2020 [27].

The findings presented in this document on price increases in the countries most affected by the pandemic refer to the harmonised price index of the Food and Non-Alcoholic Beverages category, without distinguishing between fresh and storable food. It would be interesting to distinguish between them in future research, in order to specify the measures to be taken with each of them.

Finally, it would also be interesting to investigate whether the increase in prices together with the reduction in income due to the reduction in economic activity, which was very pronounced in some groups, has been able to produce changes in the eating habits of consumers.

Author Contributions: Conceptualisation, B.G.-V. and R.M.-B.; Formal analysis, M.C.-R.; Methodology, M.C.-R.; Writing—original draft, B.G.-V., R.M.-B., and M.C.-R. All authors have read and agreed to the published version of the manuscript.

Funding: This research received no external funding.

Institutional Review Board Statement: Not applicable.

Informed Consent Statement: Not applicable.

Data Availability Statement: Population Data: https://ec.europa.eu/eurostat/databrowser/view/DEMO_GIND/default/table?lang=en (accessed on 9 December 2020). Prices: https://ec.europa.eu/eurostat/databrowser/view/PRC_FSC_IDX/default/table (accessed on 9 December 2020). COVID-19 cases: https://github.com/CSSEGISandData/COVID-19 (accessed on 9 December 2020).

Conflicts of Interest: The authors declare no conflict of interest.

References

1. Allain-Dupré, D.; Taylor, A. The Territorial Impact of COVID-19: Managing the Crisis across Levels of Government. Oecd Tackling Coronavirus (COVID-19) Contributing to a Global Effort. 2020. Available online: https://www.oecd.org/coronavirus/policy-responses/theterritorial-impact-of-covid-19-managing-the-crisis-across-levels-ofgovernment-d3e314e1 (accessed on 13 May 2021).
2. Zhang, S.X.; Marioli, F.A.; Gao, R. A second wave? What do people mean by COVID waves?—A working definition of epidemic waves. *MedRxiv* **2021**. [CrossRef]
3. Soriano, V.; Ganado-Pinilla, P.; Sanchez-Santos, M.; Gómez-Gallego, F.; Barreiro, P.; de Mendoza, C.; Corral, O. Main differences between the first and second waves of COVID-19 in Madrid, Spain. *Int. J. Infect. Dis.* **2021**, *105*, 374–376. [CrossRef]
4. Beech, H.; Rubin, A.J.; Kurmanaev, A.; Maclean, R. The Covid-19 Riddle: Why Does the Virus Wallop Some Places and Spare Others? *New York Times*, 5 May 2020.
5. World Bank. *Global Outlook: Pandemic, Recession: The Global Economy in Crisis*; World Bank Group: Washington, DC, USA, 2020.
6. ILO. La COVID—19 y el mundo del trabajo. 6ª Edición. *International Labour Organization*. 2020. Available online: https://www.ilo.org/wcmsp5/groups/public/---dgreports/---dcomm/documents/briefingnote/wcms_755917.pdf (accessed on 16 December 2020).
7. Laborde, D.; Martin, W.; Swinnen, J.; Vos, R. COVID-19 risks to global food security. *Science* **2020**, *369*, 500–502. [CrossRef] [PubMed]
8. Bayona, E. La Pandemia Aflora el Hambre: Las Peticiones de Ayuda Para Comer Crecen Hasta un 50% en las Grandes Ciudades. *Público*. 2020. Available online: https://www.publico.es/economia/vulnerables-coronavirus-pandemia-aflora-hambre-peticiones-ayuda-comer-crecen-50-grandes-ciudades.html (accessed on 12 July 2020).
9. Gursoy, D.; Chi, C.G. Effects of COVID-19 pandemic on hospitality industry: Review of the current situations and a research agenda. *J. Hosp. Mark. Manag.* **2020**, *29*, 527–529. [CrossRef]

10. Swinnen, J. Will COVID-19 Cause Another Food Crisis? An Early Review. *International Food Policy Researcho Institute-IFPRI*. 2020. Available online: https://www.ifpri.org/blog/will-covid-19-cause-another-food-crisis-early-review (accessed on 3 December 2020).
11. Bhattacharjee, A.; Jahanshah, A. The COVID–19 crisis brings spring season for translucent activity. does it result exogenous uncertainty for the entrepreneurs and bound-less commodity pricing! *SSRN Electron. J.* **2020**, 19–32. [CrossRef]
12. Mogues, T. Los Mercados de Alimentos Durante la COVID-19. *International Fund Monetary. Fiscal Affairs*. 2020. Available online: https://www.imf.org/~{}/media/Files/Publications/covid19-special-notes/Spanish/sp-special-series-on-covid-19-food-markets-during-covid-19.ashx?la=en (accessed on 14 December 2020).
13. IPSOS. El 68% de los Españoles Cree Que la vida se ha Encarecido Mucho Tras el Covid-19. *IPSOS*. 2 de julio 2020. Available online: https://www.ipsos.com/es-es/el-68-de-los-espanoles-cree-que-la-vida-se-ha-encarecido-mucho-tras-el-covid-19 (accessed on 12 September 2020).
14. Akter, S. The impact of COVID-19 related 'stay-at-home' restrictions on food prices in Europe: Findings from a preliminary analysis. *Food Secur.* **2020**, *12*, 719–725. [CrossRef] [PubMed]
15. Boyacι-Gündüz, C.P.; Ibrahim, S.A.; Wei, O.C.; Galanakis, C.M. Transformation of the food sector: Security and resilienceduring the COVID-19 pandemic. *Foods* **2020**, *10*, 497. [CrossRef]
16. JHU. Covid-10 Map. Johns Hopkins University of Medicine. Coronavirus Resource Center. 2021. Available online: https://coronavirus.jhu.edu/map.html (accessed on 30 June 2020).
17. Economist, T. Tracking covid-19 excess deaths across countries. *Economist*, 11 May 2021.
18. Eurostat. Almost 170,000 additional deaths in the UE. *Eurostat*. 2020. Available online: https://ec.europa.eu/eurostat/documents/2995521/11438257/3-19102020-BP-EN.pdf/cc763e0b-38c3-f99f-1f9c-81fa5f3d3850 (accessed on 10 January 2021).
19. Mead, D.; Ransom, K.; Reed, S.; Sager, S. The impact of the COVID-19 pandemic on food price indexes and data collection. *Mon. Labor Rev. US Bur. Labors Stat.* **2020**, *143*, 1–12.
20. Ebrahimy, E.; Igan, D.; Peria, S. The Impact of COVID-19 on Inflation: Potential Drivers and Dynamics. *International Monetary Fund*. 2020. Available online: https://www.imf.org/~{}/media/Files/Publications/covid19-special-notes/en-special-series-on-covid-19-the-impact-of-covid-19-on-inflation-potential-drivers-and-dynamics.ashx?la=en (accessed on 8 January 2021).
21. Eurostat. Harmonised Index of Consumer Prices (HICP). *Eurostat*. 2020. Available online: https://ec.europa.eu/eurostat/web/hicp/data/database (accessed on 9 December 2020).
22. Rubene, I. Recent Developments in Euro Area Food Prices. *European Central Bank*. 2020. Available online: https://www.ecb.europa.eu/pub/economic-bulletin/focus/2020/html/ecb.ebbox202005_07~{}174eeeb845.en.html (accessed on 25 July 2020).
23. Comisión Europea; Unión Europea de la Salud. Proteger la Salud de los Europeos y Responder Coelctivamente Ante las Crisis Sanitarias Transfronterizas. *Comisión Europea*. 2020. Available online: https://ec.europa.eu/info/strategy/priorities-2019-2024/promoting-our-european-way-life/european-health-union_es (accessed on 15 January 2021).
24. Hansun, S.; Charles, V.; Indrati, C.R. Revisiting the holt-winters' additive method for better forecasting. *Int. J. Enterp. Inf. Syst. IJEIS* **2019**, *15*, 43–57. [CrossRef]
25. OECD. Cities Policy Responses. Tackling Coronavirus: (COVID-19) Contributinf to a Global Effort. 2020. Available online: https://read.oecd-ilibrary.org/view/?ref=126_126769-yen45847kf&title=Coronavirus-COVID-19-Cities-Policy-Responses&_ga=2.254320782.746211555.1620810494-1342489524.1620810494) (accessed on 14 May 2021).
26. Fuest, C.; Neumeier, F.; Peichl, A. Has the reduction in value-added tax stimulated consumption? *IFO Schnelld. Digit.* **2021**, *2021*, 1–6.
27. European Commission. Coronavirus: Commission Adopts Package of Measures to Further Support the Agri-Food Sector. *European Commission*. 2020. Available online: https://ec.europa.eu/commission/presscorner/detail/en/ip_20_788 (accessed on 15 May 2021).

foods

MDPI

Article

A Detailed Characterisation of Appetite, Sensory Perceptional, and Eating-Behavioural Effects of COVID-19: Self-Reports from the Acute and Post-Acute Phase of Disease

Nora Chaaban, Alexander Teymour Zadeh Baboli Høier and Barbara Vad Andersen *

Food Quality Perception and Society, Department of Food Science, Aarhus University, Agro Food Park 48, DK-8200 Aarhus N, Denmark; ncho_on@hotmail.com (N.C.); alexanderhoier@gmail.com (A.T.Z.B.H.)
* Correspondence: barbarav.andersen@food.au.dk; Tel.: +45-30707860

Abstract: Sensory perception alterations are common in relation to COVID-19 disease, but less is known about the characteristic of the sensory alterations, and how they associate with alterations in appetite and eating behaviour. The current study aims to investigate the acute and long-term effects of COVID-19 disease on (1) the desire for food, hunger, and satiety sensations; (2) smell, taste, and flavour perception; (3) meals and intake of food types; and (4) the frequency of commonly applied strategies to tackle potential changes in appetite and sensory perception. An online survey was conducted among Danish adults (n = 102) who had experienced changes in appetite, sensory perception, and/or food-related pleasure due to COVID-19 disease. Key results include appetite-altering effects at all times during the day when suffering from COVID-19 and often associated with impaired sensory function. Severe sensory perception alterations were found, namely, for the perception of taste, ageusia > hypogeusia > hypergeusia, and for the perception of smell, anosmia > parosmia > hyposmia > hyperosmia. Eating behavioural changes included alteration in quantitative and qualitative aspects of intake. The effects were, in general, more pronounced during the acute phase of disease than during the post-acute phase. The findings illustrate the complexity by which COVID-19 affects human appetite, sensory perception, and eating behaviour, but also point to strategies to cope with these changes.

Keywords: COVID-19; appetite; sensory perception; eating behaviour; self-reports

Citation: Chaaban, N.; Høier, A.T.Z.B.; Andersen, B.V. A Detailed Characterisation of Appetite, Sensory Perceptional, and Eating-Behavioural Effects of COVID-19: Self-Reports from the Acute and Post-Acute Phase of Disease. *Foods* **2021**, *10*, 892. https://doi.org/10.3390/foods10040892

Academic Editor: Mari Sandell

Received: 26 March 2021
Accepted: 16 April 2021
Published: 19 April 2021

1. Introduction

The most common symptoms of COVID-19 disease are fever, cough, and fatigue [1,2]. Less common symptoms include sore throat and loss of smell (anosmia) and taste (ageusia), amongst other symptoms. Initially, anosmia and ageusia were not considered symptoms of COVID-19 disease. However, since the beginning of the COVID-19 pandemic, the number of patients reporting either a total or partial loss of smell and taste in connection with a SARS-CoV-2 infection has increased. Sensory dysfunction has been documented in studies utilizing (i) self-report questionnaires, namely, the National Health and Nutrition Examination Survey [3,4], the Sino-Nasal Outcome Test [5], the short-version Questionnaire of Olfactory Disorders [3,4], and questionnaires designed for the specific study in focus, e.g., [6–8], and (ii) objective testing, namely, taste-recognition test [8,9], smell-identification test [10–12] and smell-threshold test [9]. Now (April 2020), anosmia and ageusia are on the Danish health authorities' list of common long-term effects of COVID-19 affecting people of all age groups and health, and regardless of the severity of COVID-19 illness [13]. It is estimated that around 10% of those who contract COVID-19 experience long-term symptoms (lasting for more than four weeks = ongoing symptomatic COVID-19). The proportion of those who continue to show symptoms after 12 weeks (post-COVID-19 syndrome) is still not known. Since COVID-19 is still a relatively new disease, knowledge of ongoing symptomatic COVID-19 and the post-COVID-19 syndrome is needed. The majority is expected to recover on their own, although some more slowly than others and without knowing if the symptoms will be chronic for some [13].

Although the loss of smell and taste are not critical to our health per se, perception of sensory properties is closely related to well-being since sensory perception is greatly linked to eating enjoyment, e.g., [14,15], memory recall [16], mood [17–19], etc. Therefore, loss of taste and smell can have a profound impact on people's quality of life. Further, COVID-19 has been reported to be accompanied by a loss of appetite via a decreased motivation to initiate eating (desire for food) and continue eating (general hunger and enjoyment) [20]. A healthy appetite is, in general, important when recovering from diseases since nutritional choices can affect the body's ability to prevent, fight, and recover from infections [21]. Further, good nutrition can help regain strength and reduce the likelihood of developing other health-related problems. If a lack of appetite lasts for more than a couple of days, it can cause weight loss and, in the longer term, malnutrition.

Changes in appetite due to COVID-19 have been shown to be long lasting [20], posing a risk on the quality of life and increasing the risk of malnutrition, which can negatively affect recovery. Current studies of the appetite- and sensory-related effect of COVID-19 have mainly focused on studying self-reported olfaction perception [3,5,6,8] and the biological mechanisms driving olfaction dysfunction [22–24]. A detailed characterisation of the appetite- and eating-behavioural changes caused by COVID-19 will broaden our understanding of the severity of the disease from a patient perspective and can help health professionals to qualify recommendations for proper nutrition during treatment and recovery of patients.

The overall objective of this study was to investigate the effect of COVID-19 disease on self-reported appetite, sensory perception, and eating behaviour in a Danish population. Specifically, the acute and long-term effects of COVID-19 on (1) the desire for food, hunger, and satiety sensations; (2) perception of smell, taste, and flavour; and (3) meals and intake of food types. A final aim was to study the frequency of commonly applied strategies to tackle potential changes in appetite and sensory perception.

2. Materials and Methods

2.1. Participants and Recruitment

Prior to data collection, the Central Denmark Region Committees on Health Research Ethics approved the study being conducted. Data were collected via an online survey running over a period of one month, November 2020. Participants (n = 112) were recruited from several Danish COVID-19 groups on Facebook (one group for COVID-19 patients suffering from long-term effects, and two general COVID-19 information groups) and through posts on LinkedIn and Twitter. Inclusion criteria for the study were Danish-speaking adults (above 18 years old) who had experienced changes in appetite, sensory perception, and/or food-related pleasure due to COVID-19 disease. The latter criterion was chosen to facilitate a detailed characterisation of appetite, sensory perceptional, and eating-behavioural effects of COVID-19.

Participants completing the survey were at different stages of recovery from COVID-19 disease: (i) in the acute phase being currently ill, (ii) in the post-acute phase yet still suffering from long-term effects, and (iii) in the post-acute phase and fully recovered from COVID-19 disease. This paper focuses on the results from participants in the second group, i.e., those who were in the post-acute phase yet still suffering from long-term effects of COVID-19 disease. In total, 102 participants were included in the data analysis. Table 1 shows the characteristics of the included participants. The majority of the participants were diagnosed with SARS-CoV-2 infection through a throat swab (71%). Other diagnosis methods were antibody test (11%), doctor's assessment (10%), own assessment (4%), and other (5%). A study by Parma et al., focusing on the effect of COVID-19 on sensory impairments, found no difference in results when comparing the type of diagnosis, lab tests, and clinical assessments [8]. Participants mostly reported to experience the acute phase of COVID-19 to last between 1–4 weeks; less than 1 week (17%), between 1–2 weeks (33%), between

2–4 weeks (33%), 4–6 weeks (9%), 6–8 weeks (3%), more than 8 weeks (1%), and not stated (4%). In addition to changes in appetite, sensory perception, and eating behaviour, participants were also asked to report other symptoms experienced during the acute phase by evaluating a list of the most common symptoms of COVID-19 disease collected from a systematic review and meta-analysis study [2]. Among these symptoms, participants in the present study reported the following symptoms as the most common (only 5% and above are mentioned here): dizziness (5%), blurry vision (5%), headache (7%), difficulty in concentrating (8%), loss of taste (10%), fever (11%), and loss of smell (11%). The duration of the experienced long-term effects from COVID-19 was not explored in the present study.

Table 1. Participant characteristics.

Characteristics	
Total number (n)	102
Males/females	14/88
Age (years)	41 ± 12.9 (19–69) *
BMI (kg/m^2)	25.7 ± 5 (17.8–42.9) *
Educational level (min–max) **	4.7 ± 1.3 (2–6) *
Inhabitants/household (number of persons)	3 ± 1.4 (1–8) *

* Mean ± standard deviation (range). ** Educational level: (1) lower secondary, (2) higher secondary, (3) higher secondary with trainee, (4) short-length higher education, (5) medium-length higher education, and (6) long higher education.

2.2. Online Survey

The online survey focused on self-reported changes in appetite, sensory perception, and eating behaviour, as experienced in the acute and post-acute phase due to COVID-19 disease. In this study, the post-acute phase was defined as the phase in which the COVID-19 patients had not fully recovered and thereby were suffering from long-term effects, i.e., changes in appetite, sensory perception, and/or food intake. Table 2 provides a list of response variables included in the questionnaire. Throughout the survey, the participants were asked to compare their experiences during both the acute and post-acute phase with how they felt before COVID-19. By conducting this method, the effect of COVID-19 on appetite, eating behaviour, and sensory perception was explored from a subjective perspective. Appetite- and satiety-related response variables were selected from a list of mental and physical sensations developed by Murray and Vickers [25], response variables covering sensory perception were chosen to provide an overview of potential taste, orthonasal, retronasal, off-flavour, and chemesthesis perceptional alterations, eating behavioural related response variables were selected based on subjective reports of COVID-19 patients from a recent qualitative study [20], and finally, response variables regarding strategies to cope with potential changes in appetite were selected based on results from a qualitative study on the topic [20]. The response variables were evaluated by using a three- or five-point categorical scale, with the ends indicating opposite extremes depending on the type of question. The number of categories, e.g., regarding the level of detail in intensity ratings of a sensation, rely on recommendations by previous studies and an evaluation of the appropriate reflection level required of the participants. To guide this decision, participants' replies in a qualitative study were reviewed [20]. Additionally, a "do not know/not relevant" option was included. For questions with multiple possible answers, e.g., symptoms of COVID-19 during the acute phase, a "check all that apply" format was used. Furthermore, for more specific and subjective questions, such as personal experiences with certain foods, an open-reply field was used. These open-reply fields allowed the participant to elaborate on replies and allowed a check of reliability. In the case the information provided in the open-reply field indicated that the participant has misunderstood the question, the data were removed from the analysis.

Table 2. Response variables used in the online survey.

Response Variable	During the Acute Phase	During the Post-Acute Phase
Desire for food	'During the acute phase, how large was your desire for food compared to before COVID-19?'	'While you are in the post-acute phase, how large has your desire for food been recently compared to before COVID-19'
Hunger	'Indicate how COVID-19 affected following hunger sensations, x, during the acute phase (compared to before COVID-19)' x = 'desire to eat',' stomach churning', 'empty stomach feeling', 'stomach pain', 'lack of energy', 'thoughts circulating around food' and 'shaking sensation'	'Now that you are in the post-acute phase, how will you assess following hunger sensations, x, compared to before COVID-19?' x = 'desire to eat', 'stomach churning', 'empty stomach feeling', 'stomach pain', 'lack of energy', 'thoughts circulating around food' and 'shaking sensation'
Satiety	'Indicate how COVID-19 affected following satiety sensations, x, during the acute phase (compared to before COVID-19)' x = 'general satiety', 'post-meal satisfaction', 'feeling bloated', 'heavy stomach feeling', 'nausea', 'energetic' and 'difficulty breathing'	'Now that you are in the post-acute phase, how will you assess following satiety sensations, x, compared to before COVID-19?' x = 'general satiety', 'post-meal satisfaction', 'feeling bloated', 'heavy stomach feeling', 'nausea', 'energetic' and 'difficulty breathing'
Taste perception	'During the acute phase, how did you experience the intensity of the x taste?' x = 'sweet', 'salty', 'sour' and 'bitter'	'How are you experiencing the intensity of the x taste lately?' x = 'sweet', 'salty', 'sour' and 'bitter'
Retronasal odour perception	'Did COVID-19 change your ability to perceive flavours?' 'How did the changes ability of perceiving flavour affect your desire for eating?'	
Off-flavour perception	'Did COVID-19 cause any off-flavours in your mouth?' 'How did these off-flavours affect your desire for eating?'	
Orthonasal odour perception	'Did COVID-19 change your ability to perceive odours?' 'How did the changes in the ability of perceiving odours affect your desire for eating?'	
Chemesthetic perception *	'Did COVID-19 cause any changes in the sense of touch during food intake?' 'How did these feelings affect your desire for eating?'	
Quantitative food intake	The participants were asked to indicate the portion size of their daily meals (x) during the acute phase compared to before COVID-19. x = 'breakfast', 'pre-lunch snack', 'lunch', 'afternoon snack', 'dinner', 'late night snack'	The participants were asked to indicate the portion size of their current daily meals (x) compared to before COVID-19. x = 'breakfast', 'pre-lunch snack', 'lunch', 'afternoon snack', 'dinner', 'late night snack'
Qualitative food intake	The participants were asked to indicate to what extent a certain food and type of food, x, were part of their diet during the acute phase compared to before COVID-19. x = 'vegetables', 'fruits', 'bread and cereal', 'pasta, rice and potato', 'meat, meat products and poultry', 'seafood', 'dairy products', 'eggs', 'juice and soda', 'coffee and tea', 'water', 'salty snacks' and 'sweet snacks'	The participants were asked to indicate to what extent a certain food and type of food, x, are part of their diet currently compared to before COVID-19. x = 'vegetables', 'fruits', 'bread and cereal', 'pasta, rice and potato', 'meat, meat products and poultry', 'seafood', 'dairy products', 'eggs', 'juice and soda', 'coffee and tea', 'water', 'salty snacks' and 'sweet snacks'

* Defined as food-caused chemesthetic sensations in the mouth and/or gastrointestinal region.

2.3. Data Analysis

Mean and standard error of the mean (SEM) were calculated for each question, and the number of answers in each answer option was counted. Data were illustrated in either bar charts based on the mean or as a stacked bar chart showing the distribution of answers. Student's T-test was used to analyse significant differences between the acute and post-acute phases. For questions with three answer options, e.g., hunger and satiety sensations and food intake, Fisher's exact test was used to determine significant differences in frequency of use of answer options between the acute phase and the post-acute phase. Lastly, descriptive statistics were used to summarise the characteristics of the participants, which are shown in Table 1, and the weight and height were used to calculate BMI. For all analyses, XLSTAT (version 2020.5.1, Addinsoft SARL, Paris, France) was used with a significant level of α = 0.05.

3. Results

3.1. Appetite

3.1.1. Desire for Food

In the acute phase, the majority of the participants (86%) experienced a decreased to a highly decreased desire for food, compared to before COVID-19 (Figure 1). In the post-acute phase, it was found that around 57% of the participants still experienced a decreased to a highly decreased desire for food; however, some participants (37%) experienced that their desire for food returned to as before COVID-19 (Figure 1). Further, in the post-acute phase, a small percentage of the participants (6%) experienced an increased to highly increased desire for food. Generally, the desire for food was significantly higher ($p < 0.0001$) in the post-acute phase, compared to the acute phase.

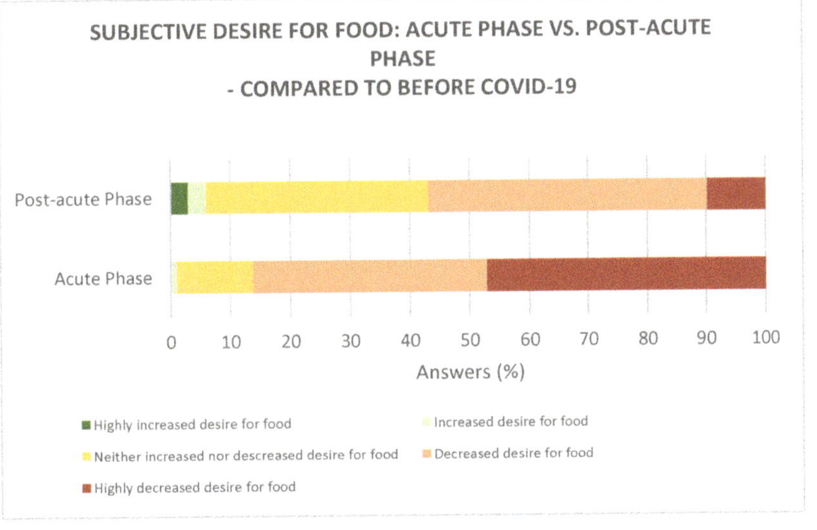

Figure 1. Subjective desire for food ratings in the acute phase ($n = 102$) and in the post-acute phase ($n = 102$), compared to before COVID-19.

Among the symptoms reported to cause changes in the *desire for food* in the two phases (acute phase/post-acute phase), the most common were loss of smell (16%/17.2%), loss of taste (16%/18.2%), food not being pleasurable (11.7%/10.6%), food tasting bad or different (9.5%/14.1%), lack of hunger sensations (11.2%/10.1%), fatigue (9.5%/8.1%) and nausea (9%/7.1%). Note that only causes above 5% are mentioned.

3.1.2. Time of Day

Overall, during the acute and post-acute phases, the majority (ranging from 70.7% to 82.7% in the acute phase and 52–80.5% in the post-acute phase) of participants experienced a decreased to a highly decreased desire for food at all times during the day, compared to before COVID-19 (Figure 2). Focusing on the meals for which participants reported the highest appetite, desire for food, was generally reported to be highest around the three main meals—in the morning, at lunchtime, and in the evening—but only during the post-acute phase. The desire for food was found to be significantly higher at any time of the day in the post-acute phase, compared to the acute phase; in the morning ($p < 0.001$), in the forenoon ($p < 0.001$), at lunchtime $p < 0.001$), in the afternoon ($p < 0.001$), in the evening ($p < 0.001$), and at late night ($p < 0.001$).

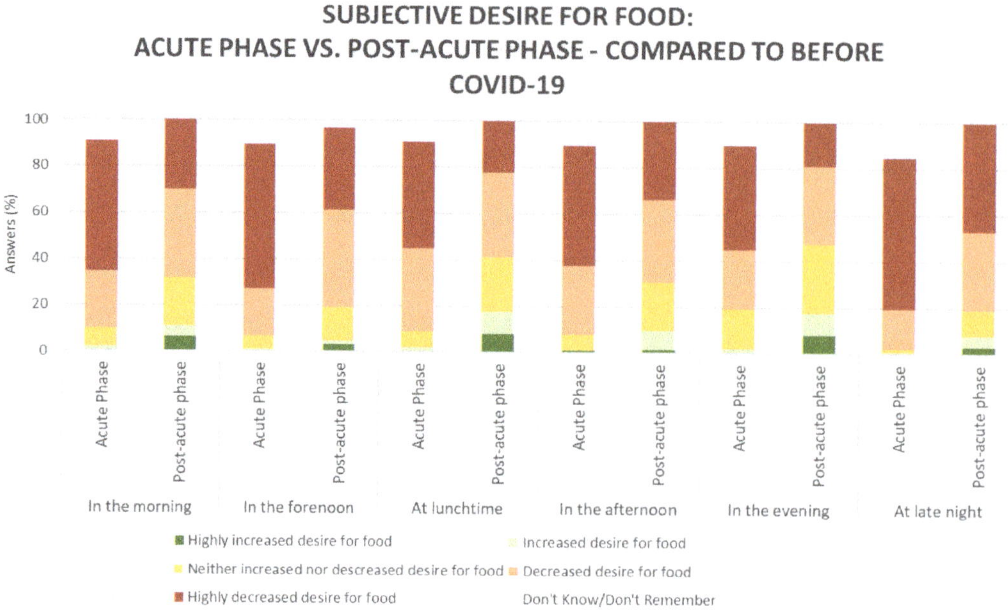

Figure 2. Subjective desire for food according to time of day during the acute phase ($n = 89$) and post-acute phase ($n = 63$) compared to before COVID-19.

3.1.3. Hunger and Satiety Sensations

The following hunger sensations were explored: desire to eat, stomach churning, empty stomach feeling, stomach pain, lack of energy, thoughts circulating around food, and shaking sensation. Regarding specific hunger sensations, the majority of participants reported 'less often' to feel a desire to eat, both during the acute phase (76%) and post-acute phase (53%) of COVID-19, compared to before COVID-19 (Figure 3), and less often to experience that their thoughts circulated around food (in the acute phase only, 58%). The sensations slowly started to return to normal in the post-acute phase. This was found since a significantly higher number of participants ($p = 0.001$, 76% vs. 53%) reported to feel a desire to eat 'less often' in the acute phase, and a significantly higher number of participants felt a desire to eat 'as often as before COVID-19' in the post-acute phase ($p < 0.0001$, 42% vs. 16%). The same tendency was found for feeling that thoughts circulated around food—a significantly higher number reported to feel the sensation 'less often' in the acute phase ($p = 0.003$, 58% vs. 36%) and a significantly higher number reported to feel the sensations 'as often as before COVID-19' in the post-acute phase ($p < 0.001$, 46% vs. 21%).

In both phases, the majority of participants (74% in the acute phase and 61% in the post-acute phase) reported 'more often' to feel a lack of energy, compared to before COVID-19. A significantly higher number of participants reported feeling a lack of energy 'as often as before COVID-19' in the post-acute phase than in the acute phase ($p = 0.004$, 28% vs. 11%), indicating that this sensation slowly returned to normal.

Regarding the physical hunger sensations, stomach churning, empty stomach feeling, stomach pain, and shaking sensation, the majority of participants reported feeling these sensations 'as often as before COVID-19' both in the acute phase (ranging from 31% to 32%) and post-acute phase (ranging from 50 to 53%). Comparing the two phases, a significantly higher number of participants reported feeling stomach churning, empty stomach, stomach pain, and shaking sensations 'as often as before COVID-19' in the post-acute phase, compared to the acute phase ($p = 0.015$, 50% vs. 32%; $p = 0.004$, 53% vs. 32%;

$p = 0.010$, 51% vs. 32%; and $p = 0.010$, 50% vs. 31%, respectively), again indicating that the majority these sensations slowly returned to normal after the acute phase of COVID-19. However, a relatively high percentage of participants (14–36%) showed difficulties remembering these sensations specifically.

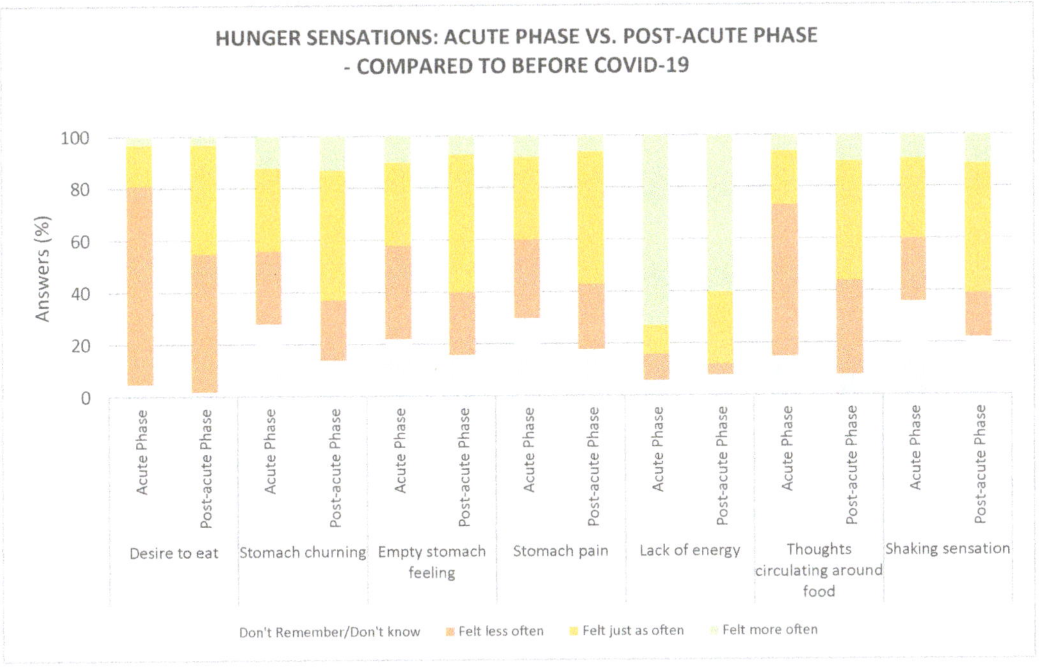

Figure 3. Hunger sensations in the acute phase ($n = 97$) and post-acute phase ($n = 100$), compared to before COVID-19.

The following satiety sensations were explored: general satiety, post-meal satisfaction, feeling bloated, heavy-stomach feeling, nausea, feeling energetic, and difficulty breathing. Regarding the satiety sensations, the majority of participants reported to 'less often' feel satisfied after consuming a meal in both the acute phase (58%) and post-acute phase (54%), compared to before COVID-19 (Figure 4). A significantly higher number of participants ($p = 0.025$) showed difficulties in remembering this sensation in the acute phase (17%), compared to the post-acute phase (6%). Additionally, the majority of the participants reported feeling energetic 'less often' in the acute phase (45%), and many reported the same during the post-acute phase (40%), compared to before COVID-19. This sensation was found to slowly return to normal since a significantly higher number of participants ($p = 0.039$) in the post-acute phase reported feeling energetic 'just as often as before COVID-19' when compared to the acute phase (43% vs. 28%). A significantly higher number of participants ($p = 0.028$) showed difficulty remembering this sensation in the acute phase (25% vs. 12%).

In both phases, participants reported feeling a general satiety 'as often as before COVID-19' (44% vs. 57%, respectively). Comparing the two phases, no significant difference was found.

Compared to before COVID-19, participants in both phases reported to 'more often' feel the following physical satiety sensations: bloated, heavy stomach, nauseous, and having difficulty in breathing. A significantly higher number of participants reported feeling 'just as often' bloated ($p = 0.047$, 51%) and having a heavy stomach (0.042, 46%) in the post-acute phase, compared to the acute phase (36% vs. 31%, respectively), indicating these sensations returning to normal. Further, a significantly higher number of participants

(p = 0.004) showed difficulty remembering the heavy-stomach sensation in the acute phase (32%), compared to the post-acute phase (14%). Regarding feeling nauseuos and having difficulty breathing, no significant difference was found between the acute and post-acute phases.

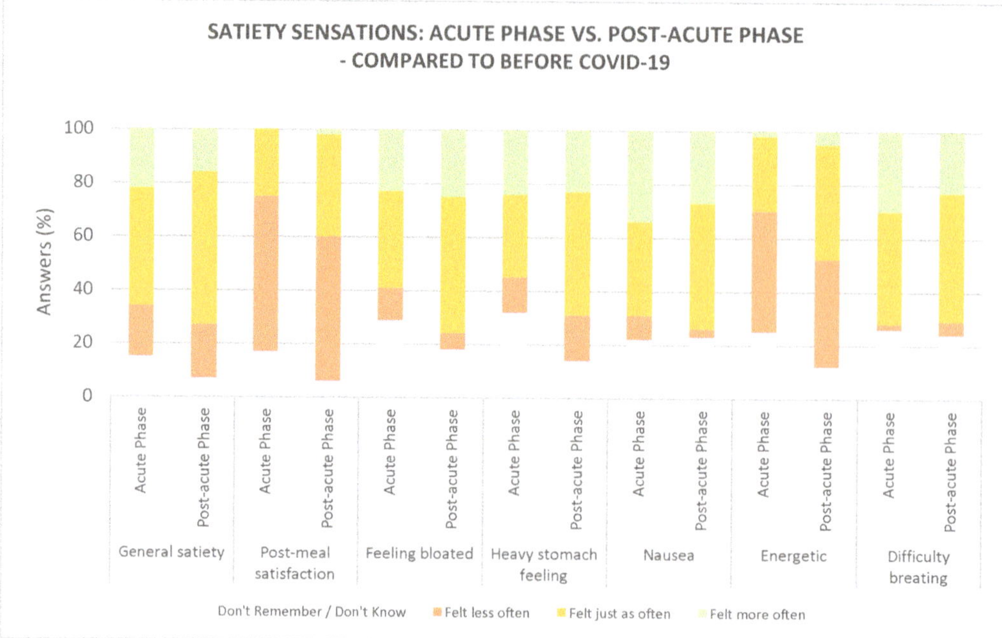

Figure 4. Satiety sensations during the acute phase (n = 102) and post-acute phase (n = 102), compared to before COVID-19.

As with the hunger sensations, a relatively high percentage of participants (12–32%) showing difficulties in remembering, especially the following sensations, was found: feeling bloated, heavy-stomach feeling, nausea, feeling energetic, and difficulty breathing.

3.2. Sensory Perception

3.2.1. Basic Taste Perception and Intake of Food with a Dominant Basic Taste

By far, the majority of the participants enrolled in the study reported changes in basic taste perception during the acute phase of COVID-19 as follows: sweet 91%, salty 89%, sour 86%, and bitter 87% (Figure 5). Among the participants who experienced changes in the ability to perceive the basic tastes, around half of the participants reported a total loss in the ability to perceive sweet (49%), salty (46.6%), sour (52.7%), and bitter (60%) tastes during the acute phase, compared to before COVID-19. The remaining participants mainly reported a 'reduced to highly reduced' ability to perceive the basic tastes, i.e., sweet (33.3%), salty (37.3%), sour (29.7%), and bitter (22.6%). During the post-acute phase, the majority reported a 'reduced to highly reduced' basic taste perception, i.e., sweet (57.2%), salty (56%), sour (44.5%), and bitter (45.3%), rather than a total loss, i.e., sweet (5.3%), salty (5.3%), sour (13.5%) and bitter (14.6%). Only a small percentage of the participants reported an 'increased to highly increased' sensitivity towards the basic tastes, which was found both in the acute phase, with sweet (8%), salty (5.32%), sour (4%), and bitter (4%), and post-acute phase, with sweet (16%), salty (5%), sour (5%), and bitter (12%). Overall, more of the participants reported a normalised ability to perceive taste in the post-acute phase, compared to the acute phase. This was reflected as a significantly higher ability to perceive sweet ($p < 0.001$), salty ($p < 0.001$), sour ($p < 0.001$), and bitter ($p < 0.001$) in the post-acute phase, compared to the acute phase.

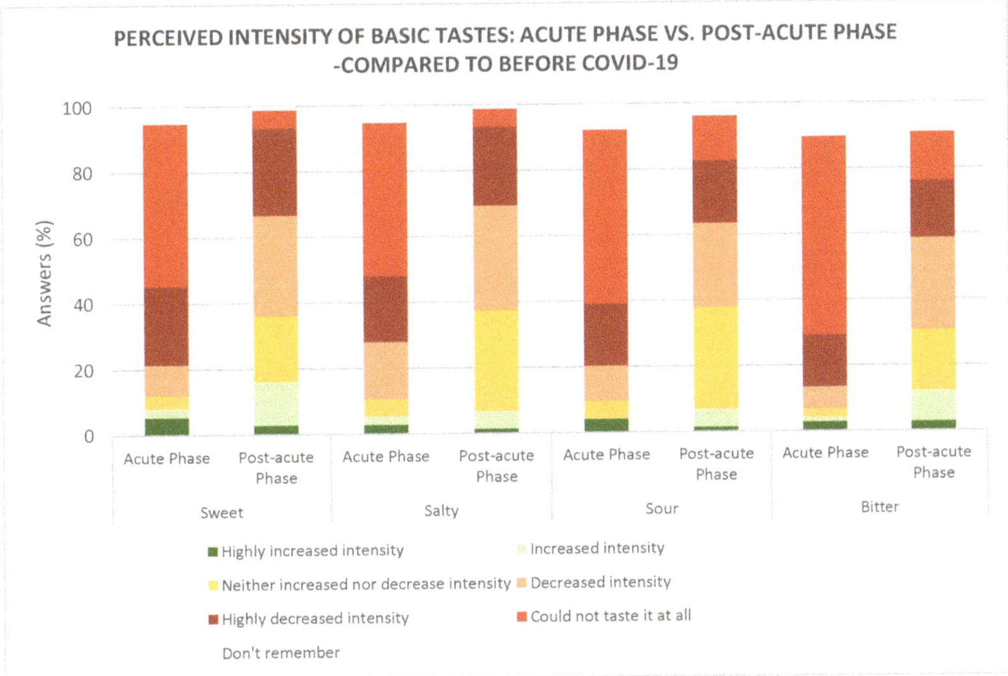

Figure 5. The perceived intensity of the basic tastes during the acute phase ($n = 75$) and post-acute phase ($n = 75$), compared to before COVID-19.

Among the participants who experienced a change in the perception of basic tastes, the intake of food with a dominant sweet, salty, sour, and bitter taste was evaluated (Figure 6). Regarding intake of food with a dominant sweet, salty, sour, and bitter taste, 58%, 32%, 28% and 34%, respectively, reported to decrease their intake of such foods, whereas 15%, 28%, 9% and 3%, respectively, increased their intake.

3.2.2. Orthonasal and Retronasal Odour Perception

The majority of the participants (92%) experienced changes in the orthonasal odour perception due to COVID-19. Note that the alteration was not specified for the acute phase and post-acute phase, respectively. Among participants reporting alterations, the majority (64%) reported a complete loss, 34% reported that the odour perception was distorted, 14% reported decreased odour perception, and 7% increased odour perception. More than half of the participants experienced changes in the retronasal odour (67%). Among these, 42% of the participants reported a complete loss of retronasal odour perception, 35% reported that food had a different flavour than usual, and 17% reported a decreased retronasal odour perception. Other (6%) reported odour-specific perception, meaning some odours were perceived retronasally, while others were not.

Participants were asked if they experienced being able to influence the ability to perceive odours orthonasally and retronasally. In both cases, the majority of participants reported 'not at all' (orthonasal: 75% and retronasal: 57%), but 14% reported being able to influence orthonasal perception, and 32% retronasal perception (ranging from a 'small' to 'high degree'). To improve the orthonasal perception, participants reported smell training and smelling known scents as the most common strategy. To improve retronasal perception, participants most often added more flavour to meals, e.g., by increasing the number of spices.

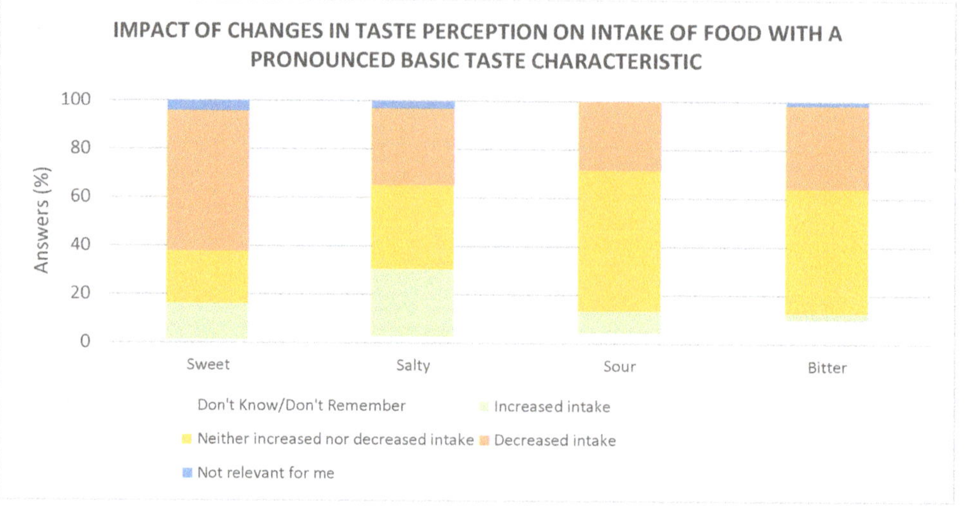

Figure 6. Intake of food with a dominant basic taste after changed perception of basic tastes during the acute phase and post-acute phase, compared to before COVID-19. Sweet (n = 74), salty (n = 72), sour (n = 67), and bitter (n = 70).

While participants experienced changes in odour perception, a reduced desire to eat was reported by the majority of participants; among 75% and 80% of the participants experiencing alterations in retronasal and orthonasal odour perception, respectively (Figure 7a). Furthermore, 51% of participants reported that changes in retronasal perception affected their food choices in a 'high to a very high degree', 21% to a 'certain degree', 19% to a 'lesser degree', while around 4% reported that altered flavour perception did 'not at all' affect food choices. A total of 39% of participants reported that changes in orthonasal odour perception affected their food choices in a 'high to a very high degree', 27% to a 'certain degree', 15% to a 'lesser degree', while around 19% reported that altered orthonasal perception did 'not at all' affect food choices (Figure 7b).

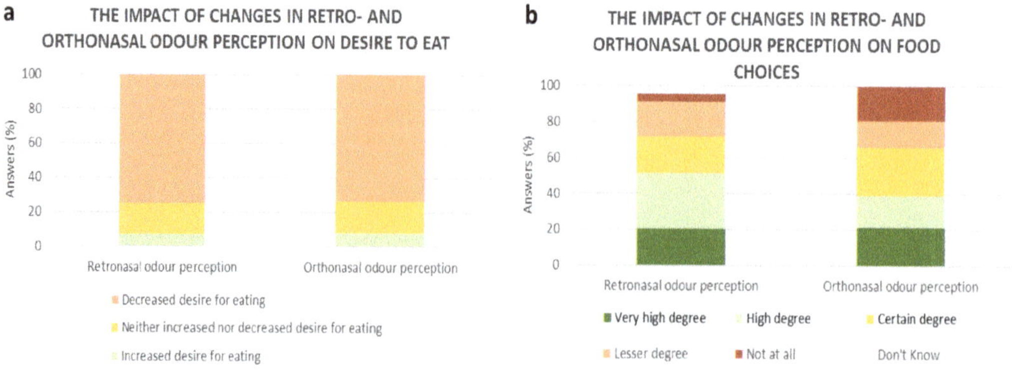

Figure 7. Participants' report on how changes in retronasal (n = 68) and orthonasal (n = 92) odour perception, respectively, impacted (**a**) desire to eat and (**b**) food choices.

3.2.3. Off-Flavour

A total of 56% of participants reported experiencing off-flavours in the terms of either metallic, rotten, smoked, and/or chemical flavours during food intake. Among these participants, 70%, 27% and 3% reported that these off-flavours 'reduced', 'neither increased nor decreased', and 'increased' their desire to eat, respectively. Furthermore, 43% of participants reported that off-flavours affected their food choices in a 'high to a very high degree', 23% to a 'certain degree', 21% to a 'lesser degree', while around 13% reported that off-flavours did 'not at all' affect food choices.

Participants were asked whether they felt they could influence the perceived intensity of off-flavours. A total of 70% reported, 'not at all' being able to affect the perceived intensity. Of the 19% who reported that they could affect the perceived intensity of off-flavours, the most often used strategies included rinsing the mouth with water, brushing teeth, chewing mint gum or liquorice, and avoiding intake of foods which provoked the off-flavour.

3.2.4. Chemesthesis

Participants were asked to evaluate whether COVID-19 caused changes in how food felt in the mouth, throat, or gastrointestinal region. Only 26 out of 102 participants reported having experienced changes in perception related to chemesthesis. From reviewing the comments of these participants, it was found that some participants misunderstood the question. Among the participants who understood the question correctly, the following changes were reported: food felt stinging on the tongue and in the throat, difficulty of food to pass through the throat—especially hard/solid foods, and food causing a burning sensation in the mouth and throat. Further, participants reported becoming more sensitive towards the texture of the food when not being able to perceive taste or odours. Especially foods with a soft texture, e.g., oatmeal, mashed potatoes, and boiled vegetables, were reported to cause nausea, and foods with a harder/crunchier texture, e.g., rye bread, apples, and pears, were reported to be more chewable and thereby more comfortable.

Altered chemesthesis during food intake caused a decreased desire for food by the majority of participants experiencing altered chemesthesis (74%). Furthermore, 52% of participants reported that altered chemesthesis affected food choices in a 'high to a very high degree', 32% to a 'certain degree', 8% to a 'lesser degree', while around 8% reported that altered chemesthesis did 'not at all' affect food choices. The majority of participants reported experiencing these changes 'most of the time' (44%).

3.3. Eating Behaviour

3.3.1. Portion Size of Main Meals and Snacks

The three main meals—breakfast, lunch, and dinner—were eaten by the majority of the participants both during the acute and post-acute phases (Figure 8). Regarding the size of the main meals, the majority reported the meals to be of a 'smaller size' in the acute phase compared to before COVID-19 (breakfast: 46%, lunch: 48%, and dinner: 70%). Comparing the two phases, it was found that the breakfast ($p = 0.02$, 46% vs. 29%) and dinner ($p < 0.001$, 70% vs. 43%) were reported to be of a 'smaller portion size' by more participants during the acute phase, compared to the post-acute phase. The breakfast, lunch, and dinner (all $p < 0.001$, 61% vs. 21%, 51% vs. 17%, 53% vs. 16%, respectively) were reported to be of 'the same size as before COVID-19' by more participants in the post-acute phase, compared to the acute phase. Regarding snack meals eaten pre-lunch, in the afternoon and late night, the majority of participants 'did not remember/did not eat' the snacks in the acute phase (pre-lunch: 71%, afternoon: 66%, late night: 66%). During the post-acute phase, more participants reported eating snacks 'of the same portion size as before COVID-19' than in the acute phase (all snack meals $p < 0.001$).

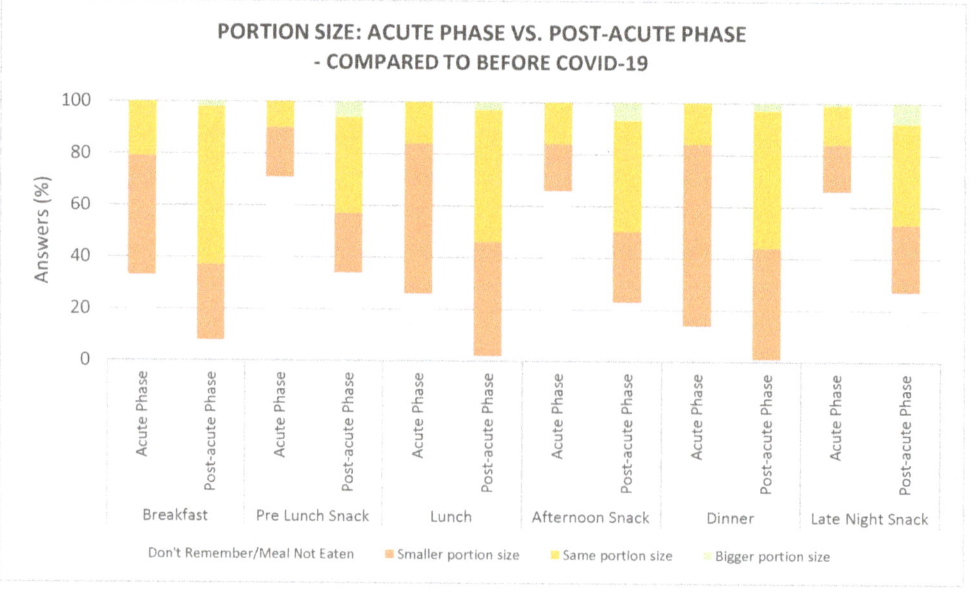

Figure 8. Portion size of meals during the acute phase (n = 102) and post-acute phase (n = 102), compared to before COVID-19.

3.3.2. Type of Diet

Vegetables, fruits, and starchy foods: During the acute phase, between 28% and 33% of participants reported vegetables, fruits, and starchy foods (bread and cereals, rice, potato, and pasta) to constitute 'a smaller proportion of the diet', compared to before COVID-19, and 43–55% reported the foods to constitute 'the same proportion of the diet', compared to before COVID-19 (Figure 9a). Comparing the two phases, a significantly higher number of participants reported vegetables (p = 0.001) and fruits (p = 0.023) to constitute 'a smaller proportion of their diet' during the acute phase (vegetables: 31% and fruit: 33%) than during the post-acute phase (vegetables: 11% and fruit: 18%). A significantly higher number of participants reported vegetables (p = 0.033) and fruits (p = 0.001) to constitute 'the same proportion of their diet' during the post-acute phase (vegetables: 68% and fruit: 67%), compared to the acute phase (vegetables: 52% and fruit: 43%). These results indicate that the intake of vegetables and fruit starts to return to normal during the post-acute phase. No significant difference was found for starchy foods between the two phases.

Meat, seafood, dairy products, and eggs: The majority of participants reported meat, meat products and poultry (42%), seafood (42%), dairy products (32%), and eggs (36%) to constitute a 'smaller proportion of the diet' during the acute phase, compared to before COVID-19 (Figure 9b). A significantly higher number of participants reported a smaller intake of meat, meat products and poultry (p = 0.017), and seafood (p = 0.001) during the acute phase, compared to the post-acute phase (25%, 19%, 19%, and 26%, respectively). For all four food categories, a significantly larger number of participants reported that the food constituted the 'same proportion of diet' during the post-acute phase, compared to the acute phase: meat, meat products and poultry (p = 0.002, 67% vs. 44%), seafood (p = 0.001, 66% vs. 42%), dairy (p = 0.030, 71% vs. 55%) and eggs (p = 0.011, 66% vs. 47%).

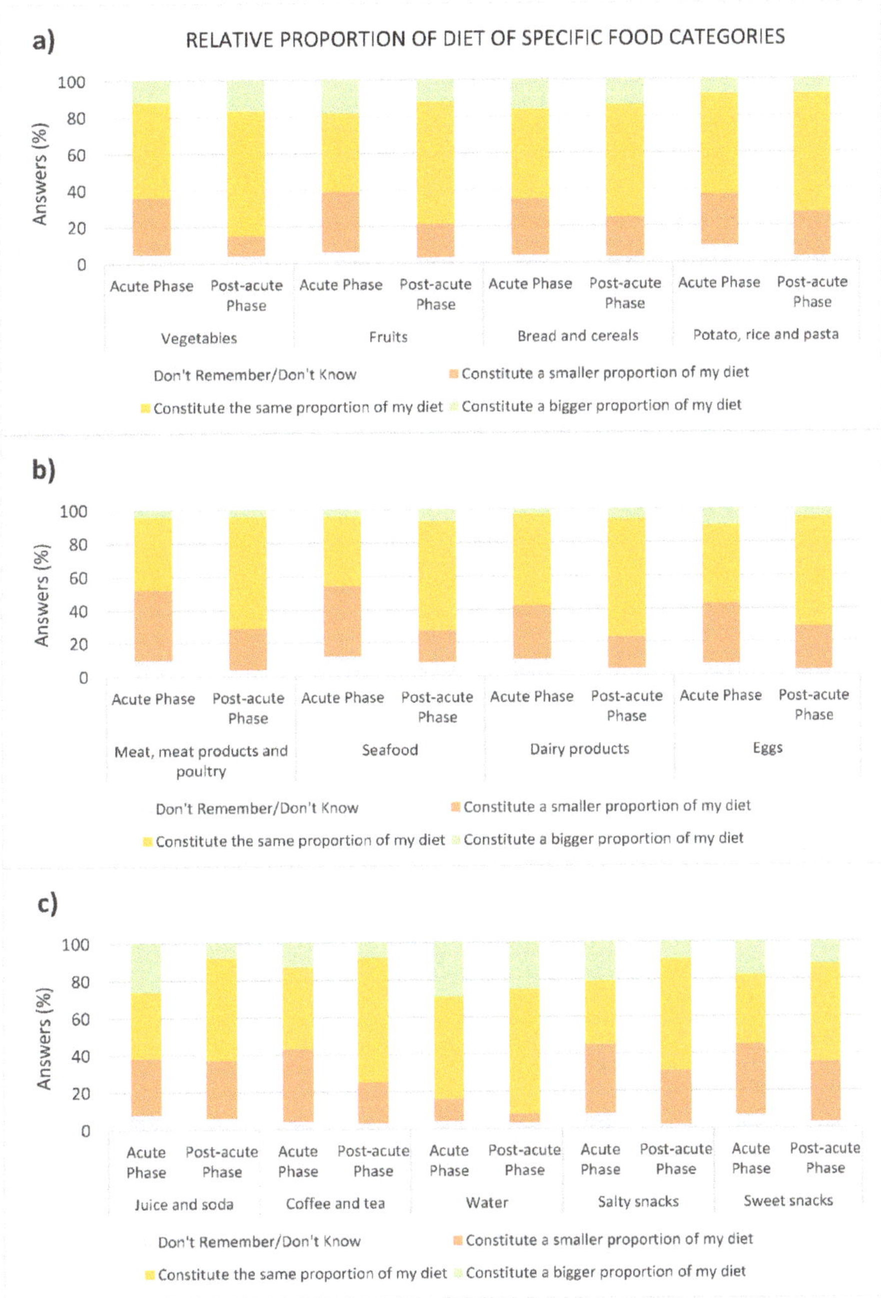

Figure 9. Type of diet during the acute phase (n = 102) and post-acute phase (n = 102), compared to before COVID-19. (**a**) Vegetables, fruits, and starchy foods; (**b**) animal products; and (**c**) drinks and snacks.

Beverages: A total of 27%, 36%, and 30%, respectively, reported juice to constitute 'a bigger', 'the same', and 'a smaller' proportion of the diet in the acute phase of COVID-19, compared to before COVID-19 (Figure 9c). More reported juice intake to be 'larger' during the acute phase, compared to the post-acute phase ($p = 0.002$, 27% vs. 9%), and more reported the intake to be 'the same' in the post-acute phase, compared to the acute phase ($p = 0.011$, 55% vs. 36%). An approximate even number of participants reported coffee/tea to make up 'a smaller' and 'the same' proportion of the diet (39% and 44%, respectively) during the acute phase. The intake started to return to normal during the post-acute phase. This was observed by more participants who reported the intake to constitute 'a smaller proportion of the diet' in the acute phase, compared to the post-acute phase ($p = 0.014$, 39% vs. 22%), and more participants reported the intake to constitute 'the same proportion of diet' during the post-acute phase ($p = 0.001$, 67% vs. 44%). The intake of water was, by the majority, 'the same' in the acute and post-acute phases (55% and 67%, respectively), compared to before COVID-19, and no significant differences were found between the two phases. A total of 30% and 26% reported that their intake of water constituted 'a bigger proportion of the diet' in the acute phase and post-acute phase, respectively.

Salty and sweet snacks: Intake of salty snacks was reported to constitute 'a bigger', 'the same', and 'a smaller' proportion of the diet, compared to before COVID-19 by 23%, 34%, and 37% of participants, respectively (Figure 9c). The number of participants who reported the intake to be 'bigger' was larger during the acute than the post-acute phase ($p = 0.038$, 23% vs. 11%), whereas the number of participants who reported their intake to be 'the same as before COVID-19' was larger during the post-acute phase ($p < 0.001$, 60% vs. 34%). The relative proportion of sweet food was reported to be 'bigger', 'the same', and 'smaller', compared to before COVID-19 by 20%, 37%, and 38% of participants, respectively. The number of participants reporting the intake to constitute 'the same' proportion of the diet was significantly higher in the post-acute phase, compared to the acute phase ($p = 0.024$, 53% vs. 37%), during which around half (53%) reported sweet food to constitute the same proportion of the diet as before COVID-19.

3.3.3. Texture, Temperature, and Preparation Method

Temperature: Regarding the temperature of food, a relatively high percentage of participants reported that warm- (43%), lukewarm- (34%), and cold food (31%) constituted a 'smaller proportion' of the diet during the acute phase, compared to before COVID-19 (Figure 10a). All three temperatures were reported to constitute the 'same proportion' of the diet by the majority of the participants during the post-acute phase (77%, 69%, and 73%, respectively). Comparing the two phases, a significantly higher number of participants reported warm- ($p < 0.001$, 43% vs. 18%) and cold food ($p < 0.001$, 31% vs. 10%) to constitute a 'smaller proportion' during the acute phase, and warm- ($p < 0.001$, 77% vs. 40%), lukewarm- ($p < 0.001$, 69% vs. 41%), and cold food ($p < 0.001$, 73% vs. 44%) to constitute 'the same proportion' in the post-acute phase, compared to the acute phase. A significantly higher number of participants ($p = 0.020$, 19% vs. 7%) showed difficulty in remembering their consumption of lukewarm food during the acute phase, compared to the post-acute phase. Comparing the three temperatures, in both phases, more people reported cold food to 'constitute a bigger proportion of my diet' (acute: 16%, post-acute: 22%) than warm- (acute: 15%, post-acute: 5%) and lukewarm food (acute: 8%, post-acute: 4%). However, also intake of warm food was found to constitute 'a larger part of the diet' in the acute phase for some participants (15%). During the post-acute phase, this number significantly dropped to 5% ($p = 0.032$).

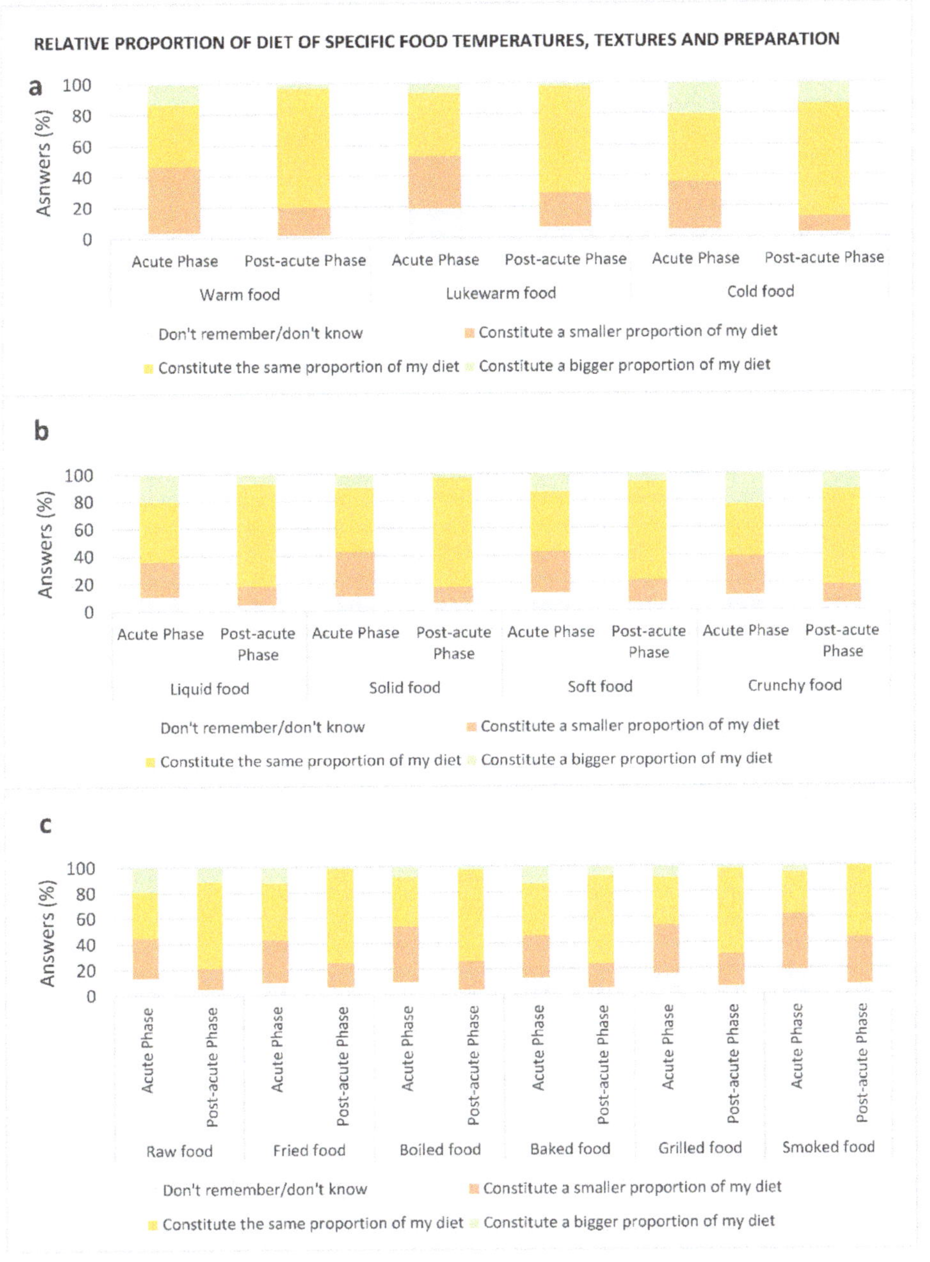

Figure 10. Type of food during the acute phase (n = 102) and post-acute phase (n = 102), compared to before COVID-19. (**a**) Temperature of food, (**b**) texture of food, and (**c**) cooking.

Texture: The majority of the participants reported the textures; liquid (44%), solid (47%), soft (44%), and crunchy (38%) food to constitute 'the same proportion' of the diet during the acute phase, compared to before COVID-19 (Figure 10b). The intake of all types of textures, was slowly returning back to normal in the post-acute phase, during which a significantly higher number of participants reported liquid ($p < 0.001$, 75% vs. 44%), solid

($p < 0.001$, 80% vs. 47%), soft ($p < 0.001$, 72% vs. 44%), and crunchy ($p < 0.001$, 70% vs. 38%) to constitute 'the same proportion' of the diet, compared to before COVID-19. Comparing and observing the texture for which most participants showed an 'increase´, it was found that intake of crunchy and liquid food was increased by the largest number of participants (24% and 21%, respectively) in the acute phase.

Preparation method: Regarding the preparation method chosen, a relatively high percentage of participants reported boiled- (43%), smoked- (43%), and grilled food (38%) to constitute 'a smaller proportion' of the diet during the acute phase, compared to before COVID-19 (Figure 10c). For, the remaining preparation methods—raw, fried, and baked food—31–33% reported the preparation methods to constitute 'a smaller proportion' of the diet. A significant higher number of participants reported the preparation method (beside smoked) to constitute a 'smaller proportion' of the diet during the acute phase, compared to the post-acute phase: raw- ($p = 0.013$, 31% vs. 16%), fried- ($p = 0.024$, 33% vs. 19%), boiled- ($p = 0.002$, 43% vs. 22%), baked- ($p = 0.036$, 33% vs. 19%), and grilled food ($p = 0.048$, 38% vs. 25%). Further, smoked- ($p = 0.028$, 5% vs. 0%) and fried food ($p = 0.016$, 12% vs. 3%) was found to 'constitute a bigger proportion' of the diet during the acute phase, compared to the post-acute phase. During the post-acute phase, the majority of participants reported all preparation methods to constitute 'the same proportion' as before COVID-19, and from the acute phase to the post-acute phase, the number of participants reporting the preparation methods to constitute 'the same proportion' as before COVID-19 increased (raw ($p < 0.0001$, 36% vs. 68%), fried ($p < 0.001$, 45% vs. 74%), boiled ($p < 0.001$, 39% vs. 72%), *baked* ($p < 0.001$, 41% vs. 69%), grilled ($p < 0.001$, 37% vs. 67%), and smoked ($p < 0.001$, 33% vs. 58%)). Comparing the preparation methods and observing the preparation method for which most participants showed an ´increase´, it was found that intake of raw- and baked food was increased by the largest number of participants (20% and 14%, respectively) in the acute phase.

3.4. Handling of Changes in Appetite, Sensory Perception, and Food Behaviour

Participants were asked to report their agreement on trying different strategies for handling changes in appetite, sensory perception, and eating behaviour (Figure 11). The list of strategies was developed based on a qualitative study with patients COVID-19, conducted prior to this online survey [20].

Overall, accepting changes in appetite and sensory perception was the strategy that was used mostly by the participants (56%) during the whole period of COVID-19 disease and recovery. While experiencing changes in appetite and/or sensory perception, the majority of participants agreed that it was important to be able to identify every ingredient in a meal (42%) and to increase their focus on senses that were well functioning (48%). From a food-behavioural perspective, the majority agreed to increase their focus on eating spicy foods (50%), and on eating healthy food (46%), whereas the majority disagreed to increase their focus on delicious foods (47%) and self-preparing foods (49%). When asked about the increased focus on crunchy food and the appearance of food, participants were equally split into three groups indicating to 'agree', 'neither agree nor disagree' and 'disagree' (compared to before COVID-19).

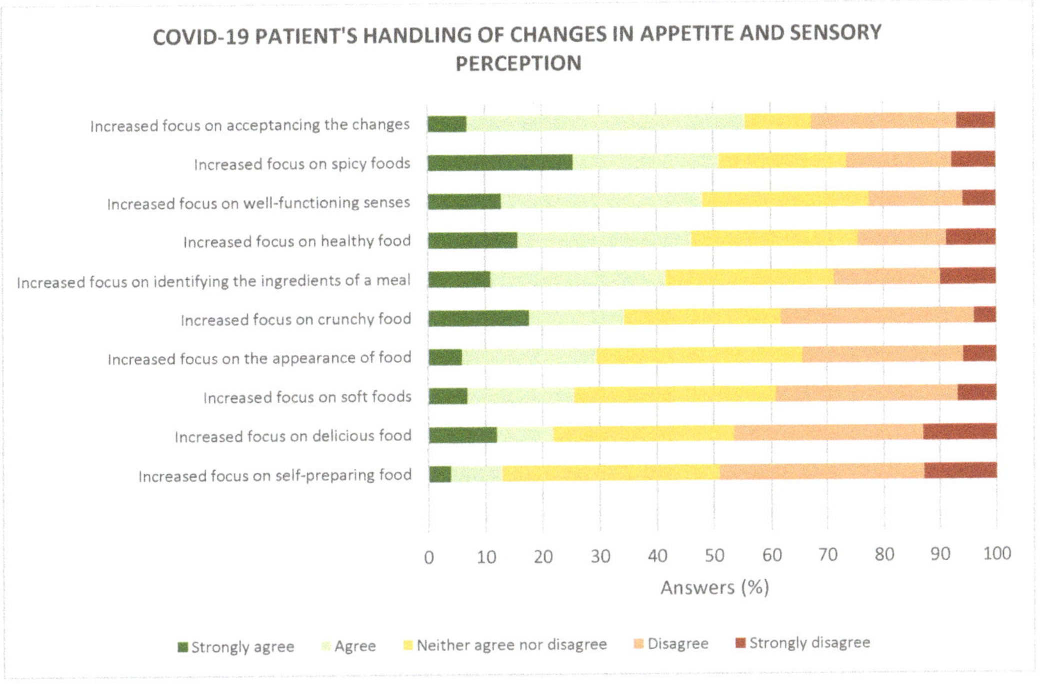

Figure 11. Participants' agreements on strategies for boosting appetite and sensory perception (n = 102).

4. Discussion

4.1. Altered Appetite

Loss of appetite is a well-documented consequence of several illnesses including various influenzas and colds [26], and recently, COVID-19 [3,20,27]. The present study presents confirmatory evidence for the effects of COVID-19 on appetite during the acute phase of the disease. Further, the study contributed to new knowledge by addressing the specific hunger and satiety sensations and by showing how the specific appetite-altering effects of COVID-19 continue into the post-acute phase of the COVID-19 disease. Among the list of symptoms presented in this online survey, lack of hunger sensations was reported to be among the main cause of reduced appetite in both the acute phase and post-acute phase of COVID-19, during which participants reported experiencing less often a desire to eat and to have thoughts circulating around food and more often feeling a lack of energy. Similar findings were found in a qualitative study on appetite among patients showing long-term effects of COVID-19 [20]. Through in-depth interviews, the COVID-19 patients expressed a lack of hunger sensations, compared to before COVID-19, and a faster fullness during the consumption of a meal resulting in reduced food intake [20]. Supporting these findings, the present study likewise found that participants more often experienced overall satiety, characterised by more often feeling bloated, heavy stomach feeling, nauseous, and difficulty in breathing, explaining the reduction in appetite.

Besides the lack of hunger sensations, participants in the current study reported that changes in chemosensory perception, i.e., alterations in taste and smell perception were among the main causes of alterations in appetite. Perception of food's sensory properties is highly linked to hunger and satiety sensations experienced during a meal and therefore play an important role in food intake control [28,29]. During the early stage of a meal, the sensory properties of food generate a positive feedback mechanism, i.e., liking of sensory properties enhance hunger and drive continued intake [30]. During the later stages of a meal, hunger sensations decline, and satiation takes over. Experiencing impairments in the ability to perceive taste and aroma properties can hinder the early stage positive feedback, resulting in a faster decrease in hunger and/or onset of satiation, experienced as a general overall satiation and lack of hunger, explaining the altered appetite by the participants in the present study. Similar results were found in a qualitative study by Høier and colleagues [20], in which COVID-19 patients expressed a lack of interest in food due to food being tasteless and not being able to smell the food while cooking to stimulate appetite. Supporting this view, a study by Merkonidis et al. [31] found that most individuals suffering from chemosensory disorders showed altered eating behaviour, reduced food intake, and/or changed food preferences. The reduced food intake was caused by a lack of sensory cues such as the smell, sight, and taste of food, which normally would motivate eating [31–33]. Further, it was found that the most common complaint among individuals with chemosensory disorders was the loss of pleasure from eating [31]. In line with these results, participants in the present study reported a diminished pleasure from eating and indicated alterations in the taste of food (to be understood as flavour) to be among the main causes.

Although feeling generally satiated, the majority of the participants, at the same time, reported feeling satisfied less often after consumption of a meal. These results can be explained by the phenomenon of sensory satisfaction [15]. Sensory satisfaction describes how the perception of sensory properties during food consumption, fulfils desires in a meal experience and can lead to a feeling of postmeal satisfaction [15,16,34,35]. Losing the ability to perceive the sensory properties of food changes the sensory experience of a meal during consumption, which can lead to the meal being perceived as less satisfying. This phenomenon is related to the hedonic aspects of food intake, which indicates that satiety is driven by both homeostatic factors and the fulfilment of hedonic desires, and the perception of sensory properties plays a causal role in satisfying these desires [35]. In the qualitative study by Høier, Chaaban, and Andersen [20], participants expressed the lack of satisfaction from eating as feeling 'unpleased senses', and for a smaller group, chemosensory dysfunction led to continued eating in order to find foods that could satisfy their sensory desires. Merkonidis et al. [31] likewise approached the association between sensory perception and intake. They found that a smaller percentage of individuals suffering from chemosensory disorders increased their food intake in the search for flavour in their meals. It can therefore be hypothesised that a lack of feeling sensory satisfied drove the increased desire for food, reported by a minority of participants in the present study.

4.2. Sensory Perception

Since the outbreak of the COVID-19 pandemic, several studies aimed to measure the prevalence and severity of chemosensory alterations among COVID-19 patients, either based on self-reporting [3,5,7,8,36] or by objective chemosensory testing [4,9,11,12,37,38]. Suffering from olfactory and gustatory dysfunction is now considered one of the most prevalent symptoms of COVID-19, ranging between 5 and 89% of the patients' complaints [36]. Data from a systematic review on COVID-19 patients showed a high global prevalence of smell (48%) and taste dysfunction (41%) and a combination of both (35%) [39]. The present study contributed to this research by characterising the chemosensory dysfunctions. It should be noted that the percentage of participants reporting chemosensory dysfunction generally is higher in the present study, compared to previous literature, due to the inclusion criteria of this study. Alterations were found for all basic taste attributes (sweet, salty, sour, and bitter) and included a total loss of basic taste perception (ageusia) and altered intensity in taste perception, in the order ageusia (total loss) > hypogeusia (decreased perception) > hypergeusia (increased perception). A study by Parma et al. likewise found impairments in two or more taste qualities among participants suffering from COVID-19 [8].

Alterations in smell perception were characterised by a total loss of smell perception (anosmia) and odours becoming altered and unpleasant (parosmia), in the order of more experiencing anosmia (total loss) > parosmia (distorted) > hyposmia (decreased perception)

> hyperosmia (increased perception). The qualitative changes in sense of smell confirm findings from Parma et al., who likewise reported anosmia, parosmia, and hyposmia as well as phantosmia. Further, in the present study, a total loss of retronasal odour perception and distorted flavour perception was found, which can be regarded as a combined dysfunctional flavor and taste perception.

Previous studies have stated that chemosensory disorders begin during the early stages of COVID-19 disease [36–38], which was confirmed by the present study. Although the loss of smell and taste function are common in acute cold, according to a study by Huart et al. [38], the effects on chemosensory dysfunction are more severe when caused by COVID-19, compared to acute colds, since both smell identification scores and taste function were found to be significantly lower in COVID-19 patients, compared to acute cold patients [38].

When reviewing existing literature, it is indicated that recovery of gustatory and olfactory disorders most often occurs within few weeks after infection, while in some cases, the recovery process will be longer [36,37]. Only participants in the post-acute phase of COVID-19 disease were included in the present study, and therefore, the time period for recovery was not the focus. However, comparing participants' self-reported perception of the basic tastes during the acute and post-acute phases, a slow recovery of taste function was found. Interestingly, more participants reported still suffering from altered ability to perceive sweet and bitter tastes, compared to salt and sour tastes, during the post-acute phase, indicating a slower recovery of sweet and bitter taste perception. Similar findings have been reported by Huart et al. [38], who investigated taste (global, sweet, sour, bitter, salty) and odour perception amongst patients recovering from COVID-19 and acute cold. Focusing on taste perception, they found worse global, sweet, and bitter perception, amongst COVID-19 recovering patients, whereas there were no differences in perception of sour and salt perception.

4.3. Eating Behaviour

COVID-19 disease is believed to put patients at risk of malnourishment since several illness-related factors including nausea, diarrhea, loss of appetite, loss of taste and smell, and stress over time can cause a reduction in food intake and the nutritional value of the diet [39,40]. As discussed previously, the majority of the participants reported a reduction in appetite during the acute phase, compared to before COVID-19. This reduction in appetite was observed at all times of the day—in the morning, in the forenoon, at lunchtime, in the afternoon, in the eve, and at late night (see Section 3.1.2). Participants further reported reductions in food intake due to COVID-19. At all times of day, main and snack meals (i.e., breakfast, pre-lunch snack, lunch, afternoon snack, dinner, and late-night snack) were reported to be of smaller portion size, or not eaten at all/did not remember. A preference towards the three main meals, i.e., breakfast, lunch, and dinner, compared to the snack meals was found, since the majority of the participants reported eating these meals, although a smaller portion size. During the post-acute phase, the desire for food was, in general, also reported to be highest around the three main meals, compared to the snack meals. This was likewise reflected in the portion size since the majority of the participants reported the main meals to be of 'the same size' as before COVID-19, although a high proportion still reported the main meals to be of a smaller size.

These results provide evidence of COVID-19 affecting the quantitative aspect of food intake and also confirm that the desire for food plays an important role in the motivation for food intake. To the authors' knowledge, this study is the first to investigate the effect of COVID-19 on food intake. Although many participants were still suffering from reduced appetite during the post-acute phase, the majority improved their food intake, which aligns with the participants' self-reports of the normalisation of their appetite. It is noticeable that even in the acute phase, participants showed a preference for the three main meals, indicating that the participants were aware of the importance of food in the process of recovery. This result brings opportunities for health professionals, such as dieticians, to focus on the main meals when pushing health-related initiatives and be aware of reduced appetite outside these meals.

In the present study, participants were asked to report on qualitative changes in their diet by evaluating if food categories during the acute and post-acute phases constituted an altered proportion of the diet. As discussed in the previous section, participants, in general, showed a reduced food intake at all times of day during the acute phase of COVID-19. This finding was reflected in an overall reduction in intake of various food categories. Most of the participants reported a normalised intake of all food categories during the post-acute phase.

During the acute phase of COVID-19, the intake of meat, seafood, eggs, coffee/tea, salty and sweet snacks was reduced. Regarding the food category 'meat, seafood, and egg', participants reported disliking these foods due to not being able to perceive the flavour, resulting in a greater awareness of the texture/consistency of this food category, which was not perceived as pleasant. Eating is a multi-sensory experience based on the perception of sensory properties of food, i.e., taste, smell, appearance, and texture. Taste (understood as flavour) has previously been found to be the main driver of hedonic eating experience [28] and more important for the hedonic aspect of food, compared to texture, smell, and appearance [41–43]. The results from the present study highlight the pronounced importance of flavour for food acceptance, in this case, meat, seafood, and egg acceptance specifically, since when taste and retronasal odour perception suffered, participants shifted focus towards the texture as the driver of food acceptance. The shift in focus towards other sensory properties of food when not being able to perceive taste and smell properties has previously been shown among individuals diagnosed with chemosensory disorders [31] and COVID-19 recovering patients [20]. The studies reported that participants obtained food-related pleasure by focusing on chemesthesis via the well-functioning sense of touch. In practice, this was achieved via the perception of the food's texture, especially by adding crunchy elements [20], and trigeminal stimulation by adding spices to food [20,44]. The increased preference for crunchy food was further reflected in the results concerning preparation methods, since an increased intake of raw food was found in the present study. Raw food is generally associated with a crunchier texture than heat-treated food. Soft textures and boiled food which generally is associated with a softer texture were reported to be less preferred during the acute phase. The present study thereby provides confirmatory evidence for focusing on chemesthesis (especially via crunchy and spicy food) to cope with taste and odour alterations, maintain food-related pleasure, and highlight actionable opportunities for individuals and health professionals interested in boosting appetite and eating enjoyment.

In the present study, sweet and salty snacks were found to constitute an altered, most often smaller, part of the diet when suffering from COVID-19, and pointed at altered basic taste perception to be (to some degree) the direct cause of altered intake of food with a dominant sweet, salty, sour, and bitter taste. Individuals experiencing chemosensory disorders commonly report changed eating behaviour [31,45], but mixed results have been reported about the association between intake quality and altered chemosensory function. One study showed a change in diet towards a more Western-style diet—high in fast food, sweet, salty, and/or fats—when suffering from olfactory dysfunction [44]. Other studies have shown a reduced intake of sweet and fatty food [46,47] and in one study also salty food [48] with a preference towards fruits and vegetables, indicating a shift towards healthier food choices. The mixed results per se, regarding changes in qualitative aspects of the diet when suffering from chemosensory dysfunction, have likewise been pointed out previously. The study by Høier, Chabaan, and Andersen [20] suggested that individuals suffering from chemosensory dysfunction can be divided into two groups—a smaller group reporting to increase intake of unhealthy food since these foods reminded them of past pleasurable experiences upon consumption of these foods, and a larger second group, reporting to focus on healthy eating since this was associated with higher mental well-being when providing the body with beneficial nutrients. The indication of two groups, showing opposite eating behavioural changes, was supported by the present study; 20% and 23% reported to increase their intake of sweet and salty snacks, respectively, and 38% and 37% reported to decrease their intake of sweet and salty snacks, respectively, as a consequence of COVID-19.

Finally, the study provides confirmatory evidence of off-flavours occurring during COVID-19 disease. Approximately half of the participants in the present study reported experiencing off-flavours, supporting previous findings by Høier et al. [20]. The off-flavours were characterised as metallic, rotten, smoked, and/or chemical flavours during food intake and were reported to lower participants' desire to eat and to some extend also to impact food choices. The characteristics of the off-flavours can explain why a relatively high percentage of participants reported smoked- (43%) and grilled food (38%) to constitute 'a smaller proportion' of the diet.

4.4. Limitations

The present study constitutes an online survey based on subjective reports. This approach offers some possibilities along with some limitations. A limitation when stating findings from the acute phase of COVID-19 is that the outcome is strongly dependent on participants' memory. Since all participants were in the post-acute phase of COVID-19 when conducting the study, retrospection was necessary but may not always be accurate. Asking multiple questions to the same response variable would allow a check of reliability. In the present study, one question per response variable was chosen under consideration of the length of the questionnaire and to avoid fatigue. The use of open-reply fields after each topic allowed the researchers to check if participants had understood the questions. Further, although the patient perspective is indeed relevant since sensations, in many cases, drive human behaviour, the findings could preferably be supported by objective means. Sensory perception could be validated, e.g., by the use of threshold tests, and eating behaviour could be validated by the use of dietary records.

Since the study aimed to provide a detailed characterisation of the appetite, sensory perceptional, and eating behavioural effects of COVID-19 disease, only participants experiencing these alterations were included in the study. Thereby, the present study does not address the prevalence of these symptoms among COVID-19 patients. Although the current study did not study nor suggest gender differences, it should be noted that the study population consisted of 88% females. A study with a bigger sample size including more males could preferably be conducted to check generalisability of the results.

4.5. Application of Findings and Suggestions for Future Research

The study brings results that can be applied by health professionals to secure eating enjoyment and thereby nutrition when suffering from altered taste and odour perception. Specifically, this article can serve as a basic information document for dietetic guidance, in which health professionals can seek information about normal occurring appetite, sensory perceptional, and eating behavioural changes due to COVID-19, along with inter-individual differences. Further, the article points to ways of applying the findings in dietary guidance. For example, a key strategy to secure eating enjoyment is to work with the drivers of pleasure. Consciously shifting focus from flavour as primary driver of consummatory pleasure towards the more well-functioning sense of touch proves to be a good strategy. In practical terms, this can be achieved via an emphasis on food textures, e.g., variation in crunchiness, by the use of raw vegetables, and/or variation in trigeminal stimulation by the use of irritants such as hot spices. However, the mental barriers and possibilities preventing/supporting these mental shifts need further study. Additionally, strategic use of healthy vs. unhealthy food to support intake and pleasure needs further exploration. Therefore, future studies should focus on how to apply these findings among COVID-19 patients, and if/how they can be broadened to other patients groups suffering from chemosensory disorders, in order to maintain appetite and secure nutrition.

5. Conclusions

This study aimed to investigate, the acute and long-term effects of COVID-19 disease on details of appetite, sensory perception, and eating behaviour. Key results revealed that although the majority reported eating behaviour and sensory perceptions to return to baseline as experienced before COVID-19, many still in the post-acute phase of the disease reported altered appetite.

The desire for food was reported severely impaired by the majority of participants during both the acute phase and post-acute phase, compared to before COVID-19. The low desire for food was, for many, associated with altered appetite and satiety sensations, and when engaging in consumption, the majority of participants were less often left with a satisfied feeling.

Basic taste- and orthonasal odour perception, were by the majority of participants, reported to be reduced due to COVID-19, and retronasal odour perception was reported to be reduced among approximately half of the participants. The changes were often characterised by a complete loss or disordered perception, along with the experience of off-flavours. Sensory alterations were reported a reduction in the desire to eat and affect food choices, yet some participants found themselves capable of improving ortho- and retronasal odour perception.

Eating focused on the three daily main meals (not snacks), both in the acute phase and post-acute phase, and meals were for many of smaller portion size. In general, during the acute phase, the majority reported alterations in the food types included in the diet, and around half reported alterations in the choice of preparation method, the texture of meals, and choice of warm–cold meals. The relative proportion of the different types of food and beverages was comparable to before COVID-19 for the majority of participants in the post-acute phase. The same tendency was found regarding the temperature of meals, preparation method, and different textures.

To cope with the changes in appetite and sensory perception, participants focused on changing their focus during eating—from a focus on taste and smell to the more well-functioning senses, e.g., touch. For many, this resulted in an increased focus on the intake of spicy, healthy, and crunchy foods.

Altogether, the findings from this study illustrate the complexity by which COVID-19 affects human appetite and sensory perception and points toward strategies to cope with these changes. Future studies could by advantage focus on validation of the results via objective measurements and application of the findings in dietary interventions for people suffering from sensory- and appetite-related impairments including COVID-19 patients.

Author Contributions: Conceptualisation, N.C., A.T.Z.B.H. and B.V.A.; methodology, N.C., A.T.Z.B.H. and B.V.A.; software, N.C. and B.V.A.; validation, N.C. and B.V.A.; formal analysis, N.C.; investigation, N.C.; resources, N.C. and A.T.Z.B.H.; data curation, N.C.; writing—original draft preparation N.C. and B.V.A.; writing—review and editing, N.C., A.T.Z.B.H., and B.V.A.; visualisation, N.C.; supervision, B.V.A.; project administration, B.V.A.; funding acquisition, B.V.A. All authors have read and agreed to the published version of the manuscript.

Funding: This research was funded by EiT Food, EiT Food ivzw, Unicenter A, Philipssite 5 bus 34, 3001 Leuven, Belgium. The KAVA grant number is: 20402.

Institutional Review Board Statement: According to the Consolidation Act on Research Ethics Review of Health Research Projects, Consolidation Act number 1083 of 15 September 2017, section 14(2) notification of questionnaire surveys or interview projects to the research ethics committee system is only required if the project involves human biological material. Ref. request: 247/2020.

Informed Consent Statement: Informed consent was obtained from all subjects involved in the study.

Data Availability Statement: The data presented in this study is available on request from the corresponding author.

Acknowledgments: The authors would like to thank the EiT Mtching COVID project partners for their interest and support of this research.

Conflicts of Interest: The authors declare no conflict of interest. The funders had no role in the design of the study; in the collection, analyses, or interpretation of data; in the writing of the manuscript, or in the decision to publish the results.

References

1. Coronavirus Disease (COVID-19). Available online: https://www.who.int/emergencies/diseases/novelcoronavirus-2019/question-and-answers-hub/q-a-detail/coronavirus-disease-covid-19#:~{}:text=symptoms (accessed on 14 January 2021).
2. Grant, M.C.; Arbyn, M.; Mohammed, Z.; McGuinness, L.; Clarke, E.L.; Wade, R.G. The prevalence of symptoms in 24,410 individuals infected by the novel coronavirus (SARS-CoV-2; COVID-19): A systematic review and meta-analysis of 148 studies from 9 countries. *PLoS ONE* **2020**, *15*. [CrossRef] [PubMed]
3. Lechien, J.R.; Chiesa-Estomba, C.M.; De Siati, D.R.; Horoi, M.; Le Bon, S.D.; Rodriguez, A.; Dequanter, D.; Blecic, S.; El Afia, F.; Distinguin, L.; et al. Olfactory and gustatory dysfunctions as a clinical presentation of mild-to-moderate forms of the coronavirus disease (COVID-19): A multicenter european study. *Eur. Arch. Otorhinolaryngol.* **2020**, *277*, 2251–2261. [CrossRef]
4. Lechien, J.R.; Cabaraux, P.; Chiesa-Estomba, C.M.; Khalife, M.; Hans, S.; Calvo-Henriquez, C.; Martiny, D.; Journe, F.; Sowerby, L.; Saussez, S. Objective olfactory evaluation of self-reported loss of smell in a case series of 86 COVID-19 patients. *Head Neck* **2020**, *42*, 1583–1590. [CrossRef]
5. Spinato, G.; Fabbris, C.; Polesel, J.; Cazzador, D.; Borsetto, D.; Hopkins, C.; Boscolo-Rizzo, P. Alterations in smell or taste in mildly symptomatic outpatients with sars-cov-2 infection. *JAMA* **2020**, *323*, 2089–2090. [CrossRef]
6. Menni, C.; Valdes, A.M.; Freidin, M.B.; Sudre, C.H.; Nguyen, L.H.; Drew, D.A.; Ganesh, S.; Varsavsky, T.; Cardoso, M.J.; Moustafa, J.S.E.-S.; et al. Real-time tracking of self-reported symptoms to predict potential COVID-19. *Nat. Med.* **2020**, *26*, 1037–1040. [CrossRef]
7. Beltran-Corbellini, A.; Chico-Garcia, J.L.; Martinez-Poles, J.; Rodriguez-Jorge, F.; Natera-Villalba, E.; Gomez-Corral, J.; Gomez-Lopez, A.; Monreal, E.; Parra-Dıaz, P.; Cortes-Cuevas, J.L.; et al. Acute-onset smell and taste disorders in the context of COVID-19: A pilot multicentre polymerase chain reaction based case–control study. *Eur. J. Neurol.* **2020**, *9*, 1738–1741. [CrossRef]
8. Parma, V.; Ohla, K.; Veldhuizen, M.G.; Niv, M.Y.; Kelly, C.E.; Bakke, A.J.; Cooper, K.W.; Bouysset, C.; Pirastu, N.; Dibattista, M. More than smell–COVID-19 is associated with severe impairment of smell, taste and, chemesthesis. *Chem. Senses* **2020**, *45*, 609–622. [CrossRef] [PubMed]
9. Vaira, L.A.; Hopkins, C.; Salzano, G.; Petrocelli, M.; Melis, A.; Cucurullo, M.; Ferrari, M.; Gagliardini, L.; Pipolo, C.; Deiana, G.; et al. Olfactory and gustatory function impairment in COVID-19 patients: Italian objective multicenter-study. *Head Neck* **2020**, *42*, 1560–1569. [CrossRef] [PubMed]
10. Vaira, L.A.; Deiana, G.; Fois, A.G.; Pirina, P.; Madeddu, G.; De Vito, A.; Babudieri, S.; Petrocelli, M.; Serra, A.; Bussu, F.; et al. Objective evaluation of anosmia and ageusia in COVID-19 patients: Single-center experience on 72 cases. *Head Neck* **2020**, *42*, 1252–1258. [CrossRef]
11. Eliezer, M.; Hautefort, C.; Hamel, A.-L.; Verillaud, B.; Herman, P.; Houdart, E.; Eloit, C. Sudden and complete olfactory loss function as a possible symptom of COVID-19. *JAMA Otolaryngol. Head Neck Surg.* **2020**, *146*, 674–675. [CrossRef]
12. Moein, S.T.; Hashemian, S.M.; Mansourafshar, B.; Khorram-Tousi, A.; Tabarsi, P.; Doty, R.L. Smell dysfunction: A biomarker for COVID-19, in 'International forum of allergy & rhinology. *Wiley Online Libr.* **2020**, *10*, 944–950.
13. Prajapati, D.P.; Shahrvini, B.; MacDonald, B.V.; Crawford, K.L.; Lechner, M.; DeConde, A.S.; Yan, C.H. Association of subjective olfactory dysfunction and 12-item odor identification testing in ambulatory COVID-19 patients, in 'International forum of allergy & rhinology. *Wiley Online Libr.* **2020**, *10*, 1209–1217.
14. Long-Term Health Effects of COVID-19. Available online: https://www.sst.dk/en/English/Corona-eng/COVID9-and-novel-coronavirus/Long-term-effects-of-COVID-19 (accessed on 14 January 2021).
15. Andersen, B.V.; Hyldig, G. Consumers' view on determinants to food satisfaction. A qualitative approach. *Appetite* **2015**, *95*, 9–16. [CrossRef]
16. Andersen, B.V.; Hyldig, G. Food satisfaction: Integrating feelings before, during and after food intake. *Food Qual. Prefer.* **2015**, *43*, 126–134. [CrossRef]
17. Harris, R. Life's a smelling success: Using scent to empower your memory and learning. *Int. J. Arimatherapy* **2005**, *15*, 58–59. [CrossRef]
18. Spinelli, S.; Monteleone, E.; Ares, G.; Jaeger, S.R. Sensory drivers of product-elicited emotions are moderated by liking: Insights from consumer segmentation. *Food Qual. Prefer.* **2019**, *78*, 1–12. [CrossRef]
19. Spinelli, S.; Jaeger, S.R. What do we know about the sensory drivers of emotions in food and beverages? *Curr. Opin. Food Sci.* **2019**, *27*, 82–89. [CrossRef]
20. Høier, A.T.Z.B.; Chaaban, N.; Andersen, B.V. Possibilities for maintaining appetite in recovering COVID-19 patients. *Foods* **2021**, *10*, 464. [CrossRef]
21. Healthy AtHome: Healthy Diet. Available online: https://www.who.int/campaigns/connecting-the-world-tocombat-coronavirus/healthyathome/healthyathome---healthy-diet (accessed on 14 January 2021).
22. Butowt, R.; von Bartheld, C.S. Anosmia in COVID-19: Underlying Mechanism and Assessment of an Olfactory Route to Brain Infection. *Neuroscientist* **2020**. [CrossRef]

23. Han, A.Y.; Mukdad, L.; Long, J.L.; Lopez, I.A. Anosmia in COVID-19: Mechanisms and Significance. *Chem. Senses* **2020**, *45*, 423–428. [CrossRef] [PubMed]
24. Li, Y.; Bai, W.; Hashikawa, T. The neuroinvasive potential of SARS-CoV2 may play a role in the respiratory failure of COVID-19 patients. *J. Med. Virol.* **2020**, *92*, 552. [CrossRef]
25. Murray, M.; Vickers, Z. Consumer views of hunger and fullness. A qualitative approach. *Appetite* **2009**, *53*, 174–182. [CrossRef]
26. Eccles, R. *Mechanisms and Symptoms of Common Cold and Flu*; Eccles, R., Weber, O., Eds.; Common cold. Birkhäuser Verlag: Basel, Switzerland, 2009; pp. 23–45.
27. Phan, L. Clinical Characteristics of COVID-19 Patients with Digestive Symptoms in Hubei, China: A Descriptive, Cross-Sectional, Multicenter Study. *Am. J. Gastroenterol.* **2020**, *115*, 766–773.
28. McCrickerd, K.; Forde, C.G. Sensory influences on food intake control: Moving beyond palatability. *Obes. Rev.* **2016**, *17*, 18–29. [CrossRef]
29. Forde, C.G. From perception to ingestion; the role of sensory properties in energy selection, eating behaviour and food intake. *Food Qual. Prefer.* **2018**, *66*, 171–177. [CrossRef]
30. Yeomans, M.R. Palatability and the Micro-structure of Feeding in Humans: The Appetizer Effect. *Appetite* **1996**, *27*, 119–133. [CrossRef] [PubMed]
31. Merkonidis, C.; Grosse, F.; Ninh, T.; Hummel, C.; Heahner, A.; Hummel, T. Characteristics of chemosensory disorders–results from a survey. *Eur. Arch. Otorhinolaryngol.* **2014**, *272*, 1403–1416. [CrossRef] [PubMed]
32. Jansen, A. A learning model of binge eating: Cue reactivity and cue exposure. *Behav. Res. Ther.* **1998**, *36*, 257–272. [CrossRef]
33. Nederkoorn, C.; Smulders, F.; Havermans, R.; Jansen, A. Exposure to binge food in bulimia nervosa: Finger pulse amplitude as a potential measure of urge to eat and predictor of food intake. *Appetite* **2004**, *42*, 125–130. [CrossRef] [PubMed]
34. Whitelock, V.; Robinson, E. Remembered meal satisfaction, satiety and later snack food intake: A laboratory study. *Nutrients* **2018**, *10*, 1883. [CrossRef] [PubMed]
35. Duerlund, M.; Andersen, B.V.; Byrne, D.V. Sensory specific desires affect and are affected by actual snack choice. *J. Food Nutr. Health* **2021**, *2*, 1–7. [CrossRef]
36. Neto, D.B.; Fornazieri, M.A.; Dib, C.; Francesco, R.C.D.; Doty, R.L.; Voegels, R.L.; Pinna, F.D.R. Chemosensory dysfunction in Covid-19: Prevalences, recovery rates, and clinical associations on large Brazilian sample. *Otolaryngol. Head Neck Surg.* **2020**, *164*, 512–518. [CrossRef]
37. Niklassen, A.S.; Draf, J.; Huart, C.; Hintschich, C.; Bocksberger, S.; Trecca, E.M.C.; Klimek, L.; Bon, S.D.L.; Altundag, A.; Hummel, T. Covid-19: Recovery from chemosensory Dysfunction. A multicentre study on smell and taste. *Laryngoscope* **2021**. [CrossRef]
38. Huart, C.; Philpott, C.; Konstantinidis, I.; Altundag, A.; Whitcroft, K.L.; Trecca, E.M.C.; Cassano, M.; Rombaux, P.; Hummel, T. Comparison of Covid-19 and common cold chemosensory dysfunction. *Rhinology* **2020**, *58*, 623–625.
39. Brugliera, L.; Spina, A.; Castellazzi, P.; Cimino, P.; Arcuri, P.; Negro, A.; Houdayer, E.; Alemanno, F.; Giordani, A.; Mortini, P.; et al. Nutritional management of COVID-19 patients in rehabilitation unit. *Eur. J. Clin. Nutr.* **2020**, *74*, 860–863. [CrossRef] [PubMed]
40. Thibault, R.; Seguin, P.; Tamion, F.; Pichard, C.; Singer, P. Nutrition of the COVID-19 in the intensive care uni (ICU): A practical guidance. *Crit Care* **2020**, *24*, 447. [CrossRef] [PubMed]
41. Moskowitz, H.R.; Krieger, B. The contribution to sensory liking to overall liking: An analysis of six food categories. *Food Qual. Prefer.* **1995**, *6*, 85–90. [CrossRef]
42. Andersen, B.V.; Brockhoff, P.B.; Hyldig, G. The importance of liking of appearance, -odour, -taste and -texture in the evaluation of overall liking. A comparison with the evaluation of sensory satisfaction. *Food Qual. Prefer.* **2019**, *71*, 228–232. [CrossRef]
43. Moskowitz, H.; Krieger, B. What sensory characteristics drive product quality? An assessment of individual differences. *J. Sens. Stud.* **1992**, *8*, 271–282. [CrossRef]
44. Stevenson, R.J.; Mahmut, M.K.; Hortsmann, A.; Hummel, T. The Aetiology of Olfactory Dysfunction and Its Relationship to Diet Quality. *Brain Sci.* **2020**, *10*, 769. [CrossRef] [PubMed]
45. Kershaw, J.C.; Mattes, R.D. Nutrition and taste and smell dysfunction. *World J. Otorhinolaryngol. Head Neck Surg.* **2018**, *4*, 3–10. [CrossRef] [PubMed]
46. Aschenbrenner, K.; Hummel, C.; Teszmer, K.; Krone, F.; Ishimaru, T.; Seo, H.-S.; Hummel, T. The influence of olfactory loss on dietary behaviors. *Laryngoscope* **2007**, *118*, 135–144. [CrossRef] [PubMed]
47. Kong, I.; Kim, S.; Prk, B.; Kim, J.; Choi, H. Olfactory dysfunction is associated with the intake of macronutrients in Korean adults. *PLoS ONE* **2016**, *11*, e0164495. [CrossRef] [PubMed]
48. Zang, Y.; Han, P.; Burghardt, S.; Knaapila, A.; Schriever, V.; Hummel, T. Influence of olfactory dysfunction on the perception of food'. *Eur. Arch. Otorhinolaryngol.* **2019**, *276*, 2811–2817. [CrossRef] [PubMed]

foods

MDPI

Article

Possibilities for Maintaining Appetite in Recovering COVID-19 Patients

Alexander Teymour Zadeh Baboli Høier, Nora Chaaban and Barbara Vad Andersen *

Food Quality Perception and Society, Department of Food Science, Aarhus University, Agro Food Park 48, DK-8200 Aarhus N, Denmark; alexanderhoier@gmail.com (A.T.Z.B.H.); ncho_on@hotmail.com (N.C.)
* Correspondence: barbarav.andersen@food.au.dk; Tel.: +45-3070-7860

Abstract: COVID-19 and sequelae thereof are known to cause chemosensory dysfunction, posing a risk for intake and adequate nutrition for recovery. The overall objective of this study was to investigate the subjective strategies for maintaining appetite applied by patients recovering from COVID-19. The study included 19 in-depth interviews, focusing on patients suffering from long-term effects of COVID-19. The results were analysed using a thematic analysis for qualitative data. Results on strategies for maintaining appetite included four key themes: (1) a focus on well-functioning senses, (2) a focus on familiar foods, (3) a focus on the eating environment, and (4) a focus on post-ingestive well-being. It was found that factors prior to, during and after food intake, as well as the context, could influence desire to eat and pleasure related to food intake. As ageusia and anosmia make characterization of food difficult, being able to recognize and memorize its flavour was important to engage in consumption. Under normal circumstances, the hedonic value of food relies predominantly on the flavour of foods. When suffering from chemosensory dysfunction, shifting focus towards the texture of food, including trigeminal stimulation during consumption, were beneficial for maintaining appetite and food-related pleasure. Furthermore, a focus on the holistic satisfying feelings of choosing healthy food, as well as a focus on other people's enjoyment during meals were reported to boost well-being around food intake. The study elaborated our understanding of the complex consequences of COVID-19, and can be applied in health promoting initiatives targeted patients recovering from COVID-19.

Keywords: COVID-19; sensory function; chemosensory dysfunction; perception; appetite; well-being; pleasure; recovery; interview

Citation: Høier, A.T.Z.B.; Chaaban, N.; Andersen, B.V. Possibilities for Maintaining Appetite in Recovering COVID-19 Patients. *Foods* **2021**, *10*, 464. https://doi.org/10.3390/foods10020464

Academic Editor: Jean-Xavier Guinard

Received: 19 January 2021
Accepted: 16 February 2021
Published: 20 February 2021

1. Introduction

Severe acute respiratory syndrome coronavirus 2 (SARS-CoV-2), also referred to as 'coronavirus disease' (COVID-19), has spread rapidly all around the world. Besides a mortality rate of COVID-19 on 2.2% as of the 8th of February 2021 [1], the long-term effects can be devastating on subjective quality of life. The long-term effects of COVID-19 vary from patient to patient, but often include chemosensory dysfunction in terms of dysosmia and dysgeusia [2]. Dysosmia is a condition affecting smell perception, and can be broadened out to the conditions; 'anosmia', which is a complete loss of the ability to detect odours, 'parosmia', which alters the odour perception, often to displeasing odours, 'hyposmia', which is decreased ability to detect odours, and 'phantosmia', which concerns spontaneously occurring odours without any triggers [3]. Dysgeusia is a condition concerning alterations of the perception of basic taste. This condition can be broadened out to; 'ageusia', which is a total loss of the ability to taste and 'parageusia', which alters the taste perception, often to displeasing tastes, and can be triggered by any or specific tastes [3].

Vaira et al. [2] found that as many as 85% of COVID-19 patients suffered from chemosensory dysfunctions in the beginning of the acute phase. Approximately 50%

of the patients showed chemosensory dysfunctions two to three weeks after infection [4], and after 60 days, 7.2% were found to suffer from chemosensory dysfunction [2].

As malnourishment for an extended period of time weakens the body's immune system, proper nutrition is important for the recovery of disease [5]. Likewise, proper nutrition is believed to be important for the recovery of COVID-19, and not least for subjective well-being. The food's sensory properties, and thereby the perception of odours, flavours, and basic tastes, are the main factors driving human motivation to eat and intake [6]. The importance of sensory properties for hedonic perception of foods and food behaviour is evident when observing the broad range of studies and models focusing on: sensory properties and acceptance, e.g., [7–9], sensory properties and preference [10], sensory properties and food behaviour [11–14], and liking as a determinant for intake [15]. A dysfunctional sensory perception therefore poses a serious risk to appetite and the nutrition needed for a fast recovery. A study of the characteristics of patients with chemosensory disorders showed that 50% of participants changed their food habits and preferences [16]. Furthermore, chemosensory dysfunction poses a significant impact on day-to-day life [17]. Affected subjects report reductions in quality of life [18–21], with reduced pleasure from eating being one of the most distressing symptoms of chemosensory loss and the main complaint from patients seeking medical attention [16].

Sensory properties are, however, not the only drivers of motivation to eat, and therefore, insight into how COVID-19 patients cope with chemosensory dysfunction in order to maintain motivation to eat and food-related pleasure, can lead to actionable knowledge for use in dietary therapy supporting recovery, and further, to advise patients on how to maintain food-related quality of life in general.

The overall aim of the present study was to investigate the subjective strategies for maintaining appetite applied by patients recovering from COVID-19. Specifically, the objectives were to study how COVID-19 affected appetite, sensory perception and food-related pleasure, and further to study, how patients cope with chemosensory dysfunction to maintain appetite and food-related pleasure.

2. Materials and Methods

The study applied one-on-one in-depth interviews, due to the type of information required by the participants. In-depth interviews allow the participants to unfold any sensitive personal matter that would not have been appropriate to discuss in a group [22], and further, this method allows the researcher to collect detailed information beyond the surface level. Up until now, the study of strategies for coping with changes in appetite due to COVID-19 has been an unexplored field of research. Therefore, a semi-structured interview guide was chosen to ensure that all predefined themes were addressed; also, the semi-structured design allowed the researcher to follow up new points raised by the participant that had not been thought of when preparing the interview guide.

Prior to data collection, the Central Denmark Region Committees on Health Research Ethics approved the study being conducted. The participants were informed that the interview was being recorded, and were also informed of their legal rights, and how their data would be used and stored. The participants gave verbal consent for the use of their replies in the research study. The interviews were conducted using Zoom or similar communication platforms depending on the preference of the participant. Telephone interviews were conducted in a few cases, when the participants were unable to partake in a video interview. The interviews were recorded using the record function in the video meeting application, or using a dedicated telephone recording application. Video interviews were preferred, in order to get a better understanding of the participants, and enabling their body language to be read.

2.1. Participants

A total of 19 interviews were conducted among a Danish population suffering from long-term effects of COVID-19, meaning that participants were in the post-acute phase of

Codid-19, yet were still showing symptoms, including changes in appetite. Participants were recruited via posts in a Danish Facebook group for people suffering from long-term effects of COVID-19. Inclusion criteria were: being over 18 years of age and suffering from long-term effects of COVID-19 including changes in appetite. All participants reported being in the post-acute phase of COVID-19 at the time of data collection. A total of 14 participants had been diagnosed with COVID-19 through antibody or swab test. The remaining five participants had been diagnosed with COVID-19 via subjective assessment, due to not being eligible for a test through the Danish health care system at the time of the disease. These five participants all showed regular long-term effects of COVID-19 symptoms, such as anosmia, ageusia, and fatigue. Across the total study sample, the average period for showing symptoms during the acute phase of the disease was reported to be 16 days, and none had been hospitalized due to COVID-19. The most frequent symptoms during the acute phase of COVID-19 included fever, ageusia, fatigue, anosmia, throat pains, headache, and difficulties breathing. In the post-acute phase, the most frequent symptoms included ageusia, anosmia, fatigue, parosmia and difficulties breathing. Other less common symptoms included headache, feeling nausea, difficulties concentrating, throat pains, parageusia and phantosmia. One participant (participant #2) reported having suffered from a blood clot earlier in life. As the present study focused on strategies for coping with changes in appetite due to COVID-19, and the blood clot incident had been years earlier in the participant's life, it was decided to include the participant in the present study. The remaining participants had not been diagnosed with diseases other than COVID-19. Participant characteristics can be seen in Table 1.

Table 1. Participant characteristics.

Characteristics	
Nationality	Danish
N_{total}	19
Gender; males, females	2 males, 17 females
Age; mean (min–max)	44 (25–66)
COVID-19 diagnose (antibody, swab, assessment)	2, 12, 5
Symptoms most often mentioned during the acute phase of COVID-19 (n) *	Fever (n = 10); Ageusia (n = 7); Fatigue (n = 7); Anosmia (n = 6), Throat pains (n = 5), Headache (n = 4); Difficulties breathing (n = 4).
Symptoms most often mentioned during the post-acute phase of COVID-19 (n) *	Ageusia (n = 14); Anosmia (n = 13); Fatigue (n = 6); Parosmia (n = 4); Difficulties breathing (n = 4)

* Symptoms reported by four or more participants are included in the list of symptoms.

2.2. Pilot Test

The interview guide was pilot-tested among four colleagues and naive participants with different ages, educational backgrounds and COVID-19 histories; one of the participants had been diagnosed with COVID-19, and the remaining three were non-diagnosed. None of the pilot participants were included in the final sample population. A pilot test is generally a recommended procedure when conducting interviews [23]. The pilot test aimed to ensure a proper flow and understanding of the questions, and to decide whether questions should be added and/or excluded. The interview guide was refined according to the feedback; questions were rephrased and words were replaced with ones that were more easily understood by all naïve participants.

2.3. Interview Protocol

The interviews lasted 45 to 90 min each and were conducted by two interviewers independently. Both had educational training in the qualitative research method. The interviewers developed the interview guide together, discussed the interview approach prior to conducting the interviews, and conducted the pilot test together. During the interviews, the interviewers followed the structure of the interview guide, the format of the questions, and discussed the interview style, in order to ensure uniformity and transparency in the interviewing style.

The interviews followed a semi-structured interview guide, a format that allowed the interviewer to pursue topics raised during the interview [24]. The interviewer firstly informed the participant about the purpose of the study. The following questions were designed to start the conversation and dialogue, including introductory, opening, transition, key and ending questions. The questions led the participants to reflect on health, appetite, sensory perception ability and food-related pleasure. It was emphasized that the questions were easy to understand, short, clear and could engage the participants in detailed elaboration. To explore the strategies affecting participants' appetite, they were firstly directly asked about how they coped with changes in appetite, sensory perception ability and food-related pleasure. If the participant reported not having consciously applied any strategies, they were asked to elaborate on concrete situations regarding every theme presented in the interview guide. This allowed the researcher to analyse for coping strategies applied subconsciously. All participants were interviewed once during the study, and at the end of the interview, the participants were thanked for their participation and assured anonymity.

2.4. Data Analysis

Researcher triangulation was conducted in order to ensure all important points from the interviews were included in a thematic analysis—the foundational analytical method. Researcher triangulation is generally recommended to overcome fundamental biases, arising when using a single researcher [25]. Different types of researcher triangulation can be applied: data triangulation, method triangulation, investigator triangulation and theory triangulation, respectively. In the current study, investigator triangulation was applied by using more than one researcher to conduct the interviews and analyse the data. Prior to the thematic analysis, the two interviewers independently and without prior discussion conducted descriptive summaries of all the interviews. The descriptive summaries served to capture a summed picture of each interview, and as such, provided the basis for the thematic analysis, together with the video material. From the summaries, the two researchers, likewise without prior discussion, clarified the themes to address the overall aim of the study. Approximately 21 h of video material and the descriptive summaries created the basis for the thematic analysis. The investigator triangulation approach ensured that the same data set was interpreted by more than one researcher; each provided their independent analysis before further comparison, which is important for decreasing bias in the analysis of data [25].

Thematic analysis is a method for identifying, analysing, interpreting and reporting patterns and themes within qualitative data [26]. The analysis included a systematic flow of moving forwards and backwards between the phases seen in Figure 1. The analysis progressed from familiarization with the data to a systematic change and reorganization of themes. The themes and concepts emerged from patterns in quotes and the descriptive summaries of the interviews. Patterns in responses and meanings related to the study aim were reflected in the chosen themes.

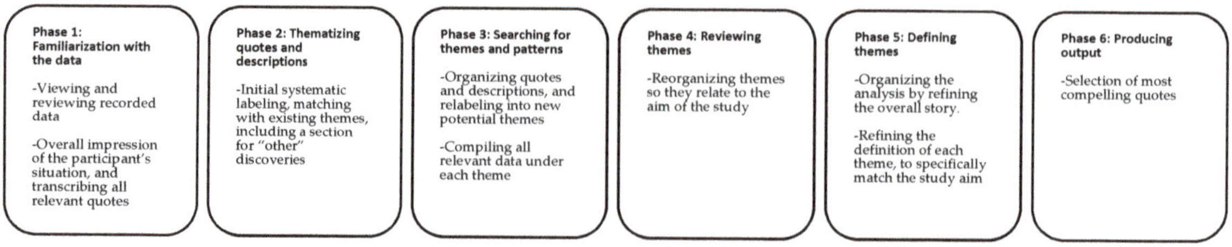

Figure 1. Overview of the phases conducted in the thematic analysis. Guidelines and adaptation from [26].

In case the verbal language did not clearly communicate a participants' feeling or intention, and the feeling or intention was found to be important for the data interpretation, participants' body language was taken into consideration. For example, if the liking of food was mentioned to be altered, but the participant did not describe the direction of the change clearly, (dis)liking could be revealed via facial expressions. Body language was mentioned as a note in the descriptive summaries and/or put in brackets after a quote. The thematic analysis was split into a descriptive part and an action-based part. Phase one and two, in Figure 1, were conducted independently, and afterwards combined and discussed in order to exclude any misinterpretations and to avoid any important points being missed. Afterwards, the two researchers conducted a thematic analysis for one part each. Towards the end of the analysis, the parts were combined and discussed, to ensure uniformity in the interpretation of the data.

3. Results

In this section, the results are divided into two sections; the first section describes the general subjective experiences of the effects of COVID-19 on appetite, sensory perception of taste, flavour and smell, and food-related pleasure, and the second section describes the subjective strategies applied by the participants to maintain their appetite and food-related pleasure. The latter section is based on the main themes that arose from the thematic analysis to maintain appetite: (1) focus on well-functioning senses, (2) focus on familiarity, (3) focus on well-being, and (4) focus on eating environment.

3.1. The Effect of COVID-19 on Appetite, Sensory Perception and Food-Related Pleasure

3.1.1. Appetite

An effect of COVID-19 on appetite was mentioned by most participants. During the acute phase of COVID-19, most participants experienced a decreased appetite:

Participant 7: Female, 59: "I had to convince myself to eat, and that was extremely difficult. The food had to be placed in front of me, and sometimes I ate, and at other times, I barely touched it."

Participant 10: Female, 35: "If I were to describe it in one word, then it would be >forced< eating."

Participant 12: Female, 40 "We had to remind each other (to eat), my husband and I, the first couple of weeks, because he did not feel like eating as well."

Besides suffering from decreased appetite during the acute phase of COVID-19, some participants were still suffering from a decreased appetite as a long-term effect of COVID-19 ($n = 13$):

Participant 5: F, 26: "I still have a reduced appetite, and I do not really feel hungry" ... "I do not experience that rumbling in the stomach as before, and the feeling of satiation come earlier than usual because I do not feel hungry to begin with."

Participant 7: F, 59: "I then started sensing that I was a bit hungry, but I did not know what I wanted (to eat). I did not feel like eating because I could not really taste or smell anything."

Participant 8: F, 42: "I do not think about food until I am really hungry"

Participant 10: F, 35: "Some days I do not feel like eating at all."

Though many participants described that they in general did not feel hungry nor a desire to eat during COVID-19, or reached satiation faster during a meal, some participants expressed the post-meal satiety sensation as 'unsatisfying' ($n = 3$). Satiation was not only depending on a physical feeling of fullness but had to include a feeling of 'pleased senses' based on the perceived sensory properties of food in order to feel fully satisfied. As a result, some participants expressed exposing themselves to food continuously in the search for 'sensory satisfaction':

Participant 19: F, 34: "When I ate something, I did not feel satisfied (caused by the lack ability to perceive the sensory properties of food). Therefore, I had a desire to eat shortly after, but I did not eat a lot of food, because I could not taste it anyway."

Participant 16: M, 48: "The less I was able to taste, the more I ate of it. (...) It was like walking around searching for something that could quench the thirst, but I was not able to find it."

Participant 22: F, 25: "You also eat to achieve the taste experience, and if it is not there, then it is like something is missing. For every bite you take, you are hoping to be able to taste something. I actually feel that more food is needed for me to feel satiated."

Participant 15: F, 33: "especially when I could not perceive any flavours, it was very difficult to tell, when I felt satiated."

3.1.2. Sensory Perception

Taste Perception

During the acute phase, several participants expressed experiencing ageusia, with a complete lack of basic taste perception [3].

Participant 8: F, 42: "I could not taste anything, neither salt, sour, sweet, bitter (during the acute phase)."

Participant 18: M, 53: "Citrus fruits, like lemon and orange that normally are very sweet or very intense in taste, these were completely tasteless."

In the post-acute phase, many participants reported suffering from ageusia, with the majority experiencing hypogeusia, a decreased basic taste perception, as basic taste perception gradually returned.

Participant 3: F, 29: "It is just thick and warm (about the sensory experience of coffee)."

Participant 6: F, 54: "It was very slow, and it started coming back gradually (about perceiving basic tastes)."

During the period where basic taste perception gradually returned, the participants expressed a need for higher concentrations of basic taste attributes to be able to perceive them:

Participant 10: F, 35: "If there is too much salt (in the food), then I am able to perceive it. If you put regular amounts, then I can't perceive it. Not at all."

Participant 12: F, 40: "It has to be extreme (amounts) before I can taste it. When I make rice, I tend to put too much salt in and the others (her family) do not like it."

Perception of taste did not necessarily return equally fast for all basic taste perceptions. Several participants experienced a higher sensitivity towards single basic taste attributes compared to others. Whenever a specific basic taste attribute was perceived in a meal, it was found to completely dominate the flavour experience:

Participant 20: F, 46: "I can taste salt, because I use a lot of it. Much more than I used to. I can barely taste sourness. I can taste sweetness."

Participant 3: F, 29: "If I ordered iced coffee, then it should not be too sweet, because then that was the only thing I could taste. It quickly became too sweet, and that's the same with sour and saltiness."

Participant 13: F, 43: "If something is pickled, it gets an extremely intense vinegar taste."

Participant 14: F, 66: "If lemon is present in food, then everything tastes sour. Lemon (sourness) takes over (the taste perception)."

The gradual return of basic taste perception was explicitly expressed by the participants, but the total time period suffering from ageusia varied between participants.

Participant 3: F, 29: "It disappeared for a week or two, and then it came back (about perceiving the basic tastes)."

Participant 19: F, 34: "I was not able to perceive basic tastes in three weeks, then it started coming back gradually, but not completely."

Participant 16: M, 48: "The first two months (basic taste) were completely gone, and then I started being able to perceive sourness, saltiness and sweetness."

Odour Perception

When reflecting on the acute phase, several participants expressed a partial or complete loss of odour perception, defined as anosmia [3], making them incapable of characterizing food items.

Participant 13: F, 43: "I tried eating chocolate cake, but it was completely tasteless (because she could not perceive the flavour of chocolate)"

Participant 7: F, 59: "I cannot tell if it is a pear or an apple, a blackberry or a blueberry."

Others described not being able to perceive flavour at all, indicating a combined basic taste and odour impairment.

Participant 10: F, 35: "The other day, I accidently made a very strong curry dressing (..) and it was like eating soft ice cream, that does not taste of anything."

Participant 15: F, 33: "I thought of eating mackerel in tomato with my oatmeal in the morning because I could not taste anything."

Participant 6: F, 54: "If you had me blind-folded, and you put food in my mouth, I would not be able to guess what food it was."

During the post-acute phase, the majority of participant experienced a partial or complete loss of odour perception:

Participant 12: F, 40: "I have always been extremely sensitive towards scents. I could throw up by the least disgusting scent, and I do not feel like that anymore. It really has to be the extremes (to be able to smell it)."

Participant 5: F, 26: "I was able to taste the sweetness (in a strawberry cake), but I could not taste the flavour of the chocolate, the strawberry and the cream."

Participant 9: F, 50: "If you can imagine eating ketchup. The only thing you can perceive is the sweetness."

Participant 22: F, 25: "I cannot taste the difference between food. I cannot taste if it is a tomato or if it is a cucumber. I can only taste the saltiness and the sourness."

In a few cases, the participants suffered from hyperosmia during the post-acute phase, defined as hypersensitive sense of smell [3].

Participant 9: F, 50: "I really like lilies, and I think they smell amazing. Sometimes I cannot handle the scent (about her sense of smell being hypersensitive)."

Participant 7: F, 59: "I am a nurse, so the smell of faeces and stuff like that make me sick. It was not like that before (because the smell is too intense)."

Participant 6: F, 54: "I made a dish with cabbage in it, that I normally liked, and suddenly I could not handle the smell of the cabbage (because the smell of cabbage was too intense)."

Some participants further experienced parosmia and phantosmia during the post-acute phase, defined as altered odour perception and the experience of odours that are not actually present, respectively [3]. The most common off-flavours mentioned were soap, metallic, rotten, smoked, perfume and chemicals:

Participant 5: F, 26: "Some scents smell different compared to before. For example, my perfume smells different than before."

Participant 2: F, 44: "I was in bed in an isolated place, and I could smell Christmas cookies. There were not any Christmas cookies (when she was in isolation at the hospital)."

Participant 8: F, 42: "There was a day, were I wanted to eat a big, delicious steak. (. . .) The taste is like how rotten animals smell."

Participant 10: F, 35: "If I eat a crunchy piece of bacon, then it tastes like soap."

Participant 13: F, 43: "I love Pepsi Max. That's my drug. It is actually terrible (that she cannot drink it anymore). I can cry about it. I know it is such a piddly thing, but now it tastes like perfume. It is like someone is spraying perfume directly in my mouth."

Participant 13: F, 43: "Salted almonds, that I also love. When I tried eating them recently, they tasted like smoke. That smoke taste is absolutely disgusting."

Participant 16: M, 48: "The taste of alcohol is very, very strongly chemical."

Participant 18: M, 53: "Everything tasted like iron."

Participant 20: F, 46: "When I eat jam, it tastes like soap."

Chemesthesis Perception

A few participants experienced an altered sense of touch during the post-acute phase of COVID-19. Specifically, they reported an altered feeling of the tongue; burning, tingling and numbing sensations. This feeling would come and go, and was mentioned by one participant to have a direct effect on the ability to perceive textures and taste:

Participant 7: F, 59: "I have days where I feel something burning, and it feel on the outer edge of my tongue. I actually have an almost constant burned tongue sensation."

Participant 12: F, 40: "It is like I have something (points at the middle of her tongue), in the middle, it is like I can only taste something if it is on the edge of my tongue (about her tongue feeling burned)."

Participant 19: F, 34: "it feels like there is a layer (on her tongue), then I cannot taste as much. But when I do not feel like that, I can taste everything."

3.1.3. Food-Related Pleasure

Many participants mentioned that the hedonic aspect of eating was severely altered due to COVID-19. Specifically, the participants expressed a reduced food-related pleasure during the post-acute phase, and eating was, by most participants, expressed to be motivated by a bodily need for energy and nutrients rather than a mental desire.

Participant 3: F, 29: "If I am a little hungry, I wait (until she become hungrier) as it does not give any feeling of pleasure. Before it was nice to eat, but now I eat to not collapse."

Participant 15: F, 33: "The only thing that matters is to go from being hungry to be satiated. Then you do not have to eat any more. It is not fun or cheering to eat anymore."

Participant 13: F, 43: "Eating is not only for the health of the body. Eating is also for pleasure, and I absolutely do not feel any pleasure of food right now. It is sheer for survival."

Participant 12: F, 40: "You start smelling food and you start saying, Yummy, this smells wonderful (before COVID-19). I am looking forward to eat. This feeling is gone, and it has really bothered me. (...) It is really sad. There is no fun about cooking anymore."

The lack of a hedonic incentive to eat was reported to result in an overall lower food intake.

Participant 3: F, 29: "I have eaten less, and I also eat less now, because it does not make sense to have a second portion. I stop eating when I am full, because there is no enjoyment in it."

Participant 5: F, 26: "For example, I do not eat an extra portion because it tastes good."

Participant 22: F, 25: "Normally the smell of food makes me hungry, and I cannot do that anymore. I do not enjoy food as much. I still eat what I need when I am hungry. It is just not the same joy around food."

Being able to perceive taste, flavour and odours, was pointed out as crucial for eating-related pleasure. Furthermore, several of the participants described that the sensory experience was not only affecting perception of the food product, but also overall quality of life.

Participant 2: F, 44: "Being able to taste these foods means a lot to me, and it affects me mentally. Food has become my enemy. It has been my best friend for so many years."

Participant 9: F, 50: "It is a serious reduction in quality of life, when you lose your ability to taste and smell."

Participant 20: F, 46: "The satisfaction from cooking and serving a meal is completely gone. That is truly sad. (. . .) it is really becoming quite a plague to cook, and it is not as nice as it used to be."

3.2. Strategies for Coping with Changes in Appetite and Maintaining Food Pleasure

3.2.1. A Focus on Well-Functioning Senses

Perception of the food's sensory properties were found to be closely related to the hedonic value of food. All of the participants who experienced dysgeusia and dysosmia changed their diet, excluding foods that tasted bad, and eating ones that they liked. Furthermore, they focused on the senses where perception was maintained, such as chemesthesis.

During the period in which the participants suffered from ageusia, parageusia, anosmia and parosmia, many participants experienced a greater appreciation of crunchy foods. Crunchy foods gave the participants some pleasure when eating, and though the level of pleasure was not perceived to the same extent as before COVID-19, a focus on the well-functioning sense of touch helped them maintain some pleasure with food intake.

Participant 6: F, 54: "It was very interesting with nuts in chocolate, they are crunchy, so there is still some satisfaction while eating it, even though it had no taste."

Participant 9: F, 50: "it was really gross to eat soft/smattered foods, when it had no taste... I like it when it is something crunchy. I like a crunchy browned surface on meat."

Participant 14: F, 66: "It is an experience to eat a carrot, because it gives my mouth something to work with. Instead of eating something that you chew in seconds, it takes a bit longer to eat a carrot."

Participant 16: M, 48: "to eat something crunchy, that's also a way to snack."

Additionally, adding a creamy component to alter the mouthfeel was appreciated by a few participants.

Participant 3: F, 29: "e.g., a salad with different veggies, nuts, seeds and some dressing to get something soft, and then it is important that the vegetables still are a bit hard on the inside."

Participant 12: F, 40: "Normally I never use milk in my coffee, but when someone offer me coffee now (after she got COVID-19), I add milk, so I get a different mouth feeling."

Several participants experienced that an increased trigeminal stimulation either via a variation in hot and cooling sensations, or an oral burning sensation made them appreciate food more, as it would compensate for ageusia and anosmia. When suffering from these conditions, the participants described that adding chili to a meal, would increase the hedonic experience.

Participant 3: F, 29: "e.g., a salad with different veggies, nuts, seeds, hot chicken so I get something hot and cold."

Participant 5: F, 26: "Now it is just the flavours and smells (that are missing). (. . .) Texture, hot food, cold food, a variation, that's delicious!"

Participant 2: F, 44: "The spicier the food, the better I like it. Then I feel like I am getting just a bit of my senses back."

Participant 8: F, 42: "I have tried to eat chili, just because it was fun to feel the burning sensation. I could not taste anything at all."

Participant 12: F, 40: "I always keep a glass of jalapeños in my fridge. I add it to my food just to get some kind of a flavour (to be understood as sensory perception)."

Participant 20: F, 46: "It gives a burning sensation. It can also help me get rid of the bad taste."

3.2.2. A Focus on Familiarity

Several participants expressed that familiarity of foods became an overly important factor for their hedonic food experience. The memory of how the food used to taste was necessary for their desire to eat and liking of foods.

Participant 7: F, 59: "One thing I have learned, is that I need to be able to remember how it taste, otherwise I will not like it."

Participant 13: F, 43: "I actually tried eating some spaghetti bolognese from a restaurant the other day. (...) It had no taste at all, but because I know that I used to like it, it still did something for me."

Participant 10: F, 35: "hash (=Biksemad in Danish, a mixed dish with potatoes, meat, vegetables) was on sale, and I thought, we will take it, because I know how it is supposed to taste. That works for me because we have eaten that before, when I was able to taste."

Participant 13: F, 43: "I cannot really taste it, but the feeling of remembering that this is good, sometimes makes me go: Mmmmh Nice! This is really good."

Several participants expressed that exposure to unfamiliar food or unexpected sensory stimuli resulted in extreme dislike.

Participant 7: F, 59: "Just a small thing, a different sound, different look, or different taste, then I instantly loose the desire to eat, and then I give up."

Participant 10: F, 35: "She then picked the wrong one, so I got sausage mix (instead of hash -a dish with potatoes, meat, vegetables) When I tried to eat it, it grew 10 times bigger (in my mouth) than it really was."

3.2.3. A Focus on the Eating Environment

In nearly all cases, the participants experienced an increased appreciation of eating with family, as it caused them to shift their focus from not being able to perceive the food's flavour to enjoying the social company. Furthermore, the social company helped the participants eat proper meals.

Participant 7: F, 59: "I enjoy it the most, if I am with others. If I am eating alone, I have to convince myself that I have to eat."

Participant 10: F, 35: "But luckily, I cook for myself and my son every day, so the point that I am cooking for someone, and eating with someone—I think that has a big impact on me."

Participant 12: F, 40: "It surely helps eating together, right? I think that was the reason that my appetite has returned. (...) If I had been alone, I would not have eaten anything, so it was really good that I had my kids and my husband."

Participant 15: F, 33: "When I lost my sense of smell and taste, it did not matter anymore—I ate because Claus, my boyfriend, and I had to. Luckily, he was here, otherwise what I ate would not have mattered to me."

Participant 20: F, 46: "Then I am just eating a piece of bread. I do not get advanced when the others are not home. So, I definitely eat less when I am alone."

Participant 22: F, 25: "I would not be cooking if I was eating alone, then I would just eat a small meal. I just enjoy it more when I am eating with someone, I automatically eat a bit more."

Even though most participants enjoyed the company of their family, some participants expressed the importance of a calm eating environment during a meal to be able to focus on the sensory properties of the meal that could vaguely be perceived. The participants felt that noise from people around them, as well as food and non-food-related odours could distract their attention, and as a result lower their appetite.

Participant 2: F, 44: "When I am eating, I need complete silence. I need to focus on what I am eating and focus my mind on the taste it used to have."

Participant 11: F, 43: "Also because we talk a lot, and my family talks a lot, and my husband and daughter talk loudly, it sometimes makes it difficult to be around, they also have to learn to lower their voices. It can affect my appetite so I have to leave the table."

Participant 13: F, 43: "At some point I had to tell her: you have to wash off your perfume, otherwise I will not be able to sit near you."

3.2.4. A Focus on Overall Well-Being

As mentioned above, the sense of taste and sense of smell had a big impact on the pleasure related to food intake. When the participants had lost their sense of taste and sense of smell, they focused on other aspects of consumption that could increase their well-being.

Some participants described how they consciously worked with adjusting expectations towards the eating experience and how this helped them not to feel disappointed.

Participant 9: F, 50: "It must come back—and if it do not, there is really nothing I can do about it."

Participant 14: F, 66: "If I am expecting to be able to taste the meal, then it is going to feel like a bummer. If I have already decided to focus on the texture, or the looks of it, it is not going to be fantastic, but kind of like what I came for—it is living up to my expectations."

Participant 14: F, 66: "I do not want to be negative about it, It is going to be some long days."

Participant 18: M, 53: "It is what it is, just going to have to wait till things turn around."

Participant 19: F, 34: "So you accept the circumstances (...) Oh well, it is what it is, you just got to wait for it to return."

Some participants found pleasure in foods they used to perceive as 'highly palatable' and ate according to what they previously liked, despite the inability to perceive the flavour. The highly palatable foods were often unhealthy foods like chocolate and chips. For this group, taste was secondary, and the feeling of giving themselves a treat was primary.

Participant 6: F, 54: "I have always had kind of a chocolate addiction. I could control it for a couple of months at a time, but would always get a fall back, and regain control after a couple of months. (. . .) I would typically fall in at Christmas, and regain control around easter. It was the same every year. When we were home sick (with COVID-19), we ate a lot of chocolate."

Participant 16: M, 48: "I ate a lot of chips, pork crisps and chocolate. It got out of hand. The less I was able to taste, the more I ate."

Participant 22: F, 25: "Well, by making some palatable meals (you bring pleasure to the meal), and not just something like a fried fish fillet on rye bread or something. That you make something a bit more delicious—more tasty."

Other participants cut down on all unhealthy foods and redirected their pleasure from primarily being driven by the food's flavour to focusing on the healthiness and the feeling of eating nutrients, which are considered good for their body. This group particularly focused on eating healthier, and cut down on all hyper caloric and unhealthy foods.

Participant 3: F, 29: "When you cannot tell the difference, you might as well just eat healthy."

Participant 13: F, 43: "I have told myself: if you cannot taste it, there is no reason to fill yourself with sugar, there is no reason for that anymore."

Participant 15: F, 33: "I thought of eating mackerel in tomato with my oatmeal in the morning because I could not taste anything. I did not get to try it though."

Participant 16: M, 48: "And I think, in some way, I have been addicted to taste, because I have significantly changed my eating habits since we found out that when we were not able to taste anything, we might as well hire a dietician to help us eat a bit healthier."

4. Discussion

Overall, the participants were able to reflect upon and describe how COVID-19 affected their appetite, sensory perception and food-related pleasure. Conversely, participants in general showed difficulties verbalizing the strategies applied. This could be related to patients coping on a subconscious level and thereby not being capable of memorizing their strategies up front when asked in an interview. Via questioning, and especially elaboration of patients' behaviour during meal preparation and intake, several strategies could be identified and interpreted (see Section 3.2).

4.1. COVID-19's Effect on Sensory Perception

Chemosensory dysfunction is common when suffering from COVID-19 [2,4], and as other studies indicate, the majority of COVID-19 patients suffered from chemosensory dysfunction, either in the acute phase, or the sequelae thereof. Specifically, olfactory and gustatory impairments are common among COVID-19 patients, both when reviewing the literature [27,28], and when looking at the participants of the present study. In the present study, the frequency of a combined olfactory and gustatory loss was more common, than suffering from gustatory or olfactory loss separately, which was also the results of a recent study [28]. Some participants experienced the condition worsening over time, observed as anosmia and ageusia developing into parosmia and parageusia. Whether this development in sensory dysfunction is a symptom of soon recovery or opposing long-term effects, cannot be concluded based on the present study.

4.2. COVID-19's Effect on Appetite

Most participants reported experiencing alterations in appetite sensations, and consequently, food intake, due to COVID-19, but the quality of this change (higher versus lower appetite and food intake) varied between the participants. The majority of participants experienced a decreased appetite and consequently decreased food intake, whereas fewer participants experienced a constant unsatisfied appetite, resulting in a constant search for food and thus increased food intake. These results support results from previous studies on patients with chemosensory disorders, where more patients reported to eat less than eat more [21,29].

4.2.1. Decreased Appetite and Intake

Common flu and common cold often correlate with reduced appetite [30]. Furthermore, a study on patients suffering from chemosensory disorders showed that a sudden onset of anosmia, as is the case with COVID-19, is related to weight loss due to lack of appetite [21]. This was likewise reported in the present study, where the participants experienced a reduced appetite during the acute phase of COVID-19, along with the symptoms fever, headache and throat pains (see Section 2.1). Specifically, participants expressed that hunger sensations in general were absent, prior to the meals, and also that a satiated state was reached faster during a meal compared to before COVID-19. Satiation, defined as a feeling of fullness evolving during a meal [31], is linked to the palatability of the food consumed based on perception of the sensory properties. During the early stages of a meal, an 'appetiser' effect had previously been found, as a response to palatable food experiences [32]. In line with this, models of appetite suggest a positive-feedback reward mechanism underlying the sensory enhancement of appetite [33]. In several studies, increases in hunger have been reported in the early stages of a meal as a response to palatable foods [32,34,35]. In addition, palatability has also been liked to significant increases in food intake in general [36–38]. At later stages during the meal, hunger sensations decline, and the sensation of satiation takes over. As COVID-19 often causes a dysfunctional sensory perception, the palatability of food is suffering. The subjective reports suggesting a faster satiation during a meal could therefore likely be related to a decreased or omitted early stage positive-feedback on appetite, resulting in a faster decrease of hunger sensations and onset of satiation. Furthermore, sensory stimuli are linked to motivation to initiate feeding [35], via value representation of the rewarding properties of food. The link between sensory stimuli, hedonics and appetite has, for example, been observed via the effect of odours on appetite stimulation, and especially via the focus on congruent foods [39–41]. As seen in the results from the present study, most of the participants suffered from ageusia and anosmia, and several of them mentioned specifically how this affected their appetite. The ability to perceive odours when cooking a meal was mentioned by several participants as a factor that, under normal circumstances, would induce appetite, but due to the dysfunctional sensory perception, this motivation to eat was not stimulated.

4.2.2. Increased Appetite and Intake

The second group of participants experienced a conditioned increase in appetite, which was expressed as 'unpleased senses', and a constant search for foods to reach satisfaction (see Section 3.1.1). This experience is related to the term 'sensory satisfaction', which is regarded as a state of contentment where sensory desires are fulfilled [42–44]. Eating is both hedonically and homeostatically driven, and the underlying hypothesis behind 'sensory satisfaction' propose that a faster hedonic fulfilment via the food's sensory properties is associated with a lower desire to continue eating, resulting in faster satiation and potentially a reduced intake [45,46]. When suffering from ageusia and anosmia, the sensory stimulation and resulting hedonic response is decreased, which can explain why these participants' search for sensory stimulation to experience the food-related pleasure they normally experienced before COVID-19. A sensory satisfying meal has previously been found to result in faster satiation, among subjects not suffering from COVID-19 [42,46]. In these two studies, sensory satisfaction increased by altering the trigeminal stimulation during the meal using cayenne pepper. Subjects exposed to the slightly spicier meal felt significantly more sensory satisfied [42], which resulted in reaching a satiated state significantly faster compared to when they ate the meal without added cayenne pepper. This could indicate that a lack of sensory satisfaction could result in an increased intake. The participants in the present study explicitly expressed their satiated sensation as 'unsatisfied', as a result of the inability to perceive taste and flavour. although the participants felt satiated, they kept eating, trying to satisfy their desire for sensory stimulation. Similar results have been found among patients suffering from chemosensory disorders [21]. Patients who ate more reported doing so in an effort to taste the food. Therefore, it can be hypothesized that patients who are able to perceive (or sometimes perceive) flavour sensations are the ones increasing their appetite and intake in order to experience the pleasure from the flavour sensation. Future studies will have to clarify this hypothesis.

Increased appetite, and consequently, increased intake, could further be related to the emotional consequences of COVID-19. Though differences were found in the present study with respect to how chemosensory dysfunction affected the participants' emotional state, all participants expressed negative emotions related to suffering from dysgeusia and dysosmia. A previous study of the effect of COVID-19 on mental and emotional state focused on the effect of quarantine and social isolation [46], but not explicitly related to sensory perception. The study found frequent reports of mental and emotional implications, such as anxiety, emotional distress, and fear during the COVID-19 pandemic [47,48]. Presence of negative emotions such as anxiety, fear and emotional distress, and lack of positive emotions such as happiness, relaxation and positive mood [49], as well as boredom [50], are all emotional states associated with increased food intake. All of the above-mentioned mental or emotional implications were either mentioned by the participants specifically, or could be interpreted from their statements. Therefore, it cannot be ruled out that the emotional effects of the COVID-19 pandemic in general also cause appetite altering effects.

4.3. *COVID-19's Effect on Food-Related Pleasure*

Food-related pleasure was mentioned directly or indirectly by all participants. Dysgeusia and dysosmia led to low food-related pleasure, and a lot of the participants expressed being surprised to experience the impact of chemosensory dysfunction on well-being.

A recent study including more than 8000 respondents across 14 countries investigated associations between food, drinks and feeling good [51]. Respondents were instructed to write down the first four words that came to mind when thinking about food and beverages and feeling good. The sensory and hedonic value of food was mentioned by 34% of the respondents, and the sub-category 'taste good' was mentioned by 24% of the respondents [51]. This study, combined with the broad range of studies and models focusing on sensory properties and acceptance, e.g., [7–9], sensory properties and preference [10], sensory properties and food behaviour, e.g., [11–14], and liking as a determinant for intake [15] highlight the close relationship between sensory properties and pleasure, and when put in the context of the present study, this explains the huge impact of sensory dysfunction on subjective well-being.

4.4. Coping Strategies

A key result from the present study was the identification of four key themes when studying the subjective strategies involved in maintaining appetite and food-related pleasure during the acute and post-acute phase of COVID-19: (1) A focus on well-functioning senses, (2) a focus on familiarity, (3) a focus on the eating environment, and (4) a focus on post-ingestive well-being.

4.4.1. A Focus on Well-Functioning Senses

The sensory food experience is related to the five sensory modalities: sight, smell, hearing, taste and touch. The importance of sensory properties and their cross-modal interactions for appetite and hedonic food appreciation is evident, but little has been done to examine the relative importance of each of the five modalities, respectively, for hedonic food appreciation; are all five modalities of equal importance, or does a single (or several) modality/modalities stand out? The few studies conducted previously all point at 'taste' (to be understood as flavour, the combined perception of aroma and taste) as being the most important sensory modality for hedonic food appreciation [52–54], indicating that consumers do not pay equal attention to all modalities. The importance of flavour was supported in the present study, but the importance of flavour per se in relation to the other modalities differed among the participants and affected subjective food-related pleasure and appetite.

Participants suffering from dysosmia and dysgeusia but who managed to shift their focus from flavour primarily driving food-related pleasure towards focusing on the sense of touch via perceptions of textures, temperatures and burning sensations, described a higher degree of food-related pleasure than the participants maintaining their focus on (lack of) the taste and odour sensations. Though the participants did not experience the same degree of food-related pleasure, as before COVID-19, a focus on the well-functioning senses, or other areas of the meal that brought pleasure, such as social eating (see Sections 3.2.3 and 4.4.3), helped the participants to maintain some degree of food-related pleasure, as it would increase the chance for the meal to meet their expectations. Inability to focus on the well-functioning sense of touch resulted in increased frustration regarding food intake, and a feeling of dissatisfaction. In studies on patients suffering from chemosensory disorders, similar changes in food habits have been reported. Often, diets were enriched with hot and spicy food [16,21,55]. Additionally, there was a trend towards choosing foods with a crunchy, crispy, smooth and creamy texture [16,21].

The ability to shift focus toward well-functioning senses can be hypothesized to depend on individual differences. The study by Andersen and colleagues [54] further found that for 42% of consumers, the taste (flavour) was the most important sensory modality for the hedonic food experience. For 19% of the subjects, appearance was most important, and for 18%, odour and texture, respectively, were most important. For only 9% of the subjects was taste the least important modality. Therefore, it might be that the participants in our study who managed to focus on the sense of touch were not primarily driven by the flavour in their appreciation of food.

4.4.2. A Focus on Familiarity

Some participants showed reluctance to eat novel foods, especially while suffering from ageusia. Being unable to perceive any tastes or flavours increased the importance of recognizing the food. Eating unfamiliar foods was associated with disgust, and made the participant incapable of continuing the meal. This reaction can be explained by the evolutionary importance of sensory perception for surviving, where the sensory perception guided humankind to avoid ingesting potential harmful foods [56]. The reaction of disgust could therefore be interpreted as an innate reaction, where unfamiliar food that has not been learned to be safe to ingest is rejected, if the sensory properties cannot be perceived and thereby help to determine, whether the meal is safe to eat [57].

4.4.3. A Focus on the Eating Situation

In the present study, many participants expressed having an increased food intake when partaking in a social eating context, for example eating dinner with their family. The participants described the increased food intake to be caused by the focus of the meal to be shifted away from the food itself, and towards the joy of eating with family. No research explains the effect of eating with others when suffering from chemosensory dysfunction, but as discussed in Section 4.2.2, shifting focus away from the flavour and taste of food, towards other aspects of eating still providing pleasure, could increase the food intake. Eating with family and friends is generally known to increase food intake [58], and under normal circumstances, the availability of food per se when eating in a social context can cause an individual to overindulge [59]. Though none of the participants in the current study overindulged while suffering from chemosensory dysfunction, this phenomenon of social eating could still drive the increased food intake reported in the present study.

4.4.4. A Focus on Post-Ingestive Well-Being

Food-related pleasure (and reward) within the appetite space has primarily focused on wanting and liking of food, prior to or during consumption, respectively. Though it has been argued that food-derived reward also depends on the mental and bodily well-being experienced after eating [44], this element has not been researched until recently [60–62]. 'Post-ingestive food pleasure' is defined as a subjective conscious sensation of pleasure and joy experienced after eating [62] and is driven by both mental and physical sensations, which can be measured via interoception. Interoception functions as a basis for self-awareness and subjective feelings, and can provide insights into the extended appetite experience.

In the present study, it was found that the participants expressed both mental and physical sensations involved in food-related pleasure, and appetite could be maintained by focusing on these both in the relation to the prior to, during and post eating experience.

Specifically, it was found that some participants found pleasure in eating unhealthy foods like chocolate and chips. For this group, the actual flavour was secondary (as it was in many cases reported to be dysfunctional), and the memory of the flavour and the feeling of giving themselves a treat was primary. Calorie-dense foods have previously been proposed as 'comfort foods' [63], and when associated with prior experiences, can hold a nostalgic or sentimental appeal [64], as in the present study. As discussed above, suffering from ageusia can have a devastating effect on an individual's well-being, and it can be the root of several negative emotions. Among the triggers of comfort food eating is the feeling of negative emotions [65], and/or the intent to remedy negative emotions [66]. Eating palatable food is in general known to release mood-enhancing chemicals in the brain [67], which can explain the desire for unhealthy food reported by some participants. Wagner et al. [63] found no difference in subjects' (n = 100) emotional state after watching an upsetting movie, as long as they were offered something to eat. The subjects were either exposed to their favourite comfort food, popcorn, a neutral snack, or nothing to eat at all [63]. Thereby, comfort eating can be interpreted as a subjective behaviour to relieve negative emotions.

In the current study, another group of participants applied the opposite strategy and focused on healthy eating as a source of pleasure. These participants lowered their expectations for the sensory experience, and focused on the holistic feeling of eating foods, which were considered nutritionally 'good for their body'. This group of participants specifically stopped eating any type of unnecessary or unhealthy foods, as the hedonic aspect related to the sensory properties could no longer be perceived. Interestingly, this group of people reported a higher food-related pleasure and appetite than the group focusing on pleasure from unhealthy foods. The quantitative element of pleasure and the difference between the participants applying a healthy versus unhealthy eating behavioural strategy can be explained in terms of (dis)confirmed expectations. Satisfaction is experienced either when expectations are confirmed, or when experiencing a positive disconfirmation (positive surprise), whereas dissatisfaction is felt when expectations are negatively disconfirmed (disappointment) [68]. Down-adjustment of expectations towards the sensory experience, can thus cause a higher likelihood of expectation confirmation compared to maintaining a focus on the flavour as it used to be perceived, which is more likely to cause a negative disconfirmation with dysfunctional sensory perception as a result of COVID-19. Healthy eating has many positive effects on the body, both mentally [69] and physically [70], and previous studies have found that healthy eating was associated with post-ingestive food satisfaction [71] and post-ingestive well-being [61].

4.5. Suggestions for Future Research

The present research increased our understanding of the subjective experience of changes in appetite, sensory perception and food-related pleasure when suffering from the long-term effects of COVID-19. Of potentially even greater importance, the research highlighted potential strategies for coping with these changes to maintain appetite and food-related pleasure. The findings are of relevance for health professionals (e.g., dieticians) working with patients suffering from COVID-19 and chemosensory disorders. In addition, the findings can potentially be extended to other patient groups showing decreased appetite and anhedonia. Therefore, a future focus should be on how to apply the findings among these patient groups.

Along with the application of the findings, the authors suggest that future research focuses on treatments for restoring taste and odour perception. One potential treatment for odour perception impairment is smell training, where patients sniff odours regularly to relearn them. There is evidence from before the COVID-19 pandemic that smell training can improve smell function in people with such impairments [72], but the effect on smell function among COVID-19 patients suffering from anosmia is unknown. In addition to the study of the effect of smell training, the wider applicability of the training procedure in a COVID-19 situation needs to be addressed. For example, it is relevant to know whether anosmic patients in all phases of the COVID-19 disease can benefit from the training, and how to manage the training without the risk of infecting others.

4.6. Limitations

From the findings of the present study, it is evident that the sensory food experience is very important for appetite; both for the motivation to engage in food intake and for the continued consumption during a meal. However, as the participants also expressed other symptoms related COVID-19 disease and long-term effects, we cannot conclude that changes in appetite and food-related pleasure are due to sensory impairments only. To be able to study this directly, it would be desirable to compare groups of COVID-19 patients with and without sensory impairments, respectively, and study effects on appetite and food-related pleasure.

A limitation when stating our findings from this qualitative study is that the outcome was strongly dependent on the replies of the participants. A total of 19 people participated, and the results might therefore not apply to the whole group of people suffering from long-term appetite-related effects of COVID-19. Furthermore, the findings are from subjective reports only, and are not backed-up by objective means.

For a total of five participants in the present study, the diagnosis with COVID-19 relied on a subjective assessment, as they were not eligible for a test through the Danish health care system at the time, when they were in the acute phase of COVID-19. Therefore, they do not hold concrete documentation for a diagnosis with COVID-19. However, all five participants mentioned having suffered from symptoms regarded common for COVID-19 during the acute phase of the disease, and were suffering from the common symptoms of long-term consequences of COVID-19, when they were enrolled in the study.

Finally, as the participants were in the post-acute phase of COVID-19 when enrolled in the study, findings regarding the acute phase rely on the memory of the participants. We know from previous studies that retrospection is not necessarily accurate. Therefore, the results regarding the acute phase of COVID-19 should preferably be confirmed in studies in which subjects are not relying on their memory to provide replies.

5. Conclusions

Changes in sensory perception, appetite and food-related pleasure are common, both in the acute and post-acute phases of COVID-19. This qualitative in-depth interview study brought a deeper understanding of the subjective experience of these changes, the effect on eating behaviour, and how to cope with the changes to maintain appetite and pleasure when recovering from COVID-19.

The study provides confirmatory evidence for an effect of COVID-19 on chemosensory functions, which include ageusia, anosmia, and parosmia as the most common during the post-acute phase of COVID-19. Chemosensory function was, via the study of dysfunction, found to be the most influential driver for food-related pleasure.

Though all participants suffered from lowered enjoyment of food-related experiences, this study points at new strategies to maintain appetite and food-related pleasure via a focus on other pleasure giving factors when suffering from chemosensory dysfunction. Four key themes were identified; a focus on well-functioning senses, a focus on familiar foods, a focus on the eating environment, and a focus on post-ingestive well-being. The wider applicability of these strategies could be advantageously applied in future studies focusing on day-to-day living with COVID-19 and diseases likewise affecting sensory perception.

Author Contributions: Conceptualization, A.T.Z.B.H., N.C. and B.V.A.; methodology, A.T.Z.B.H., N.C. and B.V.A.; validation, A.T.Z.B.H., N.C. and B.V.A.; formal analysis, A.T.Z.B.H. and B.V.A.; investigation, A.T.Z.B.H. and N.C.; resources, A.T.Z.B.H. and N.C.; data curation, A.T.Z.B.H. and N.C.; writing—original draft preparation A.T.Z.B.H. and B.V.A.; writing—review and editing, A.T.Z.B.H., N.C. and B.V.A.; visualization, A.T.Z.B.H.; supervision, B.V.A.; project administration, B.V.A.; funding acquisition, B.V.A. All authors have read and agreed to the published version of the manuscript.

Funding: This research was funded by EiT Food—EiT Food ivzw, Unicenter A, Philipssite 5 bus 34, 3001 Leuven, Belgium. The KAVA grant number is: 20402.

Institutional Review Board Statement: According to the Consolidation Act on Research Ethics Review of Health Research Projects, Consolidation Act number 1083 of 15 September 2017, section 14(2) notification of questionnaire surveys or interview projects to the research ethics committee system is only required if the project involves human biological material. Ref. request: 247/2020.

Informed Consent Statement: Informed consent was obtained from all subjects involved in the study.

Data Availability Statement: The data presented in this study are available on request from the corresponding author. The data are not publicly available due to the qualitative character of the study.

Acknowledgments: The authors would like to thank the EiT Mtching COVID project partners for their interest and supporting this research.

Conflicts of Interest: The authors declare no conflict of interest. The funders had no role in the design of the study; in the collection, analyses, or interpretation of data; in the writing of the manuscript, or in the decision to publish the results.

References

1. WHO. WHO Coronavirus Disease (Covid-19) Dashboard. 2021. Available online: https://covid19.who.int/ (accessed on 8 January 2021).

2. Vaira, L.A.; Hopkins, C.; Petrocelli, M.; Lechien, J.R.; Chiesa-Estomba, C.M.; Salzano, G.; Cucurullo, M.; Salzano, F.A.; Saussez, S.; Boscolo-Rizzo, P.; et al. Smell and taste recovery in coronavirus disease 2019 patients: A 60-day objective and prospective study. *J. Laryngol. Otol.* **2020**, *134*, 1–7. [CrossRef] [PubMed]
3. Hummel, T.; Landis, B.N.; Hüttenbrink, K.-B. Smell and taste disorders. *GMS Curr. Top. Otorhuinolaryngol. Head Neck Surg.* **2012**, *10*, Doc04.
4. Tenforde, M.W.; Kim, S.S.; Lindsell, C.J.; Rose, E.B.; Shapiro, N.I.; Files, D.C.; Gibbs, K.W.; Erickson, H.L.; Steingrub, J.S.; Smithline, H.A.; et al. Symptom Duration and Risk Factors for Delayed Return to Usual Health Among Outpatients with COVID-19 in a Multistate Health Care Systems Network. *Morb. Mortal. Wkly. Rep.* **2020**, *69*, 993–998. [CrossRef]
5. Farhadi, S.; Ovchinnikov, R.S. The Relationship between Nutrition and Infectious Diseases: A Review. *Biomed. Biotechnol. Res. J.* **2020**, *2*, 168–172. [CrossRef]
6. Boesveldt, S.; de Graaf, K. The Different Role of Smell and Taste for Eating Behavior. *Perception* **2017**, *46*, 307–319. [CrossRef] [PubMed]
7. Harper, R. The nature and importance of individual differences. In *Criteria of Food Acceptance: How Man Chooses What He Eats*; Solms, J., Hall, R.L., Eds.; Forester: Zürich, Switzerland, 1981; pp. 220–237.
8. Land, D.G. What is sensory quality? In *Sensory Quality in Food and Beverages: Definition, Measurement and Control*; Williams, A.A., Atkins, R.K., Eds.; Ellis Horwood: Chichester, UK, 1983.
9. Tuorila, H. Sensory perception as a basis of food acceptance and consumption. In *Consumer-Led Food Product Development*; MacFie, H.J.H., Ed.; Woodhead Publishing Limited: Cambridge, UK, 2007; pp. 34–59.
10. Khan, M.A. Evaluation of food selection patterns and preferences. *Crit. Rev. Food Sci. Nutr.* **1981**, *15*, 129–153. [CrossRef]
11. Cardello, A.V. The role of human senses in food acceptance. In *Food Choice, Acceptance and Consumption*; Meiselman, H.L., MacFie, H.J.H., Eds.; Blackie Academic and professional: Glasgow, UK, 1996; pp. 1–82.
12. Fürst, T.; Connors, M.; Bisogni, C.; Sobal, J.; Falk, L.W. Food choice: A conceptual model of the process. *Appetite* **1996**, *26*, 247–266. [CrossRef]
13. Connors, M.; Bisogni, C.; Sobal, J.; Devine, C. Managing values in personal food systems. *Appetite* **2001**, *36*, 189–200. [CrossRef] [PubMed]
14. Köster, E.P. Diversity in the determinants of food choice: A psychological perspective. *Food Qual. Prefer.* **2009**, *20*, 70–82. [CrossRef]
15. Drewnowski, A.; Hann, C. Food preferences and reported frequencies of food consumption as predictors of current diet in young women. *Am. J. Clin. Nutr.* **1999**, *70*, 28–36. [CrossRef] [PubMed]
16. Merkonidis, C.; Grosse, F.; Ninh, T.; Hummel, C.; Haehner, A.; Hummel, T. Characteristics of chemosensory disorders—Results from a survey. *Eur. Arch. Oto-Rhino-Laryngol.* **2015**, *272*, 1403–1416. [CrossRef]
17. Frasnelli, J.; Hummel, T. Olfactory dysfunction and daily life. *Eur. Arch. Otolaryngol.* **2005**, *262*, 231–235. [CrossRef]
18. Miwa, T.; Furukawa, M.; Tsukatani, T.; Costanzo, R.M.; DiNardo, L.J.; Reiter, E.R. Impact of olfactory impairment on quality of life and disability. *Arch. Otolaryngol. Head Neck Surg.* **2001**, *127*, 497–503. [CrossRef]
19. Temmel, A.F.; Quint, C.; Schickinger-Fischer, B.; Klimek, L.; Stoller, E.; Hummel, T. Characteristics of olfactory disorders in relation to major causes of olfactory loss. *Arch. Otolaryngol. Head Neck Surg.* **2002**, *128*, 635–641. [CrossRef]
20. Hummel, T.; Nordin, S. Olfactory disorders and their consequences for quality of life. *Acta Otolaryngol.* **2005**, *125*, 116–121. [CrossRef]
21. Keller, A.; Malaspina, D. Hidden consequences of olfactory dysfunction: A patient report series. *BMC Ear Nose Throat Disord.* **2013**, *13*, 8. [CrossRef] [PubMed]
22. DiCicco-Bloom, B.; Crabtree, B.F. Making Sense of Qualitative Research—The Qualitative Research Interview. *Med. Educ.* **2006**, *40*, 314–321. [CrossRef] [PubMed]
23. Yin, R.K. *Qualitative Research from Start to Finish*; The Guilford Press: New York, NY, USA, 2010.
24. Smith, J.A.; Osborn, M. Interpretive Phenomenological Analysis. In *Doing Social Psychology Research*; Breakwell, G.M., Ed.; The British Psychological Society: Leicester, UK; Blackwell Publishing Ltd.: Hoboken, NJ, USA, 2004; pp. 229–254.
25. Noble, H.; Heale, R. Triangulation in research, with examples. *Evid. Based Nurs.* **2019**, *22*, 67–68. [CrossRef] [PubMed]
26. Braun, V.; Clarke, V. Using Themantic analysis in psychology. *Qual. Res. Psychol.* **2006**, *3*, 77–101. [CrossRef]
27. Rojas-Lechuga, M.J.; Izquierdo-Domínguez, A.; Chiesa-Estomba, C.; Calvo-Henríquez, C.; Villarreal, I.M.; Cuesta-Chasco, G.; Bernal-Sprekelsen, M.; Mullol, J.; Alobid, I. Chemosensory dysfunction in COVID-19 out-patients. *Eur. Arch. Otorhinolaryngol.* **2020**. [CrossRef]
28. Fjaeldstad, A.W. Prolonged complaints of chemosensory loss after Covid-19. *Dan. Med. J.* **2020**, *67*, 1–11. [PubMed]
29. Aschenbrenner, K.; Scholze, N.; Joraschky, P.; Hummel, T. Gustatory and olfactory sensitivity in patients with anorexia and bulimia in the course of treatment. *J. Psychiatry Res.* **2008**, *43*, 129–137. [CrossRef] [PubMed]
30. Eccles, R. Mechanisms and symptoms of common cold and flu. In *Common Cold*; Eccles, R., Weber, O., Eds.; Birkhäuser Verlag: Basel, Switzerland, 2009; pp. 23–45.
31. Blundell, J.; Bellisle, F. *Satiation, Satiety and the Control of Food Intake: Theory and Practice*; Blundell, J., France, B., Eds.; Woodhead Publishing Limited: Cambridge, UK, 2013.
32. Yeomans, M.R. Palatability and the Micro-structure of Feeding in Humans: The Appetizer Effect. *Appetite* **1996**, *27*, 119–133. [CrossRef] [PubMed]

33. Reichelt, A.C.; Westbrook, R.F.; Morris, M.J. Integrating of rewards signalling and appetite regulation peptide in the control of food-cue responses. *Br. J. Pharmacol.* **2015**, *172*, 5225–5238. [CrossRef]
34. Hill, A.J.; Magson, D.; Blundell, J.E. Hunger and palatability: Tracking ratings of subjective experience before, during and after the consumption of preferred and less preferred food. *Appetite* **1984**, *5*, 361–371. [CrossRef]
35. Yeomans, M.R.; Gray, R.W.; Mitchell, C.J.; True, S. Independent Effects of Palatability and Within-meal Pauses on Intake and Appetite Ratings in Human Volunteers. *Appetite* **1997**, *29*, 61–76. [CrossRef] [PubMed]
36. Bobroff, E.M.; Kissileff, H. Effects of changes in palatability on food intake and the cumulative food intake curve of man. *Appetite* **1986**, *7*, 85–96. [CrossRef]
37. Guy-Grand, B.; Lenhert, V.; Doassans, M.; Bellisle, F. Type of test-meal affects palatability and eating style in humans. *Appetite* **1994**, *22*, 125–134. [CrossRef]
38. Hill, S.W. Eating responses of humans during dinner meals. *J. Comp. Physiol. Psychol.* **1974**, *86*, 652–657. [CrossRef]
39. Zoon, H.F.A.; De Gaaf, C.; Boesveldt, S. Food odours directs specific appetite. *Foods* **2016**, *5*, 12. [CrossRef]
40. Morquecho-Campos, P.; De Gaaf, C.; Boesveldt, S. Smelling our appetite? The influence of food odors on congruent appetite, food preferences and intake. *Food Qual. Prefer.* **2020**, *85*, 1–7. [CrossRef]
41. Ferriday, D.; Brunstrom, J.M. How does food-cue exposure lead to larger meal sizes? *Br. J. Nutr.* **2008**, *100*, 1325–1332. [CrossRef]
42. Andersen, B.V.; Byrne, D.V.; Bredie, W.L.P.; Møller, P. Cayenne pepper in a meal: Effect of oral heat on feelings of appetite, sensory specific desires and well-being. *Food Qual. Prefer.* **2017**, *60*, 1–8. [CrossRef]
43. Cornil, Y.; Chandon, P. Pleasure as a Substitute for Size: How Multisensory Imagery Can Make People Happier with Smaller Food Portions. *J. Mark. Res.* **2016**, *53*, 847–864. [CrossRef]
44. Møller, P. Satisfaction, satiation, and food behaviour. *Curr. Opin. Food Sci.* **2015**, *3*, 59–64. [CrossRef]
45. Andersen, B.V. Sensory Factors in Food Satisfaction. An Understanding of the Satisfaction Term and a Measurement of Factors Involved in Sensory- and Food Satisfaction. Ph.D. Thesis, Technical University of Denmark, Kgs. Lyngby, Denmark, 2014; 175p.
46. Møller, P. Gastrophysics in the brain and body. *Flavour* **2013**, *2*, 2–4. [CrossRef]
47. Canet-Juric, L.; Andrés, M.L.; del Valle, M.; Lópes-Morales, H.; Poó, F.; Galli, J.I.; Yerro, M.; Urquijo, S. A Longitudinal Study on the Emotional Impact Cause by the COVID-19 pandemic Quarantine on General Population. *Front. Psychol.* **2020**, *11*, 1–17. [CrossRef]
48. Kleinberg, B.; van der Vegt, I.; Mozes, M. Measuring Emotions in the Covid-19 Real World Worry Dataset. *arXiv* **2020**, arXiv:2004.04225.
49. Devonport, T.J.; Nicholls, W.; Fullerton, C. A systematic review of the association between emotions and eating behaviour in normal and overweight adult populations. *J. Health Psychol.* **2019**, *24*, 3–24. [CrossRef] [PubMed]
50. Abramson, E.E.; Stinson, S.G. Boredom and eating in obese and non-obese individuals. *Addict. Behav.* **1977**, *2*, 181–185. [CrossRef]
51. Sulmont-Rossé, C.; Drabek, R.; Almli, V.L.; van Zyl, H.; Silva, A.P.; Kern, M.; McEwan, J.A.; Ares, G. A Cross-cultural perspective of feeling good in the context of foods and beverages. *Food Res. Int.* **2019**, *115*, 292–301. [CrossRef]
52. Moskowitz, H.; Krieger, B. What sensory characteristics drive product quality? An assessment of individual differences. *J. Sens. Stud.* **1993**, *8*, 271–282. [CrossRef]
53. Moskowitz, H.; Krieger, B. The contribution of sensory liking to overall liking: An analysis of sex food categories. *Food Qual. Prefer.* **1995**, *6*, 83–90. [CrossRef]
54. Andersen, B.V.; Brockhoff, P.B.; Hyldig, G. The importance of liking of appearance, -odour, -taste and -texture in the evaluation of overall liking. A comparison with the evaluation of sensory satisfaction. *Appetite* **2019**, *71*, 228–232. [CrossRef]
55. Aschenbrenner, K.; Hummel, C.; Teszmer, K.; Krone, F.; Ishimaru, T.; Seo, H.S.; Hummel, T. The influence of olfactory loss on dietary behaviors. *Laryngoscope* **2008**, *118*, 135–144. [CrossRef]
56. Reed, D.R.; Knaapila, A. Genetics of taste and smell: Poisons and Pleasures. *Prog. Mol. Biol. Transl. Sci.* **2010**, *94*, 213–240. [CrossRef]
57. Pliner, P.; Pelchat, M.; Grabski, M. Reduction of neophobia in humans by exposure to novel foods. *Appetite* **1993**, *20*, 111–123. [CrossRef] [PubMed]
58. Ruddock, H.K.; Brunstrom, J.M.; Vartanian, L.R.; Higgs, S. A systematic review and meta-analysis of the social facilitation of eating. *Am. J. Clin. Nutr.* **2019**, *110*, 842–861. [CrossRef]
59. Herman, C.P. The social facilitation of eating or the facilitation of social eating? *J. Eat. Disord.* **2017**, *5*. [CrossRef]
60. Duerlund, M.; Andersen, B.V.; Byrne, D.V. Dynamic Changes in Post-Ingestive Sensations after Consumption of a Breakfast Meal High in Protein or Carbohydrate. *Food* **2019**, *8*, 413. [CrossRef]
61. Duerlund, M.; Andersen, B.V.; Grønbeck, M.S.; Byrne, D.V. Consumer reflections on post-ingestive sensations. A qualitative approach by means of focus group interviews. *Appetite* **2019**, *142*, 104350. [CrossRef]
62. Duerlund, M.; Andersen, B.V.; Wang, K.; Chan, R.C.K.; Byrne, D.V. Post-Ingestive Sensations Driving Food Pleasure: A Cross-Cultural Consumer Study Comparing Denmark and China. *Foods* **2020**, *9*, 617. [CrossRef]
63. Wagner, H.S.; Ahlstrom, B.; Redden, J.P.; Vickers, Z.; Mann, T. The myth of comfort food. *Health Psychol.* **2014**, *33*, 1552–1557. [CrossRef]
64. Locher, J.L.; Yoels, W.C.; Maurer, D.; Van Ells, J. Comfort foods: An exploratory journey into the social and emotional significance of food. *Food Foodways* **2005**, *13*, 272–297. [CrossRef]

65. Dubé, L.; LeBel, J.L.; Lu, J. Affect asymmetry and comfort food consumption. *Physiol. Behav.* **2005**, *86*, 559–567. [CrossRef] [PubMed]
66. Evers, C.; Marijn Stok, F.; de Ridder, D.T. Feeding your feelings: Emotion Regulation Strategies and Emotional Eating. *Pers. Soc. Psychol. Bull.* **2010**, *36*, 792–804. [CrossRef]
67. Magnen, J.L. *Hunger*; Cambridge University Press: Cambridge, UK, 1986.
68. Deliza, R.; Macfie, H.J.H. The generation of sensory expectation by external cues and its effect on sensory perception and hedonic ratings: A review. *J. Sens. Stud.* **1996**, *11*, 103–128. [CrossRef]
69. Firth, J.; Gangwisch, J.E.; Borsini, A.; Wootton, R.E.; Mayer, E.A. Food and mood: How do diet and nutrition affect mental wellbeing? *BMJ* **2020**, *369*, m2382. [CrossRef] [PubMed]
70. Ezzati, M.; Riboli, M.D. Behavioral and dietary risk factors for noncommunicable diseases. *N. Engl. J. Med.* **2013**, *369*, 954–964. [CrossRef]
71. Andersen, B.V.; Hyldig, G. Consumers' view on determinants to food satisfaction. *A qualitative approach. Appetite* **2015**, *95*, 9–16. [CrossRef] [PubMed]
72. Boesveldt, S.; Postma, E.M.; Boak, D.; Welge-Luessen, A.; Schöpf, V.; Mainland, J.D.; Martens, J.; Ngai, J.; Duffy, V.B. Amosmia—A clinical review. *Chem. Senses* **2017**, *7*, 513–523. [CrossRef] [PubMed]

foods

MDPI

Article

COVID-19 Lockdown and Self-Perceived Changes of Food Choice, Waste, Impulse Buying and Their Determinants in Italy: QuarantEat, a Cross-Sectional Study

Alessandro Scacchi [1], Dario Catozzi [1], Edoardo Boietti [1], Fabrizio Bert [1,2,*] and Roberta Siliquini [1,2]

1 Department of Public Health Sciences and Paediatrics, University of Torino, 10126 Torino, Italy; alessandro.scacchi@unito.it (A.S.); dario.catozzi@unito.it (D.C.); edoardo.boietti@unito.it (E.B.); roberta.siliquini@unito.it (R.S.)
2 City of Health and Science of Turin, Azienda Ospedaliero-Universitaria, 10126 Torino, Italy
* Correspondence: fabrizio.bert@unito.it

Abstract: Data about self-perceived food choice (FC) changes and their determinants during COVID-19 lockdowns are limited. This study investigated how the Italian lockdown affected self-perceived food purchases (FP), occurrence of impulse buying (IB), household food waste production (HFWP) and their determinants. A web-based cross-sectional survey was distributed in May 2020, collecting an opportunistic sample of the Italian population. A total of 1865 (70% females) people were enrolled, the median age was 29 (IQR 16.0). Most of the sample increased overall FP (53.4%), food consumption (43.4%), reduced HFWP (53.7%) and halved the prevalence of IB (20.9%) compared to the period before the lockdown (42.5%). Baking ingredients, fresh vegetables, fresh fruit and chocolate had the largest sales increase by individuals, while bakery products, fresh fish and salted snacks purchases highly decreased. Increased FP was associated with the occurrence of IB (adjOR 2.48, $p < 0.001$) and inversely associated with not having worked during lockdown (adjOR 0.71, $p = 0.003$). Multivariable logistic regressions revealed occurrence of IB was associated with low perceived dietary quality (adjOR 2.22, $p < 0.001$), resulting at risk, according to the Emotional Overeating Questionnaire (EOQ, adjOR 1.68, $p < 0.001$), and inversely associated with decreased HFWP (adjOR 0.73, $p < 0.012$). Reduced HFWP was associated with higher perceived dietary quality (adjOR 2.27, $p < 0.001$) and negatively associated with low score at WHO-5 Well-Being Index (adjOR 0.72, $p = 0.002$). The Italian lockdown highly affected FC behaviours, leading to positive and sustainable habits towards food purchase and consumption. Public health interventions are needed to keep these new positive effects and avoid negative consequences in case of future lockdowns.

Keywords: lockdown; COVID-19; coronavirus; food choice; food purchase; food waste; impulse buying; food consumption; mental health; emotional eating

Citation: Scacchi, A.; Catozzi, D.; Boietti, E.; Bert, F.; Siliquini, R. COVID-19 Lockdown and Self-Perceived Changes of Food Choice, Waste, Impulse Buying and Their Determinants in Italy: QuarantEat, a Cross-Sectional Study. *Foods* **2021**, *10*, 306. https://doi.org/10.3390/foods10020306

Academic Editor: Derek V. Byrne
Received: 2 November 2020
Accepted: 29 January 2021
Published: 2 February 2021

1. Introduction

On 21 February 2020, the first case of indigenous SARS-CoV2 infection in Italy was reported. A few days later, the lockdown was established in some provinces of northern Italy [1,2]. On 9 March 2020, the Italian Government decided for a stringent containment measure of lockdown on the entire national territory [3]. This measure was effective in flattening the epidemic curve and bought valuable time, allowing for the number of intensive care beds to be nearly doubled before the National Health System reached maximum capacity [4]. During lockdown, people could leave their homes only for primary activities such as work in key sectors of industry, care and services, physical exercise, medical care or food shopping. On 3 May, the government declared the end of the first phase of the lockdown by introducing a series of less restrictive anti-contagion rules [5].

The global pandemic of COVID-19 has caused radical changes in the structure of people's daily routines in most of the countries around the world, including the way people buy food, that has changed dramatically [6]. In the weeks immediately preceding the Italian lockdown, people began to panic-buy and stockpile essential and non-perishable products such as water, gloves, carbohydrate-rich staples (e.g., bread, pasta), canned food, hand sanitisers, and even toilet paper [6]. On a national scale, in March, during the acute phase of the lockdown, a +17% of grocery sales was reported, reaching almost EUR 6 billion, EUR 860 million more than the same period during the previous year. Purchase choices were mostly directed toward the stocking of non-perishable foods, in order to face potential scarcity situations. The increase in purchases affected moreover pasta, UHT (ultra-high temperature) milk, canned fish, flours and eggs, frozen foods, cold cuts and parmesan, and water [7]. Neighbourhood shops were preferred over hypermarkets, due to large queues and proximity [8]. A similar trend affected online shopping, reaching virtual overcrowding and service outages [9]. The major increase in purchases occurred in South Italy, despite being the least affected territory by COVID-19 [8].

These data are not surprising. In the literature, indeed, it is well known that during home confinement people tend to increase their food intake [10,11]. A quarter of the Italian population consumed more food and one third increased time spent cooking at home [12], while an Italian study showed that half of the sample felt anxious about their eating habits, consumed comfort food and were inclined to increase food intake to feel better [13]. Furthermore, during lockdown the perception of weight gain was observed in almost half of an Italian sample and young people resulted having a higher adherence to the Mediterranean diet [14]. Another study, conducted in Poland, reported that during quarantine people ate more snacks [15]. In particular, those with a high BMI (body mass index) tended to introduce less vegetables, fruit and beans in their diet, while a greater amount of alcohol and tobacco consumption was reported [15]. The reported big changes in food purchase and consumption habits, such as the increased reuse of leftovers, could have affected the production of household food waste, as reported in a Tunisian study [16].

However, evidence about changes of food choice, household food waste production and their associated factors during lockdown in Italy is poor. Existing studies have been carried out on limited samples or have collected data for short periods, in the primeval phase of the lockdown. Hence, it is important to increase our knowledge on the self-reported change of habits that occurred during home confinement, to encourage proactive strategies in view of potential future lockdown measures and to keep any new positive behaviours toward maintaining a sustainable and healthy lifestyle in the future.

The aim of this study was to investigate, during lockdown, how Italian people have perceived the change of their food purchases and eating habits and what are the factors associated with the self-perceived increase in food purchases, occurrence of impulse buying and household food waste production. To date, this is the first study investigating the impact of the lockdown on these habits in a national sample.

2. Materials and Methods

2.1. Study Design and Questionnaire

The QuarantEat study investigated how a sample of Italian inhabitants was affected by the lockdown in terms of self-perceived variations of food purchase, food consumption habits, physical activity levels and how home confinement impacted on mental well-being as well as on the presence of emotional overeating.

An online survey was developed using the Uniquest (LimeSurvey) platform, which was made available by the University of Turin. Our questionnaire was spread among the Italian population through a web link shared by institutional social media pages and the personal accounts of researchers. This procedure led to the enrolment of an opportunistic sample of citizens. The survey was spread a few days after the end of the Italian lockdown, starting from May 6th, in order to highlight the effects of the whole home confinement experience on people's habits and behaviours. The enrolment ended on the 31 of May, some weeks after the end of the lockdown.

The research protocol was approved by the Ethical Commission of University of Turin (prot. no. 197989). Inclusion criteria were: age equal or older than 18 years, living in Italy during lockdown period, being able to give informed consent to enrolment in the study in Italian. Before starting the questionnaire, each participant was shown a brief written summary including the aims of the research project, and finally each of them confirmed the enrolment to the study declaring their informed consent.

The questionnaire consisted of 40 questions, divided into 6 sections: socio-demographic assessment, physical activity, food purchase habits, food consumption behaviours, mental well-being evaluation and occurrence of emotional overeating. Two validated tests were included: the 5-item World Health Organization Well-Being Index (WHO-5) questionnaire and the Emotional Overeating Questionnaire-5 (EOQ-5). The full version of the questionnaire, translated into English, is available as a Supplementary Materials.

The socio-demographic section included personal data (age, gender, smoking status, relationship status, offspring) and a variety of items regarding the living environment, such as housing place, the presence of a backyard, cohabitation, geographical context (region of Italy) and the working condition during home confinement. Self-reported height and weight were included to calculate BMI. Regions of Italy were later gathered in three geographical areas as advised by National Institute of Statistics (ISTAT): North, Centre, South and Isles. Physical activity (PH) habit was investigated by asking if exercises were practiced during and before lockdown.

Overall self-perceived food consumption changes during lockdown were assessed, as well as the quality of diet and food waste, in terms of subjective increased, decreased or unvaried during lockdown compared to the period before. In addition, diet was investigated, intended as every eating regime with the purpose of body control (weight loss or gain, high protein diets), or medical reasons (due to allergies and food intolerance). Finally, we asked if, during lockdown, on an everyday basis, time spent cooking increased, decreased, or remained unchanged.

The place of food purchase was investigated (supermarket, discount, market, neighbourhood shop, online shop, home delivery), along with shopping frequency in terms of overall times leaving home for buying food per week (the Italian government suggested to go shopping no more than once per week) [17]. Impulse buying behaviour was assessed by asking if any sense of guilt or unnecessary purchase occurred after grocery shopping during lockdown, and if it ever happened before the lockdown. Finally, we proposed a list of 50 foods asking whether their purchase increased, decreased or unchanged during lockdown, as well as if it has never been bought.

To evaluate the impact of home confinement on mental health and psychological well-being in people living in Italy immediately after the lockdown, a section of the survey included the WHO-5 questionnaire, validated in Italian language and used worldwide in research [18]. It can be used as a sensitive and specific screening tool for risk of depression. This questionnaire contains five non-invasive statements about feelings during the last 14 days. A WHO-5 cut-off score of $\leq$50 is recommendable for screening for clinical depression [18].

To evaluate the occurrence of Emotional Overeating during lockdown as a coping mechanism, a section of the survey included the EOQ-5 questionnaire, validated in Italian language [19]. The EOQ-5 is a brief, valid and reliable 5-item self-report that measures the frequency of overeating behaviour in response to five negative emotions (anxiety, sadness, loneliness, tiredness and anger) during the last 28 days. A cut-off score of 2 points identifies individuals at risk for binge eating disorders. Higher EOQ-5 scores are associated with higher risk of binge eating, lower mental well-being, and lower mindful eating [19].

2.2. Statistical Analysis

Overall descriptive analyses were performed for the most prominent variables, showing frequencies for categorical variables and medians and interquartile range (IQR) for scalar variables since the normality Shapiro–Wilk test proved a non-normal distribution for age and shopping frequency. Data were also shown divided in a geographical fashion; Chi-squared test or nonparametric Mann–Whitney or Kruskal–Wallis Tests were performed.

Logistic regression analysis was performed to evaluate determinants of three prominent phenomena, highlighted by collected data and supported by evidence: increased food purchase, occurrence of impulse buying and reduction in household food waste production.

While the vast majority of variables were included in the models unchanged, for analytical purposes some of them were aggregated: for example, education level was dichotomised, aggregating university degree and post-doc studies into high level and the remaining values as middle-low level. WHO-5 and EOQ-5 scores were also dichotomised based on validated threshold values.

The selection of independent variables included into the regression models was achieved with a stepwise backwards method, in which three covariates were protected from exclusion: age, gender and education, since their potential exclusion in the final models could have led to highly biased outcomes. Results were expressed as adjusted odds ratios (AdjOR) and their 95% confidence intervals (95% CI). The statistical significance threshold was set at $p < 0.05$. The software employed for the analysis was IBM SPSS Statistics (Version 25.0). Cases with missing values were excluded from logistic regressions (listwise deletion) and retained in the descriptive analysis (pairwise deletion).

3. Results

3.1. Participant Characteristics

A total of 2524 individuals began the online survey, and 1923 of them completed every item displayed. Fifty-eight records were excluded due to inclusion/exclusion criteria: 26 of them revealed to be aged 17 years old or younger, while 32 people spent the lockdown period outside the country, reducing the number of eligible records to 1865. An analysis of completion time revealed median duration was 9:25 min (IQR 4:05); since no record could be highlighted as an outlier, none of them were discarded.

Among the sample, 69.9% of participants were female, and the median age was 29 (IQR 16.0), and almost half of them lived in northern Italy (49.7%). People in our sample living in the northern regions most commonly resulted in being women (p-value = 0.021) and older (p-value < 0.001). Almost an equal number of responders stated having reached the educational level of high school (43.1%) and university degree (42.1%), but with an important geographical variability (p-value = 0.006).

The majority of our sample resulted in living with a partner or family (81.8%), with 11.7% living alone and 6.5% with one or more roommate(s). In addition, this variable resulted in an uneven geographical distribution, with fewer people living alone or with cohabitants in the south (p-value = 0.002)

Regarding housing, 60.8% of respondents live in a flat or apartment, and 32.9% in an independent house. Living in an independent house was more common in the south (p-value = 0.005). One third of our sample, regardless of housing, stated to have in use a private yard or garden.

Only 56.7% of our sample actually worked during lockdown, with a maximum of 62.6% in the north and a minimum of 44.4 in south (p-value<0.001). There was a similar geographical distribution for healthcare workers, with a 20.3% in the northern regions, 17.7% in the centre and 14.2% in southern ones (p-value = 0.040), and an overall representation of 18.3%. Approximately one-fifth (21.9%) of the sample stated to regularly smoke, slightly more in the south (22.3%, p-value = 0.033).

The WHO-5 survey revealed a significant number of respondents (42.4%) potentially at risk of depression development, and 50.7% displayed the occurrence of a significant number of episodes of emotional overeating. None of these scores show asymmetric geographical distribution.

In addition, more than half of our sample claimed to have practiced physical activity during lockdown and as many as 76.2% of participants followed some kind of dietary regimen. Additional descriptive data are provided in Table 1.

Table 1. Participant characteristics stratified by geographical area: Descriptive and Chi-square analysis.

Variables		Median [IQR] or n (%) All (*n* = 1865)	North (*n* = 927)	Centre (*n* = 593)	South and Isles (*n* = 345)	*p*-Value
Geographical area	North Centre South and Isles	927 (49.7) 593 (31.8) 345 (18.5)				
Age		29 [16.0]	29 [17.0]	29 [15.0]	27 [16.0]	**<0.001**
Gender	Female Male Non-Binary	1304 (69.9) 558 (29.2) 3 (0.2)	679 (73.2) 246 (26.5) 2 (0.2)	394 (66.4) 199 (33.6) 0 (0.0)	231 (67.0) 113 (32.8) 1 (0.3)	**0.021**
Relationship status *Missing = 5*	Into stable relationship or married Single/divorced/widow	1194 (64.2) 666 (35.8)	600 (64.9) 324 (35.1)	387 (65.5) 204 (34.5)	207 (60.0) 138 (40.0)	0.193
Education level	Primary/Middle Sch. High School University degree Post-graduate ed.	87 (4.6) 803 (43.1) 785 (42.1) 190 (10.2)	45 (4.9) 397 (42.8) 387 (41.7) 98 (10.6)	16 (2.7) 246 (41.5) 265 (44.7) 66 (11.1)	26 (7.5) 160 (46.4) 133 (38.6) 26 (7.5)	**0.008**
Living condition	Not alone Alone	1647 (88.3) 218 (11.7)	794 (85.7) 133 (14.3)	538 (90.7) 55 (9.3)	315 (91.3) 30 (8.7)	**0.002**
Offspring	No Yes	1441 (77.3) 424 (22.7)	709 (76.5) 218 (23.5)	468 (78.9) 125 (21.1)	264 (76.5) 81 (23.5)	0.508
Housing *Missing = 1*	Room Flat Independent house	117 (6.3) 1133 (60.8) 615 (32.9)	45 (4.9) 572 (61.7) 308 (33.2)	55 (9.3) 359 (60.5) 179 (30.2)	17 (4.9) 202 (58.6) 123 (35.7)	**0.005**
Yard/garden	Yes No	618 (33.1) 1247 (66.9)	313 (33.8) 614 (66.2)	181 (30.5) 412 (69.5)	124 (35.9) 221 (64.1)	0.200
Working during lockdown	Working Not working	773 (41.4) 1093 (58.6)	431 (46.5) 496 (53.5)	237 (40.0) 356 (60.0)	104 (30.1) 241 (69.9)	**<0.001**
Healthcare worker	Yes No	342 (18.3) 1523 (81.7)	188 (20.3) 739 (79.7)	105 (17.7) 488 (82.3)	49 (14.2) 296 (85.8)	**0.040**
Smoke habit	Yes No	409 (21.9) 1456 (78.1)	182 (19.6) 745 (80.4)	150 (25.3) 443 (74.7)	77 (22.3) 268 (77.7)	**0.033**
WHO-5 Well-being	≤50 >50	1074 (57.6) 791 (42.4)	512 (55.2) 415 (44.8)	348 (58.7) 245 (41.3)	214 (62.0) 131 (38.0)	0.075
EOQ-5	At risk Not at risk	920 (49.3) 945 (50.7)	435 (46.9) 492 (53.1)	299 (50.4) 294 (49.6)	186 (53.9) 159 (46.1)	0.070
BMI *Missing = 10*	*Underweight* *Normal* *Overweight* *Obese*	118 (6.4) 1273 (68.6) 366 (19.7) 98 (5.3)	65 (7.0) 638 (69.0) 173 (18.7) 49 (5.3)	39 (6.6) 391 (66.5) 131 (22.3) 27 (4.6)	14 (4.1) 244 (71.3) 62 (19.2) 22 (6.4)	0.203
Sport during lockdown	*Yes* *No*	1220 (65.4) 645 (34.6)	612 (66.0) 315 (34.0)	391 (65.9) 202 (34.1)	217 (62.9) 128 (37.1)	0.553
Being on a diet during lockdown	*Yes* *No*	444 (23.8) 1421 (76.2)	218 (23.5) 709 (76.5)	143 (24.1) 450 (75.9)	83 (24.1) 262 (75.9)	0.958

Abbreviations: IQR, interquartile range; N, number; Who-5, 5-item World Health Organization Well-Being Index (WHO-5) questionnaire; EOQ-5, Emotional Overeating Questionnaire-5 (EOQ-5); BMI, body mass index. In order to enhance readability, *p*-values < 0.05 are shown bolded.

3.2. Food Purchase Habits

Regarding food purchase habits (Table 2), the majority of our sample increased food purchases (53.4%), while 7.2% reduced them. Food consumption increased in 43.4% of the respondents. A similar size of the sample (46.5%) improved the perceived nutrition quality, while one quarter (26.6%) worsened it. More than half respondents (53.4%) reduced household food waste production. A similar amount of people (55.1%) increased time spent cooking at home, but with smaller increments in the south (p-value < 0.001).

People went shopping once or less per week in 68.9% of cases, and the mean was 1.51 (SD 1.31) occasions. A significant geographical difference was found for this behaviour, reaching its peak in southern Italy and Islands (p-value = 0.003). Most grocery shopping was made directly by the respondents (81%), showing significant geographical differences: 83.3% in the north, 81.5% in the centre and 74.8% in the south (p-value = 0.003).

Impulse buying before the lockdown occurred in 42.5% of the sample, while a strong reduction happened during lockdown; just the 20.9% of respondents occurred in this behaviour, halving its prevalence by 51%.

Delivered food was chosen by 16.6% of the sample, showing a strong North–South geographical gradient, from 18.8% to 11.9% (p-value = 0.011). 0.001).

Table 2. Food purchase and consumption habits stratified by geographical area: Descriptive and Chi-square analysis.

Variables		Median (IQR) or n (%) All (n = 1865)	North (n = 927)	Centre (n = 593)	South and Isles (n = 345)	p-Value
Food purchase	Decreased	134 (7.2)	54 (5.8)	50 (8.4)	30 (8.7)	0.090
	Unvaried	735 (39.4)	359 (38.7)	229 (38.6)	147 (42.6)	
	Increased	996 (53.4)	514 (55.4)	314 (53.0)	168 (48.7)	
Food consumption	Decreased	237 (12.7)	132 (14.2)	65 (11.0)	40 (11.6)	0.130
	Unvaried	818 (43.9)	402 (43.4)	270 (45.5)	138 (40.0)	
	Increased	810 (43.4)	393 (42.4)	258 (43.5)	167 (48.4)	
Perceived nutrition quality	Less healthy	502 (26.9)	230 (24.8)	161 (27.2)	111 (32.2)	0.079
	Unvaried	495 (26.6)	441 (47.6)	270 (45.5)	157 (45.5)	
	Healthier	868 (46.5)	256 (27.6)	162 (27.3)	77 (22.3)	
Household food waste production	Decreased	1002 (53.7)	500 (53.9)	315 (53.1)	187 (54.2)	0.933
	Unvaried	800 (42.9)	399 (43.0)	256 (43.2)	145 (42.0)	
	Increased	63 (3.4)	28 (3.0)	22 (3.7)	13 (3.8)	
Time spent cooking	Decreased	184 (9.9)	82 (8.8)	53 (8.9)	49 (14.2)	**<0.001**
	Unvaried	654 (35.0)	305 (32.9)	206 (34.7)	143 (41.4)	
	Increased	1027 (55.1)	540 (58.3)	334 (56.3)	153 (44.3)	
Grocery shopping	Personally	1513 (81.1)	772 (83.3)	483 (81.5)	258 (74.8)	**0.003**
	Someone for me	352 (18.9)	155 (16.7)	110 (18.5)	87 (25.2)	
N° trips for shopping	1/week or less	1285 (68.9)	669 (72.2)	400 (67.5)	216 (62.6)	**0.003**
	>1 per week	580 (31.1)	258 (27.8)	193 (32.5)	129 (37.4)	
	mean (SD)	1.51 (1.3)	1.42 (1.2)	1.55 (1.3)	1.65 (1.4)	**0.001**
Impulse buying during lockdown	No	1476 (79.1)	740 (79.8)	464 (78.2)	272 (78.8)	0.752
	Yes	389 (20.9)	187 (20.2)	129 (21.8)	73 (21.2)	
Impulse buying before lockdown	No	1073 (57.5)	554 (59.8)	326 (55.0)	193 (55.9)	0.147
	Yes	792 (42.5)	373 (40.2)	267 (45.0)	152 (44.1)	
Delivery food	No	1556 (83.4)	753 (81.2)	499 (84.1)	304 (88.1)	**0.011**
	Yes	309 (16.6)	174 (18.8)	94 (15.9)	41 (11.9)	
Food shops *	Supermarket	1635 (87.7)	806 (86.9)	543 (91.6)	297 (86.1)	0.171
	Small shops	711 (38.1)	452 (48.8)	231 (39.0)	128 (37.1)	0.846
	Discount market	306 (16.4)	133 (14.3)	110 (18.5)	63 (18.3)	0.057
	Online shops	277 (14.9)	172 (18.6)	74 (12.5)	31 (9.0)	**<0.001**
	Market	162 (8.7)	100 (10.8)	55 (9.3)	7 (2.0)	**<0.001**

* For this question, multiple answers were allowed. In order to enhance readability, p-values < 0.05 are shown bolded.

Finally, a majority of purchases were made in supermarkets (86.7%), while 14.9% of the respondents used online shopping, with great geographical diversities (p-value < 0.001), as well as for market purchases, occurring in 8.7% of the sample and being largely more common in the north (p-value <

3.3. Food Purchase Trends

Regarding food purchase trends, a selection of the most increased and decreased foods is shown in Figure 1. Baking products and fresh healthy food had the largest sales increase by individuals (flour and yeast +63.2%, eggs +48.4% fresh vegetables +41.2%, fresh fruits +39.0%) as well as chocolate (+26.4%) as indulgence food. Large increases affected pasta and UHT milk, too. The largest individual purchase decreases affected bakery products (pizza delivery −29.4%, ice-cream and cakes −21.7%, bread −18.3%), highly perishable foods (fresh fish −28.2%) and salted snacks (−18.4%). The complete list of purchases is available as a Supplementary Materials.

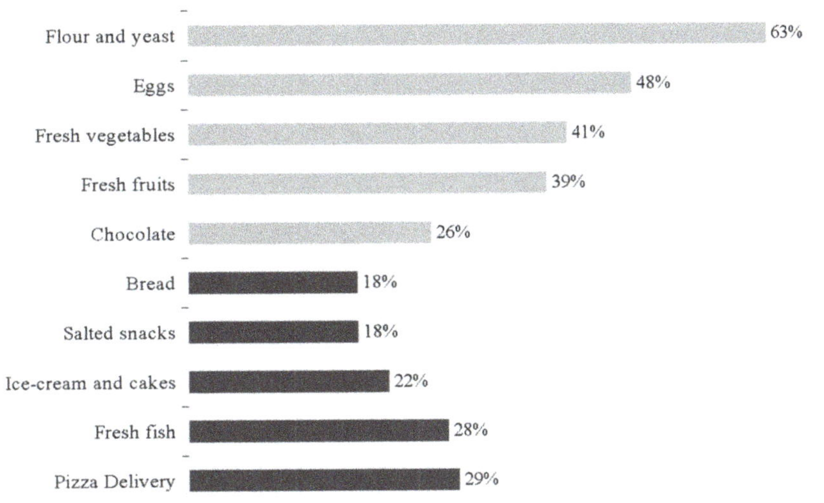

Figure 1. Top five increased (light grey) and top five decreased (dark grey) foods purchased.

3.4. Determinants of Changes of Food Purchase, Household Food Waste Production and Occurrence of Impulse Buying

Multivariable analysis final models are shown on Table 3. Due to only a small number of participants identifying themselves as "Non-binary" gender (n = 3), this category was unable to be analysed and eventually removed for logistic regression model.

The first model was designed to find associated factors of increased food purchase among population, and the strongest one resulted to be the occurrence of impulse buying (adjOR 2.48, p-value < 0.001) followed by increased time spent cooking (adjOR 2.12, p-value < 0.001), presence of offspring (adjOR 1.76, p-value = 0.0101), perceived nutrition quality (less healthy adjOR 1.66, p-value < 0.001; healthier adjOR 1.29, p-value = 0.033), while protective factors were being single (adjOR 0.78, p-value = 0.028), not having worked during lockdown (adjOR 0.71, p-value = 0.003) and younger age (adjOR 0.98, p-value = 0.002).

The occurrence of impulse buying during lockdown was positively associated with increased food purchase (adjOR 2.72, p-value < 0.001), low perceived quality of diet (adjOR 2.22, p-value < 0.001), living alone (adjOR 1.89, p-value = 0.002), resulting overweight (adjOR 1.44, p-value = 0.024), time spent cooking (decreased adjOR 1.58, p-value = 0.039; increased adjOR 1.36, p-value = 0.034) high score in EOQ-5 survey (adjOR 1.68, p-value < 0.001) and a low score in WHO-5 questionnaire (adjOR 1.73, p-value < 0.001), while the only protective covariate resulted to be a decrease in household food waste production (adjOR 0.73, *p*-value = 0.012).

Table 3. Multivariable analysis: determinants of self-perceived changes of food purchase, household food waste production and occurrence of impulse buying.

Variables		Increased Food Purchase		Impulse Buying		Decreased Household Food Waste Production	
		p-Value	OR (IC 95%)	*p*-Value	OR (IC 95%)	*p*-Value	OR (IC 95%)
Age		**0.002**	0.98 (0.97–0.99)	0.115	0.99 (0.98–1)	**0.006**	**1.01 (1–1.02)**
Gender	Female		Ref		Ref		Ref
	Male	0.055	0.81 (0.65–1)	**0.219**	0.83 (0.63–1.11)	**0.002**	**0.59 (0.47–0.74)**
Education	High		Ref		Ref		Ref
	Med-Low	0.644	1.05 (0.85–1.31)	0.302	0.88 (0.68–1.13)	0.602	0.94 (0.76–1.17)
Sentimental status	Not single		Ref				Ref
	Single	**0.028**	0.78 (0.62–0.97)			0.074	0.82 (0.66–1.02)
Offspring	No		Ref				
	Yes	**0.001**	1.76 (1.25–2.47)				
Cohabitation	Yes				Ref		
	No			**<0.002**	1.89 (1.32–2.71)		
Working during lockdown	Yes		Ref				Ref
	No	**0.003**	0.71 (0.57–0.89)			**0.024**	1.3 (1.03–1.62)
Smoking habit	No		Ref				Ref
	Yes	0.069	1.25 (0.98–1.58)			**<0.001**	1.8 (1.42–2.29)
BMI score	Normal				Ref		Ref
	Underw.			0.318	0.76 (0.44–1.3)	**0.009**	0.71 (0.48–1.06)
	Overw.			**0.024**	1.44 (1.05–1.96)	**0.017**	1.37 (1.06–1.77)
	Obese			0.367	1.27 (0.75–2.15)	**0.034**	1.63 (1.04–2.57)
Time spent cooking	Unvaried		Ref		Ref		Ref
	Decreased	0.192	0.79 (0.55–1.13)	**0.039**	1.58 (1.02–2.45)	**0.056**	1.41 (0.99–2)
	Increased	**<0.001**	2.12 (1.71–2.61)	**0.034**	1.36 (1.02–1.8)	**<0.001**	1.52 (1.23–1.88)
Perceived nutrition quality	Unvaried		Ref		Ref		Ref
	Less Healthy	**<0.001**	1.66 (1.3–2.12)	**<0.001**	2.22 (1.68–2.93)	**0.001**	1.37 (1.08–1.75)
	Healthier	**0.033**	1.29 (1.02–1.63)	**0.196**	0.8 (0.57–1.12)	**<0.001**	2.27 (1.77–2.9)
EOQ Score	Not at risk				Ref		
	At risk			**<0.001**	1.68 (1.29–2.19)		
Dietary regimen during lockdown	No						Ref
	Yes					**0.043**	0.79 (0.62–0.99)
WHO-5 Score	> 50				Ref		Ref
	≤ 50			**<0.001**	1.73 (1.32–2.27)	**0.002**	0.72 (0.59–0.89)
Household food waste production	Unvar. or increased				Ref		
	Decreased			**0.012**	0.73 (0.57–0.93)		
Impulse buying	No		Ref				Ref
	Yes	**<0.001**	2.48 (1.91–3.22)			**0.013**	0.73 (0.57–0.94)
Food purchase	Unvaried				Ref		Ref
	Decreased			**0.223**	1.4 (0.82–2.4)	**0.011**	1.69 (1.13–2.53)
	Increased			**<0.001**	2.72 (2.05–3.62)	**0.221**	1.14 (0.92–1.41)

Each column refers to a binary logistic regression model. Empty boxes refer to variables excluded using stepwise backward selection. Abbreviations: IQR, interquartile range; N, number; Who-5, 5-item World Health Organization Well-Being Index (WHO-5) questionnaire; EOQ-5, Emotional Overeating Questionnaire-5 (EOQ-5); BMI, body mass in-dex. In order to enhance readability, *p*-values < 0.05 are shown bolded.

The last model estimates the determinants of decreased household food waste production, finding in healthier perceived nutrition quality the strongest positive predictor (adjOR 2.27, *p*-value < 0.001) and in male gender the strongest negative predictor (0.59, *p*-value = 0.002). Additionally age (adjOR 1.01 per year, *p*-value = 0.006), not working during lockdown (adjOR 1.30, *p*-value = 0.024), smoking (adjOR 1.80, *p*-value < 0.001), BMI different than normal (underweight adjOR 0.71, *p*-value = 0.009; overweight adjOR 1.37, *p*-value = 0.017, obese adjOR1.63, *p*-value = 0.034), time spent cooking (adjOR 1.52, *p*-value < 0.001), WHO-5 score below 50 points (adjOR 0.72, *p*-value = 0.002), impulse buying (adjOR 0.73, *p*-value = 0.013), decreased food purchased (adjOR 1.69, *p*-value = 0.011) had statistically significant results.

4. Discussion

Home confinement during lockdown caused strong self-perceived changes in the food purchasing habits and behaviours of Italian residents. The majority of our sample perceived to have increased overall food purchase, food consumption and improved diet quality, reducing household food waste production, increasing time spent cooking at home and halving the prevalence of impulse buying.

Most of our sample followed Italian government suggestions about shopping frequency [17], limiting it to once or less per week, as found in other studies [16,20]. The lowering of shopping frequency was possible by concentrating most purchases in one time and at one place. Indeed, most food purchases occurred in supermarkets, as shown in other studies (64.3% and 75.8%, main frequencies) [16,20]. A big group of purchases were made in neighbourhood shops, where a +40% of sales was registered in April 2020, compared to last year, as well as a +23% compared to March 2020 [21], whilst a remarkable percentage of purchases were made online. The severe restrictions to movements and the presence of long queues out of supermarkets could have discouraged many customers, causing a shift of choice from hyper/supermarkets toward online or small neighbourhood shops. For these reasons, 27.6% of the customers changed their trusted store during lockdown [22]. Expectedly, a minimum number of purchases was performed in street markets, due to strong limitation of them or closure during lockdown.

The large increase in baking products purchases (flour/yeast, eggs, butter and fresh cheese) reflected the increase in self-production and consumption of foods such as pizza, homemade desserts and bread [14] that many people experimented with during lockdown [20,21,23]. Moreover, there was an increase in Google searches for recipes and baking [20]. Conversely, among the most decreased food purchases in our sample were delivered pizza, bread, ice cream and cakes. These foods could have been prepared at home instead of being bought. Actually, most of the sample increased time spent cooking, as an attempt to face boredom for the interruption of the work routine [14], less availability of out-of-home food, up to entertainment of children at home [24], resulting in a positive effect of home confinement. Indeed, home cooking is a healthy habit, related to better dietary quality, lower adiposity and greater adherence to Dietary Approaches to Stop Hypertension (DASH) and Mediterranean diets [25].

The recourse to foods for coping with stress and anxiety could have caused the increase in purchases of chocolate and biscuits [20,23,26]. Indeed, chocolate is also consumed as a stress relief, causing improvement of mood, but at the same time is related with emotional eating [27]. Furthermore, during lockdown there was an augmented prevalence of sleeping disturbances, depressive and anxiety symptoms in Italy [28]. Our results confirmed this trend, indeed the lockdown impacted on the mental health of a critically high portion of our sample. Almost half had a score of ≤50 in the WHO-5 questionnaire, resulting in low mental well-being and being at high risk of depression development [18]. Moreover, there was a high occurrence of emotional overeating in most of our sample during home confinement period, leading to a pathologic relation with nutrition, as an enormous palliative response to negative feelings. Our findings raise the need for public health interventions to take care of these people and to block the development of heavier mental health issues that can last after the pandemic as psychological aftermaths.

The rise of purchases of shelf-stable foods is typically associated with emergencies and uncertain times, even suggested by the government in the USA [29]. High increases of UHT long shelf-life milk and pasta purchases were reported in our samples and, in several articles [8,21,26], giving witness to their purchase was heavily affected by the psychological impact of the pandemic on the occurrence of "panic buying" during lockdown [30].

Interestingly, cheap price and the large amount of spare time to be spent cooking at home could have contributed to the high increase in purchases of fresh vegetables and fruits, flour and eggs, confirming similar upward trends for basic ingredients found in the literature [20,24,31]. By contrast, we observed a decrease in purchases of ready-to-eat vegetables, as already found [8,21], suggesting increased attention was spent transforming raw food into dishes, therefore limiting the purchase of ready-made products. This trend could have improved the diet quality of our sample, since the daily consumption of fruits and vegetables has become the main tool for prevention of cardiovascular disease, from the public health viewpoint worldwide [32]. Moreover, a high consumption of fruit and vegetables, if kept over time, could be related to lower frailty risk [33] and inversely associated with the risk of cardiovascular disease [34] and mortality [35].

However, a strong decrease in fresh fish purchases occurred in our sample, since it is one of the most perishable foods, characterised by a short shelf life and usually sold in street markets, which were mostly limited or closed during lockdown. Taken together, these factors could have led to this decrease, as shown in Spain [20]. The Italian annual per capita consumption of fish was estimated by European Commission at about 30.9 kg in 2017 [36]. We expect a reduction trend by 2020 that could have health consequences, if maintained in the future. Indeed, evidence confirms the salutary effects of fish consumption on the prevention of coronary artery disease, stroke and dementia [32] while showing an inverse association with the risk of all-cause mortality [37].

During lockdown, most people increased the overall amount of food purchases, while an increase in food sales during lockdown was reported in April 2020 (+18% compared to the same period in 2019, +3% compared to the previous month) [21] and during the entire lockdown [23]. Panic buying and the increase in purchases occurred also during past epidemics such as severe acute respiratory syndrome (SARS) [38]. During lockdown, stockpiling and sudden increases in purchases of food and even toilet paper have caused several problems to the retail sector all around the world, increasing concerns about shortages of non-perishable food products, contributing to the indirect, socioeconomic strong impact of coronavirus on sane people [30]. Indeed, the occurrence of impulse buying was related to an increase in food purchase in our sample, as well as having worked during lockdown or having children. Workers usually ate food cooked out of home in their workplace, but during lockdown bars and restaurants were closed, so they had to face new habits, increasing the amount of food purchases accounting for the introduction of their work meals. On the counterpart, similar mechanisms occurred in families with children, resulting in an increase in food consumed (and previously purchased) at home instead of school canteens, which were closed during the lockdown period. Finally, we observed a relation between a perceived change in diet quality and increased food purchase, resulting in an increased consumption of healthier or unhealthier foods. A study found that, during lockdown, both healthy and unhealthy foods recorded an increase in buying: unhealthy foods were purchased more often to cope with stress and improve the mood, whereas healthy foods were purchased extensively considering the aim of keeping healthy and in shape despite the lockdown-related restrictions, resulting in both cases in a change of perception of diet quality [20].

Most of the sample reduced household food waste production, confirming recent findings [14,16] about decrease in food waste production and increased use of the leftover food during lockdown. Similar behaviours in different samples toward food waste production indicate that their drivers are likely to be similar in many cultures [39], thereby the suitability of adopting means to reduce food waste from one country to another can be explored, as it is possible to learn from the experience of other countries. The rising leftover-use routines have shown to be strong contributors to food waste reduction [39] closely followed by shopping routines. During lockdown, shopping frequency in Italy strongly decreased [40], confirming our findings, and potentially affecting household food waste production. Indeed, a negative impact of frequency of food shopping on household food waste quantities was found [41], even in an Italian sample [42]. We found a relation between not working during lockdown and reduction in household food waste production that could be explained by the lack of out-of-home meals (that before lockdown were typically some form of gatherings in restaurants, pubs or cafes) by students and an increased attention towards the economic impact of waste by general population having lost their job during this period. A relation between out-of-home meals frequency and food waste production was found in the literature [43,44]. Interestingly, the occurrence of impulse buying was related to a non-reduction in household food waste production, confirming that impulsive purchases and buying foods that are not intended to be bought can affect food waste behaviours [45,46]. The relation between being on a diet and food waste reduction confirmed the negative impact of unplanned meals shopping on household food waste production [45,46]. Indeed, people on a diet follow a planned meals routine, resulting in a precise and well-organised shopping list and behaviour.

Shopping experience has deeply changed during lockdown. Supermarkets set a maximum number of inside customers, causing big queues up to 2 h [47]. Supermarkets reduced their opening hours and working days, closing "non-essential goods" sectors [48]. Therefore, customers might have felt less time available and pressure to shop quickly [6]. These factors are thought to have a role in enhancing impulse buying [49]. Interestingly, the occurrence of impulse buying during lockdown in our sample halved its prevalence, compared to the period before. The Italian government advice to reduce shopping frequency and to buy only necessary goods might have encouraged the extensive use of shopping lists among the population. Moreover, lockdown-related job insecurity may have played a role in restricting unnecessary purchases. People with worsened diet quality, low psychological well-being or occurrence of emotional overeating could have bought and consumed more indulgence and junk food as a coping strategy for the stressful situation, feeling then guilty. The purchase and consumption of these foods (rich in fats, sugars and calories) could explain the relation of these conditions with the occurrence of impulse buying and sense of guilt after purchase.

5. Strengths and Limitations

To date, this is the first study investigating both food choice and factors associated with increased food purchase, occurrence of impulse buying and reduction in household food waste production during lockdown among the Italian general population. Moreover, the investigation was performed a few days after the end of lockdown, in order to highlight well-established effects of the whole confinement period on our sample, instead of partial investigations on different lockdown phases, potentially underestimating different behaviours that could have come out in the last weeks. Our sample was large and composed of people from every region of the country, leading the extensive data to take into account the national perspective. Finally, validated tests were adopted to assess mental well-being and the occurrence of emotional overeating, resulting in a valuable occasion of investigation of mental health and nutrition issues during the COVID-19 pandemic in Italy. However, our study has some limitations. The online spread of the survey led to an opportunistic sampling. Moreover, females accounted for 70% of our sample. Nevertheless, similar gender prevalence was observed in different studies conducted online during the lockdown period [11,14]. Food purchases and perceived change of habits were assessed in a qualitative fashion, without being given the opportunity to further explore their connections. In addition, the self-reporting of items could have represented itself a limitation in terms of quality of data (e.g., lack of memory, over/under-reporting). Finally, due to the cross-sectional design of the study, it was not possible to infer causal relationships between variables.

6. Implications and Conclusions

Overall, the effects of lockdown on population appeared to be both positive and negative, depending on the context. Food purchase, consumption and household food waste production in our sample were affected mostly in a desirable way, as evidenced by results. More efforts for public health interventions are needed to keep these new habits in the future, leading to positive behaviours toward achieving a sustainable and healthy lifestyle. Conversely, the lockdown appeared to affect heavily on mental health among a critically high portion of our sample, resulting in low psychological well-being, higher risk of depression and occurrence of emotional overeating as a possible coping strategy. Since the adoption of new lockdowns in the future cannot be excluded, policymakers should take into consideration this public health perspective, since for these people, a new containment measure could cause further negative effects on their physical and mental health. Moreover, our results can suggest strategies to the food retail sector about food categories that should be primarily provided in the case of new confinements, such as starchy foods, eggs, fresh fruits and vegetables, dairy products, considering that many issues occurred during the first lockdown regarding stockpiling and scarce food supplies in shops. Finally, the pandemic encouraged the adoption of online grocery purchase to the Italian population, offering a modern and low-risk shopping method. These services should be strengthened, especially in the southern regions of Italy, in order to make providers more resilient and prepared to satisfy an increased demand for service in the critical period to come, characterised by social distancing and home working.

Supplementary Materials: The following are available online at https://www.mdpi.com/2304-8158/10/2/306/s1, Document S1: Full regression outcomes; Document S3: Brief version of the questionnaire used for the survey; Image S2: Purchases per individual category.

Author Contributions: Conceptualisation, A.S., D.C. and F.B.; methodology, A.S., D.C. and F.B.; formal analysis, A.S., D.C., E.B. and F.B.; investigation, A.S., D.C., E.B. and F.B.; resources, R.S.; data curation, D.C.; writing—original draft preparation, E.B., D.C. and A.S.; writing—review and editing, A.S., D.C., F.B. and R.S.; visualisation, A.S. and D.C.; supervision, F.B. and R.S. All authors have read and agreed to the published version of the manuscript.

Funding: This research received no external funding.

Institutional Review Board Statement: The study was conducted according to the guidelines of the Declaration of Helsinki, and approved by the Ethics Committee of University of Turin (protocol code 197989 approved in May 2020).

Informed Consent Statement: Informed consent was obtained from all subjects involved in the study.

Data Availability Statement: The datasets generated and/or analysed during the current study are not publicly available due data are not public but are available from the corresponding author on reasonable request.

Conflicts of Interest: The authors declare no conflict of interest.

References

1. Gazzetta Ufficiale DECRETO DEL PRESIDENTE DEL CONSIGLIO DEI MINISTRI 23 febbraio 2020. Available online: https://www.gazzettaufficiale.it/eli/id/2020/02/23/20A01228/sg (accessed on 24 October 2020).
2. Gazzetta Ufficiale DECRETO DEL PRESIDENTE DEL CONSIGLIO DEI MINISTRI 1 marzo 2020. Available online: https://www.gazzettaufficiale.it/eli/id/2020/03/01/20A01381/sg (accessed on 24 October 2020).
3. Gazzetta Ufficiale DECRETO DEL PRESIDENTE DEL CONSIGLIO DEI MINISTRI 9 marzo 2020. Available online: https://www.gazzettaufficiale.it/eli/id/2020/03/09/20A01558/sg (accessed on 24 October 2020).
4. Vicentini, C.; Bordino, V.; Gardois, P.; Zotti, C.M. Early assessment of the impact of mitigation measures on the COVID-19 outbreak in Italy. *Public Health* **2020**, *185*, 99–101. [CrossRef]

5. Gazzetta Ufficiale DECRETO DEL PRESIDENTE DEL CONSIGLIO DEI MINISTRI 10 aprile 2020. Available online: https://www.gazzettaufficiale.it/eli/id/2020/04/11/20A02179/sg (accessed on 24 October 2020).
6. Martin-Neuninger, R.; Ruby, M.B. What Does Food Retail Research Tell Us About the Implications of Coronavirus (COVID-19) for Grocery Purchasing Habits? *Front. Psychol.* **2020**, *11*. [CrossRef]
7. IRI COVID-19: LA SPESA PER LARGO CONSUMO NELLA FASE ACUTA DELLA CRISI. Available online: https://www.iriworldwide.com/it-it/insights/publications/covid-19-la-spesa-per-largo-consumo-nella-fase-acu (accessed on 24 October 2020).
8. ISMEA Emergenza COVID–19: Rapporto sulla domanda e l'offerta dei prodotti alimentari nelle prime settimane di diffusione del virus. Available online: http://www.ismea.it/flex/cm/pages/ServeBLOB.php/L/IT/IDPagina/11018 (accessed on 24 October 2020).
9. Altroconsumo Fare La Spesa Online Ai Tempi Del Coronavirus: Non Pochi Problemi. Available online: https://www.altroconsumo.it/alimentazione/fare-la-spesa/news/coronavirus-spesa-online (accessed on 24 January 2021).
10. Haddad, C.; Kheir, M.B.; Zakhour, M.; Haddad, R.; Al Hachach, M.; Sacre, H.; Salameh, P. Association between eating behavior and quarantine/confinement stressors during the Coronavirus disease 2019 outbreak. *J. Eat. Disord.* **2020**. [CrossRef] [PubMed]
11. Cancello, R.; Soranna, D.; Zambra, G.; Zambon, A.; Invitti, C. Determinants of the lifestyle changes during covid-19 pandemic in the residents of northern italy. *Int. J. Environ. Res. Public Health* **2020**, *17*, 6287. [CrossRef] [PubMed]
12. Coldiretti Istat, 2 kg di Peso Per il Lockdown a Tavola. Available online: https://www.coldiretti.it/salute-e-sicurezza-alimentare/istat-2-kg-di-peso-per-il-lockdown-a-tavola (accessed on 24 October 2020).
13. Di Renzo, L.; Gualtieri, P.; Cinelli, G.; Bigioni, G.; Soldati, L.; Attinà, A.; Bianco, F.F.; Caparello, G.; Camodeca, V.; Carrano, E.; et al. Psychological aspects and eating habits during covid-19 home confinement: Results of ehlc-covid-19 italian online survey. *Nutrients* **2020**, *12*, 2152. [CrossRef] [PubMed]
14. Di Renzo, L.; Gualtieri, P.; Pivari, F.; Soldati, L.; Attinà, A.; Cinelli, G.; Cinelli, G.; Leggeri, C.; Caparello, G.; Barrea, L.; et al. Eating habits and lifestyle changes during COVID-19 lockdown: An Italian survey. *J. Transl. Med.* **2020**, *18*, 229. [CrossRef] [PubMed]
15. Sidor, A.; Rzymski, P. Dietary choices and habits during COVID-19 lockdown: Experience from Poland. *Nutrients* **2020**, *12*, 1657. [CrossRef] [PubMed]
16. Jribi, S.; Ben Ismail, H.; Doggui, D.; Debbabi, H. COVID-19 virus outbreak lockdown: What impacts on household food wastage? *Environ. Dev. Sustain.* **2020**, *22*, 3939–3955. [CrossRef]
17. Ministero della Salute Come fare la spesa ai tempi del Covid-19 e Rispettare le Corrette Prassi di Igiene Alimentare. Available online: http://www.salute.gov.it/portale/news/p3_2_1_1_1.jsp?lingua=italiano&menu=notizie&p=null&id=4299 (accessed on 24 October 2020).
18. Topp, C.W.; Østergaard, S.D.; Søndergaard, S.; Bech, P. The WHO-5 well-being index: A systematic review of the literature. *Psychother. Psychosom.* **2015**, *84*, 167–176. [CrossRef]
19. Casu, G.; Gremigni, P.; Masheb, R.M. Emotional overeating questionnaire: A validation study in Italian adults with obesity, overweight or normal weight. *Eat. Weight Disord.* **2019**. [CrossRef]
20. Laguna, L.; Fiszman, S.; Puerta, P.; Chaya, C.; Tárrega, A. The impact of COVID-19 lockdown on food priorities. Results from a preliminary study using social media and an online survey with Spanish consumers. *Food Qual. Prefer.* **2020**, *86*, 104028. [CrossRef] [PubMed]
21. ISMEA Emergenza COVID–19: 2° Rapporto Sulla Domanda e L'offerta dei Prodotti Alimentari Nell'emergenza Covid-19. Available online: http://www.ismea.it/flex/cm/pages/ServeBLOB.php/L/IT/IDPagina/11017 (accessed on 24 October 2020).
22. Centro di Ricerche su Retailing e Trade Marketing dell'Università Cattolica Come Cambia il Consumatore Dopo il Covid. Available online: https://www.cattolicanews.it/come-cambia-il-consumatore-dopo-il-covid (accessed on 24 October 2020).
23. Bertoletti, C. Nielsen Le Vendite in Gdo Durante il Lockdown: +4,2% Con Liberi Servizi al Traino. Gdoweek. Available online: https://www.gdoweek.it/le-vendite-in-gdo-durante-il-lockdown-42-con-liberi-servizi-al-traino/ (accessed on 24 October 2020).
24. Zhao, A.; Li, Z.; Ke, Y.; Huo, S.; Ma, Y.; Zhang, Y.; Zhang, J.; Ren, Z. Dietary diversity among chinese residents during the COVID-19 outbreak and its associated factors. *Nutrients* **2020**, *12*, 1699. [CrossRef] [PubMed]
25. Mills, S.; Brown, H.; Wrieden, W.; White, M.; Adams, J. Frequency of eating home cooked meals and potential benefits for diet and health: Cross-sectional analysis of a population-based cohort study. *Int. J. Behav. Nutr. Phys. Act.* **2017**, *14*, 109. [CrossRef] [PubMed]
26. Bracale, R.; Vaccaro, C.M. Changes in food choice following restrictive measures due to Covid-19. *Nutr. Metab. Cardiovasc. Dis.* **2020**, *30*, 1423–1426. [CrossRef] [PubMed]
27. Macht, M.; Mueller, J. Immediate effects of chocolate on experimentally induced mood states. *Appetite* **2007**, *49*, 667–674. [CrossRef] [PubMed]
28. Gualano, M.R.; Lo Moro, G.; Voglino, G.; Bert, F.; Siliquini, R. Effects of COVID-19 lockdown on mental health and sleep disturbances in Italy. *Int. J. Environ. Res. Public Health* **2020**, *17*, 4779. [CrossRef] [PubMed]
29. Ready Suggested Emergency Food Supplies. Available online: https://www.ready.gov/food (accessed on 24 October 2020).
30. Nicola, M.; Alsafi, Z.; Sohrabi, C.; Kerwan, A.; Al-Jabir, A.; Iosifidis, C.; Agha, M.; Agha, R. The Socio-Economic Implications of the Coronavirus Pandemic (COVID-19): A Review. *Int. J. Surg.* **2020**, *78*, 185–193. [CrossRef]

31. Batlle-Bayer, L.; Aldaco, R.; Bala, A.; Puig, R.; Laso, J.; Margallo, M.; Vázquez-Rowe, I.; Antó, J.M.; Fullana-i-Palmer, P. Environmental and nutritional impacts of dietary changes in Spain during the COVID-19 lockdown. *Sci. Total Environ.* **2020**, *748*, 141410. [CrossRef]
32. Román, G.C.; Jackson, R.E.; Gadhia, R.; Román, A.N.; Reis, J. Mediterranean diet: The role of long-chain ω-3 fatty acids in fish; polyphenols in fruits, vegetables, cereals, coffee, tea, cacao and wine; probiotics and vitamins in prevention of stroke, age-related cognitive decline, and Alzheimer disease. *Rev. Neurol. (Paris)* **2019**, *175*, 724–741. [CrossRef]
33. Kojima, G.; Avgerinou, C.; Iliffe, S.; Jivraj, S.; Sekiguchi, K.; Walters, K. Fruit and Vegetable Consumption and Frailty: A Systematic Review. *J. Nutr. Heal. Aging* **2018**, *22*, 1010–1017. [CrossRef]
34. Zhan, J.; Liu, Y.J.; Cai, L.B.; Xu, F.R.; Xie, T.; He, Q.Q. Fruit and vegetable consumption and risk of cardiovascular disease: A meta-analysis of prospective cohort studies. *Crit. Rev. Food Sci. Nutr.* **2017**, *57*, 1650–1663. [CrossRef]
35. Wang, X.; Ouyang, Y.; Liu, J.; Zhu, M.; Zhao, G.; Bao, W.; Hu, F.B. Fruit and vegetable consumption and mortality from all causes, cardiovascular disease, and cancer: Systematic review and dose-response meta-analysis of prospective cohort studies. *BMJ* **2014**, *349*, g4490. [CrossRef]
36. European Commission Consumption of Fisheries and Aquaculture Products (2017). Available online: https://ec.europa.eu/fisheries/6-consumption_en (accessed on 24 October 2020).
37. Wan, Y.; Zheng, J.; Wang, F.; Li, D. Fish, long chain omega-3 polyunsaturated fatty acids consumption, and risk of all-cause mortality: A systematic review and dose-response meta-analysis from 23 independent prospective cohort studies. *Asia Pac. J. Clin. Nutr.* **2017**, *26*, 939–956. [CrossRef] [PubMed]
38. Ding, H. Rhetorics of Alternative Media in an Emerging Epidemic: SARS, Censorship, and Extra-Institutional Risk Communication. *Tech. Commun. Q.* **2009**, *18*, 327–350. [CrossRef]
39. Stancu, V.; Haugaard, P.; Lähteenmäki, L. Determinants of consumer food waste behaviour: Two routes to food waste. *Appetite* **2016**, *96*, 7–17. [CrossRef] [PubMed]
40. ISMEA Emergenza COVID–19: 3° Rapporto Sulla Domanda e L'offerta Dei Prodotti Alimentari Nell'emergenza Covid-19. Available online: http://www.ismea.it/flex/cm/pages/ServeBLOB.php/L/IT/IDPagina/11116 (accessed on 24 October 2020).
41. Giordano, C.; Alboni, F.; Cicatiello, C.; Falasconi, L. Do discounted food products end up in the bin? An investigation into the link between deal-prone shopping behaviour and quantities of household food waste. *Int. J. Consum. Stud.* **2019**, *43*, 199–209. [CrossRef]
42. Jörissen, J.; Priefer, C.; Bräutigam, K.R. Food waste generation at household level: Results of a survey among employees of two European research centers in Italy and Germany. *Sustainability* **2015**, *7*, 2695–2715. [CrossRef]
43. Ponis, S.T.; Papanikolaou, P.A.; Katimertzoglou, P.; Ntalla, A.C.; Xenos, K.I. Household food waste in Greece: A questionnaire survey. *J. Clean. Prod.* **2017**, *149*, 1268–1277. [CrossRef]
44. Breda, M.; Hong-Bo, L. 'Waste not, want not': Exploring green consumers' attitudes towards wasting edible food and actions to tackle food waste. *Br. Food J.* **2017**, *119*, 2519–2531. [CrossRef]
45. Stefan, V.; van Herpen, E.; Tudoran, A.A.; Lähteenmäki, L. Avoiding food waste by Romanian consumers: The importance of planning and shopping routines. *Food Qual. Prefer.* **2013**, *28*, 375–381. [CrossRef]
46. Bravi, L.; Francioni, B.; Murmura, F.; Savelli, E. Factors affecting household food waste among young consumers and actions to prevent it. A comparison among UK, Spain and Italy. *Resour. Conserv. Recycl.* **2020**, *153*, 104586. [CrossRef]
47. Il Fatto Quotidiano Coronavirus, a Milano Lunghe Code Davanti Ai Supermercati: In Fila Dalle 6.30 Del Mattino. Attese anche di 2 ore. Available online: https://www.ilfattoquotidiano.it/2020/03/21/coronavirus-a-milano-lunghe-code-davanti-ai-supermercati-in-fila-dalle-6-30-del-mattino-attese-anche-di-2-ore-le-immagini/5744340/ (accessed on 24 October 2020).
48. Nardi, V. La Spesa ai Tempi del Coronavirus: Supermercati Chiusi la Domenica, Orari Ridotti e Vendita Solo di Beni di Prima Necessità. Il fatto alimentare. Available online: https://ilfattoalimentare.it/coronavirus-supermercati-orari.html (accessed on 24 October 2020).
49. Hausman, A. A multi-method investigation of consumer motivations in impulse buying behavior. *J. Consum. Mark.* **2000**, *17*, 403–419. [CrossRef]

foods

MDPI

Review

Trends in Coffee and Tea Consumption during the COVID-19 Pandemic

Fabio Castellana [1], Sara De Nucci [1], Giovanni De Pergola [2], Martina Di Chito [2], Giuseppe Lisco [3], Vincenzo Triggiani [3], Rodolfo Sardone [1,*] and Roberta Zupo [1]

[1] Unit of Data Sciences and Technology Innovation for Population Health, National Institute of Gastroenterology "Saverio de Bellis", Research Hospital, Castellana Grotte, 70124 Bari, Italy; fabio.castellana@irccsdebellis.it (F.C.); sara.denucci@irccsdebellis.it (S.D.N.); zuporoberta@gmail.com (R.Z.)
[2] Unit of Geriatrics and Internal Medicine, National Institute of Gastroenterology "Saverio de Bellis", Research Hospital, Castellana Grotte, 70124 Bari, Italy; gdepergola@libero.it (G.D.P.); martinadichito1999@gmail.com (M.D.C.)
[3] Section of Internal Medicine, Geriatrics, Endocrinology, and Rare Disease, Interdisciplinary Department of Medicine, School of Medicine, University of Bari, 70124 Bari, Italy; g.lisco84@gmail.com (G.L.); vincenzo.triggiani@uniba.it (V.T.)
* Correspondence: rodolfo.sardone@irccsdebellis.it

Abstract: Over the last two years, many countries have enforced confinement to limit both the spread of COVID-19 and the demand for medical care. Confinement has resulted in a disruption of work routines, boredom, depression, and changes in eating habits, among them consumption of coffee and tea. Following six databases, we examined articles tracking consumption of these beverages. Out of 472 articles, including 23 beverage entries, 13 matched our criteria. While no clear trend in coffee consumption during the coronavirus pandemic emerged (7 of 13 studies indicated an increase, accounting for 53.8%), tea consumption clearly increased (70% versus 30%). Considering the global health emergency continuum, more research is needed to better understand the paths underlying food choices and the ways those changes may influence health outcomes, including those related to COVID-19 disease.

Keywords: dietary behavior; COVID-19; beverage consumption; coffee; tea

Citation: Castellana, F.; De Nucci, S.; De Pergola, G.; Di Chito, M.; Lisco, G.; Triggiani, V.; Sardone, R.; Zupo, R. Trends in Coffee and Tea Consumption during the COVID-19 Pandemic. *Foods* **2021**, *10*, 2458. https://doi.org/10.3390/foods10102458

Academic Editor: Derek V. Byrne

Received: 7 September 2021
Accepted: 11 October 2021
Published: 15 October 2021

1. Introduction

The coronavirus pandemic still poses a worldwide public health challenge, with 216,229,741 cases confirmed since the World Health Organization (WHO) declared a state of global emergency in March 2020 [1]. At present, Europe is experiencing increased infections due to social mixing, summer travel, family reunions, and looser social restrictions. Set against this backdrop, a recent report by the WHO director for Europe warns of three conditions that could contribute to a new wave of hospitalizations and excess deaths before the fall: new variants, incomplete vaccine adherence (63% of people are still reluctant to undergo vaccination), and increased social mixing. By way of prevention, most governments have imposed varying degrees of self-isolation and nationwide lockdowns to curb spread of the virus. Staying-at-home has meant digital education, smart working, social isolation, job insecurity, and limited outdoor and gym activity-in short, a dramatic change in lifestyles [2]. Moreover, quarantine and distancing from families have led to a cluster of negative psychological implications, including confusion, anger, and depression due to frustration, boredom, inadequate information, and financial loss [3,4]. This burden of unpleasant feelings, combined with limitations at multiple levels, has prompted substantial changes in lifestyle, triggering a shift in eating habits in terms of a reduced control over food intake and quality. These changes include escalations in carbohydrate intake, the frequency of snacking, and home cooking during confinement [5]. Often home cooking entails a higher consumption of homemade cakes, bread, and pizza, all sharing a critical glycemic load, which could affect weight. In most cases, people ate "comfort" foods to reduce accumulated stress, relying on the biological effect of serotonin on mood.

Beverage choices also contribute to daily calorie intake and hydration, particularly nervine beverages such as coffee and tea, in view of their known potential to promote psychological well-being beyond the mere nutritional aspect [6]. Tea affects psychopathological condition (e.g., reduced anxiety), cognition (e.g., benefits in memory and attention), and brain function (e.g., activation of working memory). Yet such benefits are not attributable to a single constituent, and better gains yielded by the synergy of caffeine and L-theanine are reported as compared to their separate administration [7]. As for coffee, much scientific attention has been paid to its association with mood and emotion. One cup of coffee every four hours improves mood. Low to moderate doses of caffeine (two to five cups of coffee per day) have been shown to improve hedonic tone (the degree of pleasantness or unpleasantness associated with a given state) and reduce anxiety [8,9].

Against this background, it seemed useful to further probe the influence of COVID-19 confinement on consumption of these beverages in order to consider possible health implications, related not only to mental well-being but to overall health.

2. Methods

The present work is a narrative review article. We searched the US National Library of Medicine (PubMed), Medical Literature Analysis and Retrieval System Online (MEDLINE), EMBASE, Scopus, Ovid, and Google Scholar to find original articles covering dietary variables during the COVID-19 pandemic, selecting studies examining changes in coffee and tea consumption (see Table 1). Given the novelty of the topic and the short timeframe surrounding COVID-19 pandemic research, no skimming was applied to the study population, design, or setting. For the same reason, no age range was applied to the study population. Two investigators (RZ, SD), independently and in duplicate, searched for papers, screened titles and abstracts of the retrieved articles, reviewed entire texts, and identified articles for inclusion in this study. Of particular interest were original articles investigating dietary habits, particularly coffee and tea consumption, during COVID-19 via online or telephone questionnaires and those utilizing accounts of weekly or monthly food intake. Of note, reports offering a snapshot of coffee and tea consumption during the period of COVID-19 confinement were excluded in favor of studies comparing the consumption of the two beverages before and after the COVID-19 pandemic. The investigators tracked (1) such general information as study design, setting, sample size and demographics (age and gender), country, method of dietary assessment, and dietary exposure; and (2) principal results of nutritional surveys (changes in trends and frequency of coffee and tea consumption). Data were reported separately for coffee and tea consumption with respect to increase or decrease. Data were cross-checked by a third researcher (FC) to remove discrepancies and resolve disagreements.

Table 1. Search strategy to be used in the US National Library of Medicine (PubMed) and Medical Literature Analysis and Retrieval System Online (MEDLINE) and adapted to the other sources, according to selected descriptors.

Strategy	Descriptors Used
# 1	(diet*[tiab]) OR (feeding*[tiab])) OR (habit*[tiab]) OR (dietary lifestyle*[tiab]) OR (drinking habit*[tiab]) OR (beverage*[tiab]) OR (dietary habit*[tiab]) OR (dietary [tiab]) OR (dietary pattern*[tiab]) OR (dietary behavior*[tiab]) OR (food*[tiab]) OR (food habit*[tiab]) OR (eating habit*[tiab]) OR (coffee[tiab]) OR (tea[tiab])
# 2	(change*[tiab]) OR (modification*[tiab]) OR (alteration*[tiab]) (different*[tiab]) OR (difference*[tiab])
# 3	(SARS-CoV 2[tiab]) OR (COVID 19[tiab]) OR (severe acute respiratory syndrome coronavirus 2[tiab])
# 4	(Review) or (systematic review) or (narrative review) or (meta-analysis)
# 5	#1 AND #2 AND #3 NOT #4

3. Results

The search strategy was updated to 31 August 2021, and yielded 472 results. Of these, 13 were found pertinent to our objective and selected for analysis [10–22]. Figure 1 shows a flowchart of the literature screening process. Each of these 13 reports utilized online or telephone questionnaires about dietary habits covering a cross-section of the population. All studies were based on community sampling, and respondents were older than eighteen years. Overall, these studies included 435,616 subjects, more prevalently female (70.2%, n = 305,802 versus 29.8%, n = 129,816). The majority of the studies were European (9/13, 70%), with a minority of American and Asian (2/13, 15% each). Only one report had a multicenter setting [18], reporting data from Poland, Austria, and the UK. Table 2 covers details of the design (cohort or cross-sectional), sample size (n) and gender ratio (%), minimum age and mean (SD) or age range, setting (community or hospital), and country.

Though only 10 of the 13 selected studies investigated both coffee and tea as beverages, we analyzed a total of 23 beverage entries. It should be noted that 7 of these 10 studies considered the combined consumption of both tea and coffee, rather than treating each individually. Of the 23 entries, 13 considered coffee consumption (7/13, n = 54%) and the remaining 9, tea (9/13, n = 46%). As regards coffee, 54% of studies indicated an increase, while the remaining 46% reported a decrease in consumption. By contrast, in 70% of studies (7 or 10) tea consumption was reported as having increased while in 30% of studies (3 of 10) is was reported as having decressed. Only one report analyzed coffee consumption by variety, i.e., whether American or Arabic [11].

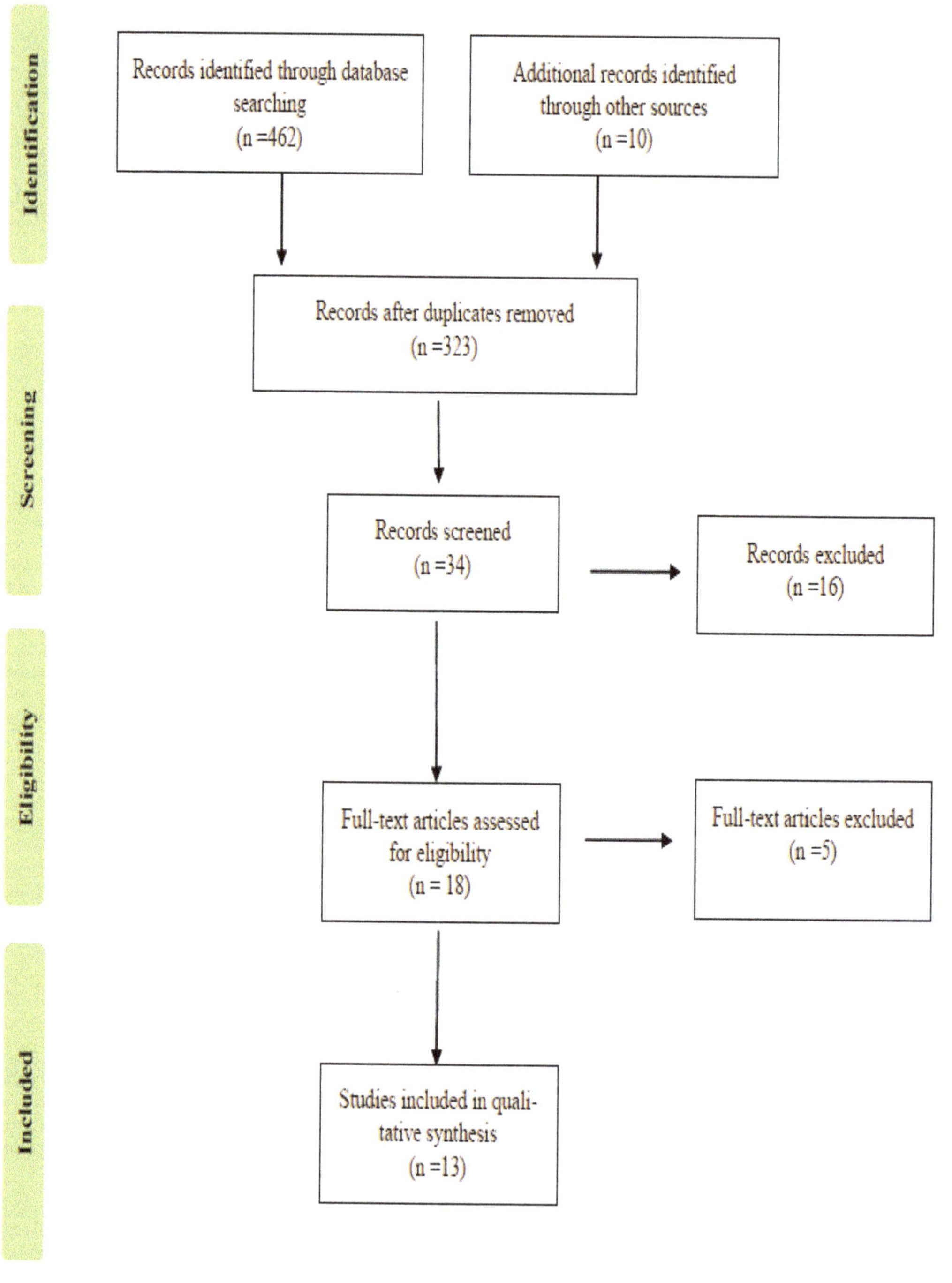

Figure 1. Flow diagram of literature screening process.

Table 2. Descriptive characteristics of included studies.

Authors, Year [Ref.]	Diet Exposure	Diet Assessment Tool	Design	*n*	Sex	Age	Setting	Country	Results	Summary of Findings
Bin Zarah A et al., 2020 [10]	Coffee, Tea	Questionnaire	Cross-sectional	3133	19.8%M 79.4% F	18+ years	Community	America (USA)	About 10% declared a reduction in frequency of consumption, whereas 31.1% an increase.	Participants reported a higher consumption of coffee and tea drinks.
Husain W et al., 2020 [11]	Coffee, Tea			415	31.3%M 68.7%F	18+ years		Asia (Kuwait)	1–2 cups/day of American coffee: 41.4% (before) vs. 33% (during) 3–4 cups/day of American coffee: 8.7% (before) vs. 5.5% (during) 1–2 cups/day of arabic coffee: 15.4% (before) vs. 15.5% (during) 5–6 cups/day of arabic coffee: 10.6% (before) vs. 7.7% (during) 1–2 cups/day of tea: 31.6% (before) vs. 33% (during) 5–6 cups/day of tea: 1.7% (before) vs. 2.9% (during)	Decreased consumption of American and Arabic coffee during the confinement, versus increase in tea consumption.
Błaszczyk-Bębenek E et al., 2020 [12]	Coffee, Tea			312	35.9%M 64.1%F	18+ years		Europe (Poland)	Hot beverage consumption frequencies: 1–3 times/month: 3.2% (before) vs. 3.5% (during) Once a week: 1.3% (before) vs. 1.6% (during) A few times a week: 5.1% (before) vs. 7.1% (during) Once a day: 16.0% (before) vs. 16.3 (during)	Hot beverages such as black coffee, herbal, or fruit tea were chosen most frequently by respondents during the confinement.
Grabia M et al., 2020 [13]	Coffee			124	17%M 83%F	17–45 years		Europe (Poland)	30% reported an increase in frequency of coffee consumption while 13% reported a reduction.	Increased coffee consumption during the confinement.

Table 2. *Cont.*

Authors, Year [Ref.]	Diet Exposure	Diet Assessment Tool	Design	*n*	Sex	Age	Setting	Country	Results	Summary of Findings
Sánchez-Sánchez E et al., 2020 [14]	Coffee, Tea			1073	27.2%M 72.8%F	16+ years		Europe (Spain)	Frequency of coffee or tea consumption: 7.89% (before) vs. 6.48% (during)	Decreased coffee or tea consumption during the confinement.
Đogaš Z et al., 2020 [15]	Coffee			3027	20.3%M 79.7%F	18+ years		Europe (Croatia)	Frequency of coffee consumption (cups/day): All participants: 2.1 ± 1.0 (before) vs. 2.1 ± 1.1 (during) Males: 2.4 ± 1.2 (before) vs. 2.0 ± 1.2 (during) Females: 2.1 ± 1.0 (before) vs. 2.1 ± 1.1 (during)	Croatian males drank fewer cups of coffee during confinement.
Di Renzo L et al., 2020 [16]	Coffee, Tea			3533	23.9%M 76.1%F	12–86 years		Europe (Italy)	Hot beverage consumption frequency increased by more than 20% during confinement.	Greater than 20% increase in the consumption of hot drinks
Luo Y et al., 2021 [17]	Coffee, Tea			2272	18.3%M 83.4%F	18+ years		Asia (China)	Coffee, tea, and water consumption increased by 29.3%.	Hot beverage consumption increased during confinement.
Skotnicka M et al., 2021 [18]	Coffee, Tea			1071	43.6%M 56.4%F	18+ years		Europe (Poland, Austria, UK)	Frequency of coffee consumption (cups/day): Poland: 76.90% (before) vs. 76.17% (during) Austria: 62.61% (before) vs. 62.32% (during) UK: 54.34% (before) vs. 49.84% (during) Frequency of tea consumption (cups/day): Poland: 57.74% (before) vs. 60.93% (during) Austria: 54.11% (before) vs. 58.36% (during) UK: 64.31% (before) vs. 74.59% (during)	More frequent consumption of tea and less frequent consumption of coffee during confinement.

Table 2. *Cont.*

Authors, Year [Ref.]	Diet Exposure	Diet Assessment Tool	Design	*n*	Sex	Age	Setting	Country	Results	Summary of Findings
Celorio-Sardà R et al., 2021 [19]	Coffee, Tea			321	20.2%M 79.8%F	18+ years		Europe (Spain)	There was a decrease in coffee and tea consumption during confinement reported by 56.7% of the study sample.	Decreased coffee and tea consumption during confinement.
Izzo L et al., 2021 [20]	Coffee			1519	28.4%M 71.6%F	0+ years		Europe (Italy)	Coffee consumption increased for 64.8% of participants.	Increased coffee consumption during confinement.
Mitchell E.S. et al., 2021 [21]	Coffee			381,564	16.6%M 83.4%F	18+		America (USA)	Caffeinated beverages such as tea and coffee decreased in the proportion of users aged 18–35 years (−2.3%), but only marginally decreased in users aged 35 years and older.	Decreased tea and coffee consumption during confinement.
Deschasaux-Tanguy M et al., 2021 [22]	Coffee, Tea			37,252	47.7%M 52.3%F	18+ years		Europe (France)	Frequency of tea consumption during confinement: Increased for 19.5%, for decreased 4.3%. Frequency of coffee consumption during confinement: Increased for 13.5%, decreased for 8.4%	Increased coffee and tea consumption during confinement.

4. Discussion

We reviewed existing literature on changes in coffee and tea consumption driven by confinement during the COVID-19 pandemic with respect to coffee and tea consumption, in view of their known impact on psychological well-being. We found no clear trend in coffee consumption, while there was a clear increase in tea consumption. However, looking at coffee consumption, it should be considered that our methodological setting only included original reports comparing the period of confinement to the previous time. However, based on the literature search, we found a survey from Poland reporting the highest frequency of coffee consumption (88.9%) among adults aged 45+ but referring only to COVID-19 confinement, with no comparison to previous habits [23]. The high frequency of coffee consumption recorded in this study suggests an increasing consumption of this beverage during the pandemic.

The unclear findings on increased coffee consumption may be understood from both social and psychological perspectives. On the one hand, people who are used to drinking coffee in family contexts on a daily basisalso enjoy coffee in social situations. Especially among adolescents, drinking coffee is a way to spend time with friends and improve one's mood [24,25]. Moreover, compared to adults, adolescents are particularly oriented toward the upgraded social image they can project by consuming caffeinated beverages. Such an aspect, if read in a pandemic key, would imply a drop in consumption, as social and community events were curtailed during COVID-19 confinement. On the other hand, from a purely psychological and emotional point of view, coffee is a good source of energy and may be used to improve mood, fight drowsiness, and enhance cognitive function [26–28]. This second point, read in a pandemic key, could explain increased coffee consumption, in light of the widespread smart working scenario and the distressed mood caused by the pandemic itself.

As for consumption of tea, findings suggest a clear increase in consumption compared to tea drinking before COVID-19.Tea is usually linked to routine and ritualized household consumption. Tea is historically instrumental in bringing the family closer together and provides a platform for sharing. In contrast, coffee consumption needed to be considered in a social, aesthetic and emotional context. Setting aside the social context, therefore, the increased consumption of tea should be understood in emotional and family-related settings. From this perspective, this beverage It has long been associated with mood and performance enhancements, such as a greater relaxation and concentration. Though tea contains many bioactive compounds, but its benefits with respect to attention, mood, and the cognitive sphere have generally been attributed to two of its components, namely, caffeine and theanine [29]. A cup of tea provides 35–61 mg of caffeine and 4.5–22.5 mg of theanine. A substantial body of research suggests that L-theanine exerts anti-stress effects in response to acute stress challenges via the inhibition of cortical neuron excitation. On the other hand, caffeine found in coffee has been reported to improve performance and memory, reduce errors in performing tasks, accelerate cognitive processing, and improve mood [30,31]. Moreover, caffeine improves concentration and attention by eliminating distractors and improving focus, which is the reason why it has the potential to improve vigilance and reaction time [8]. Yet, in pandemic settings, this feature may be read as a driver of increased tea consumption, driven by higher levels of stress and confusion induced by the epidemic situation and the dissemination of home-based smart working [32].

5. Conclusions

The lack of a clear trend in coffee consumption as the result of the COVID-19 pandemic calls for further investigation. Moreover, potential health implications should not be overlooked, especially since caffeine consumption may directly or indirectly promote bronchodilation, interfere in the process of immunomodulation, and hinder viral intracellular transcription while undergoing COVID-19 infection [33]. Furthermore, reflecting a discomfited mood and socially confining setting, we found a marked increase in tea consumption.

6. Limitations

More studies are needed to expand on these findings and examine coffee and tea consumption separately. The limited number of studies included is a major limitation of the present review, weakening the completeness and generality of findings, despite their being quite representative of the European population. Secondly, some of the selected studies examined the combined consumption of tea and coffee, rather than taking each individually. Nevertheless, this preliminary research provides food for thought. Lastly, though coffee and tea stand out as the most popular beverages worldwide, we know the cluster of nervine beverages is much wider, including cocoa, cola drinks, guarana, and maté, all of which have a tonic and stimulating effect on the central nervous system, due to the presence of natural alkaloids (for example caffeine, theophylline, theobromine, etc.)

Author Contributions: Conceptualization: R.Z.; Methodology: F.C.; Validation: G.D.P.; Investigation: R.Z., F.C., M.D.C. and G.L.; Writing—Original Draft Preparation: R.Z. and S.D.N.; Writing—Review and Editing, G.D.P. and V.T.; Visualization, R.S.; Supervision, G.D.P. and R.S. All authors have read and agreed to the published version of the manuscript.

Funding: This research received no external funding.

Conflicts of Interest: The authors declare no conflict of interest.

References

1. Available online: https://covid19.who.int/ (accessed on 3 September 2021).
2. Bennett, G.; Young, E.; Butler, I.; Coe, S. The Impact of Lockdown during the COVID-19 Outbreak on Dietary Habits in Various Population Groups: A Scoping Review. *Front. Nutr.* **2021**, *8*, 53. [CrossRef]
3. Brooks, S.K.; Webster, R.K.; Smith, L.E.; Woodland, L.; Wessely, S.; Greenberg, N.; Rubin, G.J. The Psychological Impact of Quarantine and How to Reduce It: Rapid Review of the Evidence. *Lancet* **2020**, *395*, 912–920. [CrossRef]
4. Lozupone, M.; La Montagna, M.; Di Gioia, I.; Sardone, R.; Resta, E.; Daniele, A.; Giannelli, G.; Bellomo, A.; Panza, F. Social Frailty in the COVID-19 Pandemic Era. *Front. Psychiatry* **2020**, *11*, 577113. [CrossRef]
5. Zupo, R.; Castellana, F.; Sardone, R.; Sila, A.; Giagulli, V.A.; Triggiani, V.; Cincione, R.I.; Giannelli, G.; De Pergola, G. Preliminary Trajectories in Dietary Behaviors during the COVID-19 Pandemic: A Public Health Call to Action to Face Obesity. *Int. J. Environ. Res. Public Health* **2020**, *17*, 7073. [CrossRef]
6. Socała, K.; Szopa, A.; Serefko, A.; Poleszak, E.; Wlaź, P. Neuroprotective Effects of Coffee Bioactive Compounds: A Review. *Int. J. Mol. Sci.* **2020**, *22*, 107. [CrossRef]
7. Mancini, E.; Beglinger, C.; Drewe, J.; Zanchi, D.; Lang, U.E.; Borgwardt, S. Green Tea Effects on Cognition, Mood and Human Brain Function: A Systematic Review. *Phytomedicine* **2017**, *34*, 26–37. [CrossRef] [PubMed]
8. Nehlig, A. Is Caffeine a Cognitive Enhancer? *J. Alzheimer's Dis.* **2010**, *20* (Suppl. 1), S85–S94. [CrossRef] [PubMed]
9. Haskell, C.F.; Kennedy, D.O.; Milne, A.L.; Wesnes, K.A.; Scholey, A.B. The Effects of L-Theanine, Caffeine and Their Combination on Cognition and Mood. *Biol. Psychol.* **2008**, *77*, 113–122. [CrossRef]
10. Bin Zarah, A.; Enriquez-Marulanda, J.; Andrade, J.M. Relationship between Dietary Habits, Food Attitudes and Food Security Status among Adults Living within the United States Three Months Post-Mandated Quarantine: A Cross-Sectional Study. *Nutrients* **2020**, *12*, 3468. [CrossRef] [PubMed]
11. Husain, W.; Ashkanani, F. Does COVID-19 Change Dietary Habits and Lifestyle Behaviours in Kuwait: A Community-Based Cross-Sectional Study. *Environ. Health Prev. Med.* **2020**, *25*, 61. [CrossRef] [PubMed]
12. Błaszczyk-Bębenek, E.; Jagielski, P.; Bolesławska, I.; Jagielska, A.; Nitsch-Osuch, A.; Kawalec, P. Nutrition Behaviors in Polish Adults before and during COVID-19 Lockdown. *Nutrients* **2020**, *12*, 3084. [CrossRef]
13. Grabia, M.; Markiewicz-Żukowska, R.; Puścion-Jakubik, A.; Bielecka, J.; Nowakowski, P.; Gromkowska-Kępka, K.; Mielcarek, K.; Socha, K. The Nutritional and Health Effects of the COVID-19 Pandemic on Patients with Diabetes Mellitus. *Nutrients* **2020**, *12*, 3013. [CrossRef] [PubMed]
14. Sánchez-Sánchez, E.; Ramírez-Vargas, G.; Avellaneda-López, Y.; Orellana-Pecino, J.I.; García-Marín, E.; Díaz-Jimenez, J. Eating Habits and Physical Activity of the Spanish Population during the COVID-19 Pandemic Period. *Nutrients* **2020**, *12*, 2826. [CrossRef]
15. Đogaš, Z.; Lušić Kalcina, L.; Pavlinac Dodig, I.; Demirović, S.; Madirazza, K.; Valić, M.; Pecotić, R. The Effect of COVID-19 Lockdown on Lifestyle and Mood in Croatian General Population: A Cross-Sectional Study. *Croat. Med. J.* **2020**, *61*, 309–318. [CrossRef]
16. Di Renzo, L.; Gualtieri, P.; Pivari, F.; Soldati, L.; Attinà, A.; Cinelli, G.; Leggeri, C.; Caparello, G.; Barrea, L.; Scerbo, F.; et al. Eating Habits and Lifestyle Changes during COVID-19 Lockdown: An Italian Survey. *J. Transl. Med.* **2020**, *18*, 229–344. [CrossRef] [PubMed]

17. Luo, Y.; Chen, L.; Xu, F.; Gao, X.; Han, D.; Na, L. Investigation on Knowledge, Attitudes and Practices about Food Safety and Nutrition in the China during the Epidemic of Corona Virus Disease 2019. *Public Health Nutr.* **2021**, *24*, 267–274. [CrossRef]
18. Skotnicka, M.; Karwowska, K.; Kłobukowski, F.; Wasilewska, E.; Małgorzewicz, S. Dietary Habits before and during the COVID-19 Epidemic in Selected European Countries. *Nutrients* **2021**, *13*, 1690. [CrossRef] [PubMed]
19. Celorio-Sardà, R.; Comas-Basté, O.; Latorre-Moratalla, M.L.; Zerón-Rugerio, M.F.; Urpi-Sarda, M.; Illán-Villanueva, M.; Farran-Codina, A.; Izquierdo-Pulido, M.; Vidal-Carou, M.D.C. Effect of COVID-19 Lockdown on Dietary Habits and Lifestyle of Food Science Students and Professionals from Spain. *Nutrients* **2021**, *13*, 1494. [CrossRef]
20. Izzo, L.; Santonastaso, A.; Cotticelli, G.; Federico, A.; Pacifico, S.; Castaldo, L.; Colao, A.; Ritieni, A. An Italian Survey on Dietary Habits and Changes during the COVID-19 Lockdown. *Nutrients* **2021**, *13*, 1197. [CrossRef]
21. Mitchell, E.S.; Yang, Q.; Behr, H.; Deluca, L.; Schaffer, P. Adherence to Healthy Food Choices during the COVID-19 Pandemic in a U.S. Population Attempting to Lose Weight. *Nutr. Metab. Cardiovasc. Dis.* **2021**, *31*, 2165–2172. [CrossRef]
22. Deschasaux-Tanguy, M.; Druesne-Pecollo, N.; Esseddik, Y.; de Edelenyi, F.S.; Allès, B.; Andreeva, V.A.; Baudry, J.; Charreire, H.; Deschamps, V.; Egnell, M.; et al. Diet and Physical Activity during the Coronavirus Disease 2019 (COVID-19) Lockdown (March-May 2020): Results from the French NutriNet-Santé Cohort Study. *Am. J. Clin. Nutr.* **2021**, *113*, 924–938. [CrossRef] [PubMed]
23. Sidor, A.; Rzymski, P. Dietary Choices and Habits during COVID-19 Lockdown: Experience from Poland. *Nutrients* **2020**, *12*, 1657. [CrossRef] [PubMed]
24. Ludden, A.B.; O'Brien, E.M.; Pasch, K.E. Beliefs, Behaviors, and Contexts of Adolescent Caffeine Use: A Focus Group Study. *Subst. Use Misuse* **2017**, *52*, 1207–1218. [CrossRef] [PubMed]
25. Costa, B.M.; Hayley, A.; Miller, P. Young Adolescents' Perceptions, Patterns, and Contexts of Energy Drink Use. A Focus Group Study. *Appetite* **2014**, *80*, 183–189. [CrossRef]
26. Abalo, R. Coffee and Caffeine Consumption for Human Health. *Nutrients* **2021**, *13*, 2918. [CrossRef]
27. Haskell-Ramsay, C.F.; Jackson, P.A.; Forster, J.S.; Dodd, F.L.; Bowerbank, S.L.; Kennedy, D.O. The Acute Effects of Caffeinated Black Coffee on Cognition and Mood in Healthy Young and Older Adults. *Nutrients* **2018**, *10*, 1386. [CrossRef]
28. Nehlig, A. Effects of Coffee/caffeine on Brain Health and Disease: What Should I Tell My Patients? *Pract. Neurol.* **2016**, *16*, 89–95. [CrossRef]
29. Bryan, J. Psychological Effects of Dietary Components of Tea: Caffeine and L-Theanine. *Nutr. Rev.* **2008**, *66*, 82–90. [CrossRef]
30. Camfield, D.A.; Stough, C.; Farrimond, J.; Scholey, A.B. Acute Effects of Tea Constituents L-Theanine, Caffeine, and Epigallocatechin Gallate on Cognitive Function and Mood: A Systematic Review and Meta-Analysis. *Nutr. Rev.* **2014**, *72*, 507–522. [CrossRef]
31. Sargent, A.; Watson, J.; Topoglu, Y.; Ye, H.; Suri, R.; Ayaz, H. Impact of Tea and Coffee Consumption on Cognitive Performance: An fNIRS and EDA Study. *NATO Adv. Sci. Inst. Ser. E Appl. Sci.* **2020**, *10*, 2390. [CrossRef]
32. Mari, E.; Lausi, G.; Fraschetti, A.; Pizzo, A.; Baldi, M.; Quaglieri, A.; Burrai, J.; Barchielli, B.; Avallone, F.; Giannini, A.M. Teaching during the Pandemic: A Comparison in Psychological Wellbeing among Smart Working Professions. *Sustain. Sci. Pract. Policy* **2021**, *13*, 4850. [CrossRef]
33. Romero-Martínez, B.S.; Montaño, L.M.; Solís-Chagoyán, H.; Sommer, B.; Ramírez-Salinas, G.L.; Pérez-Figueroa, G.E.; Flores-Soto, E. Possible Beneficial Actions of Caffeine in SARS-CoV-2. *Int. J. Mol. Sci.* **2021**, *22*, 5460. [CrossRef] [PubMed]

MDPI
St. Alban-Anlage 66
4052 Basel
Switzerland
Tel. +41 61 683 77 34
Fax +41 61 302 89 18
www.mdpi.com

Foods Editorial Office
E-mail: foods@mdpi.com
www.mdpi.com/journal/foods

www.ingramcontent.com/pod-product-compliance
Lightning Source LLC
Chambersburg PA
CBHW061444210926
51850CB00008B/1606
9783036550176